培文书系·心理学系列

HISTORY & SYSTEMS OF PSYCHOLOGY

心理学的历史与体系

第 6 版

〔美〕James F. Brennan 著

北京大学出版社
PEKING UNIVERSITY PRESS

北京市版权局著作权合同登记图字:01-2004-2654 号

图书在版编目(CIP)数据

心理学的历史与体系(第 6 版)/(美)布伦南(Brennan, J. F.)著. —影印本. —北京:北京大学出版社,2004.7

(心理学系列)

ISBN 7-301-07368-2

I. 心… II. 布… III. 心理学史—世界—高等学校—教材—英文 IV. B84—091

中国版本图书馆 CIP 数据核字(2004)第 033876 号

书 名: 心理学的历史与体系(第 6 版)
著作责任者: [美] James F. Brennan 著
责 任 编 辑: 张慧欣 孙明卉
标 准 书 号: ISBN 7-301-07368-2/C·0268
出 版 者: 北京大学出版社
地 址: 北京市海淀区中关村北京大学校内 100871
网 址: http://cbs.pku.edu.cn 电子信箱:pw@pup.pku.edu.cn
电 话: 邮购部 62752015 发行部 62750672 编辑部 58874097 58874098
印 刷 者: 三河市欣欣印刷有限公司
发 行 者: 北京大学出版社
经 销 者: 新华书店
850 毫米×1168 毫米 16 开 23.75 印张 408 千字
2004 年 7 月第 1 版 2006 年 9 月第 3 次印刷
定 价: 40.00 元

出 版 说 明

培文书系心理学英文影印系列旨在面向心理学领域的师生和广大心理科学的从业者和爱好者，推介国外心理学领域的英文原版专著和教材，使我国读者能够接触到原汁原味的第一手资料。

需要重申的是，作者本人的有些观点和结论尚需商榷，有些甚至是不可取的，为此提请读者加以甄别。书中的观点均不代表出版社观点。

北京大学出版社

2004年7月

简明目录

Writers and Events to the Renaissance

B.C. | A.D.

600 500 400 300 200 100 | 100 200 300 400 500 600 700 800 900 1000 1100 1200 1300 1400 1500

Jesus Christ Crusades Renaissance
Naturalist Stoics Cicero Great Schism Reformation
Biological Epicureans Plotinus Pope Gregory St. Francis Copernicus
Mathematical St. Paul St. Jerome U. of Paris
Eclectic St. Augustine R. Bacon
European Cultures
Anaxagoras Justinian Aquinas
Socrates
Plato
Aristotle Julius Caesar Constantine Charlemagne
Alexander Augustus
Zarathustra Shih Huang-ti Mohammed
Cyrus Abbasid Caliphate
Asian Cultures
Darius III Abu ibn Sina
Upanishads
Buddha
Lao-tze Mo Ti Yang Chu
Confucius Mencius

Ancient Era **Christian Era**

Intellectual Developments: Post-Renaissance to pre-Twentieth Century

1600 1700 1800 1900

Science
Galileo Newton Lagrange J. Müller Darwin Golgi/Cajal
Kepler Académie des Sciences Priestley Rolando Helmholtz
Halley Franklin Flourens
Euler Linnaeus Young Purkinje

1600 1700 1800 1900

Philosophy

Spinoza Comte
Descartes French Sensationalism Biran James
Locke Early British Empiricism Later British Empiricism

Psychology

Weber/Fechner Galton
Wundt APA
Brentano

Twentieth-Century Psychology: Major Representative Figures

1900 1910 1920 1930 1940 1950 1960 1970 1980 1990 2000

Binet Calkins Washburn Ladd-Franklin Spearman Psychonomic Society Soc. for Neuroscience APS
James Munsterberg Cattell McDougall Angell
Peirce Hall Dewey Thorndike Carr
Wertheimer Koffka Köhler Lewin
Freud Adler Jung Horney
Sullivan Fromm Neuroscience
Sechenov Bekhterev Pavlov Watson Weiss Holt Hunter Lashley Konorski Cognitive Science
Guthrie Tolman Hull Luria Skinner Mowrer
Brunswik Spence Miller
Camus Sartre Jaspers
Husserl Heidegger Buber Binswanger
Merleau-Ponty Maslow May
Bühler Rogers
Piaget Chomsky Sperry
Eysenck
Mead Seligman

Contents

Preface

My prefatory remarks for this edition cover the same ground as in the prior editions—namely, that this text is written as an introduction to psychology's past, grounded firmly in the intellectual history of Western civilization. Psychology emerged as a scientific discipline within the context of the intellectual history of Western Europe. The progression of ideas that led to the post-Renaissance development of empirical science allowed psychology to assume its present diverse form. Accordingly, the scope of contemporary systems of psychology may be best understood in terms of the evolution of Western thought from the time of antiquity. This book contains a historical perspective on the intellectual development of Western civilization, which gradually focuses on the emergence of psychology as an independent, recognized scientific enterprise.

Chapters 1–11 introduce the major themes of psychological inquiry initially considered by early Greek scholars and subsequently modified by Christian and Islamic writers. As modern science grew out of the Renaissance, the place of psychological inquiry became a source of controversy that resulted in competing philosophical models of the nature of psychology. These models are organized along characteristic national trends of psychological views proposed by scholars in France, Britain, and Germany. The tremendous advances of the empirical disciplines, which culminated in the nineteenth century, led to the articulation of the formal study of psychology in the 1870s by Wundt and Brentano.

Chapters 12–16 deal with the major systems of psychology in the twentieth century and into the current century: the American functional movement, Gestalt psychology, psychoanalysis, behaviorism, and the third force movement. Chapter 17 concludes this survey of the systems with an outline of trends within the more contemporary, post-system period of psychology's development. In the 20 years of work on the five previous editions of this project, the database of psychology has seemed to grow exponentially. The disciplinary content of psychology has been diffused to various allied fields. Cognitive science and neuroscience have matured and brought psychology into intimate contact with research trends derived from other disciplines. This development is obviously difficult to capture in a book of this nature, yet justifies even more the need for understanding the historical background of psychology.

I would like to thank those who have taken the time with previous editions of this work to offer suggestions for improvement and clarification. I especially want to thank my colleague Dr. Michael Riccards, President of Fitchburg State College, for his continued support during the various iterations of this project. I must also thank the many students who, over the years, helped me to express my ideas and always ignited the spark that made teaching psychology so much fun.

I would like to acknowledge the helpful comments of the following people, who served as reviewers for the publisher: Mary Ballou, Northeastern College; Greg Bohemier, Culver-Stockton College; and Lori Van Wallendael, University of North Carolina at Charlotte.

For their ongoing help and support, I am grateful to my wife, Maria, and my family. My daughters, Tara and Mikala, and their respective husbands, Craig and Adam, have been and continue to be a source of consistent support and inspiration over the years devoted to this project and to other academic demands. Our grandsons, Sam and Luke, now add a note of respite and joy to our lives. My family is my life, and their patience with me and this project merits far more than a dedication.

James F. Brennan

1

Introduction: Past for Present

A cursory glance at contemporary psychology reveals startling diversity. Psychology seems to mean many things to many people. In everyday life the word *psychology* has a variety of meanings with mentalistic, behavioristic, or abnormal implications. The popular media seem to reinforce this perception. For example, we often hear the words *psychological*, *psychiatric*, and *psychoanalytic* equated and used interchangeably. We often read or see research results on smoking or drug hazards conducted by psychologists but described as medical research. Or we see instances where a psychologist, using "armchair" methodology, responds with profound advice in a newspaper to a reader in distress. Nor does the college-level introductory course to psychology necessarily dispel the confusion. Those who have taken such courses may have dim, confused recollections of IQ tests, dogs salivating, hierarchies of anxiety, the Oedipus complex, figure-ground reversals, rats running through a maze, heart rate control, peer group influence, and so on. Similarly, listing the range of positions held by psychologists does not resolve the confusion. We find psychologists in hospitals and community mental health centers, in advertising and industry, in government and the military, and in universities.

Whereas the diversity of modern psychology is a source of bewilderment, psychology's range of study is justifiably broad. As a formal, independent discipline studied and taught in universities, psychology has been in existence for only a century.

However, we should recognize that people have been "psychologizing" since they first began to wonder about themselves. The long history of theories and models of psychology slowly evolved, mostly within philosophy, until the nineteenth century, when the methodological spirit of science was applied to the study of psychology and the formal discipline of psychology appeared in Western intellectual institutions.

The emergence of psychology as a formal discipline takes us to the problem of science. Generally, *science* is defined as the systematic acquisition of knowledge. However, from a more narrow perspective, the acquisition of knowledge is limited to observations validated by our senses. That is, we must see, hear, touch, taste, or smell events to confirm their existence as scientific data. This type of science is called *empiricism*, and its most controlled application is called the *experimental method*, in which variables are manipulated and measured. Over a century ago this more narrow, empirical definition of science linked up with a nineteenth-century model of what psychology should study to form the discipline of psychology. Yet neither at that time nor during the last hundred years did that form of psychology win universal acceptance. Some scholars argued for a different model of psychology, a broader definition of science, or both. Thus, psychology's long past, coupled with more recent differences of opinion about the form that the discipline of psychology should take, resulted in the heterogeneous discipline we study today.

Although the variety of opinions about psychology can be confusing, it can also be a source of excitement. Psychology is a young, unsettled, and often unwieldy discipline that has a highly stimulating subject matter to investigate—human activity. The purpose of studying psychology's history is to help remove the confusion caused by the diversity of psychology. By using this diversity as a resource rather than a hindrance, our understanding of psychology's development makes contemporary psychology richer for us. There are other reasons to study the history of psychology. Knowledge of the past, per se, is certainly worthwhile and beneficial in providing perspectives. Furthermore, the study of psychology's history may help illuminate some of the questions that have concerned scholars through the ages. However, the most pressing reason to study the history of psychology may be to understand the basis of its present diversity.

APPROACHES TO HISTORICAL INVESTIGATION

In their examination of the past, historians have proposed structures, or models, within which events may be categorized, correlated, and explained. For example, the preeminent historian of psychology E. G. Boring (1950) contrasted the *great man* and *Zeitgeist* models as they applied to the history of psychology. Expressed succinctly, the *great man* view holds that historical progress occurs through the actions of great persons who are able to synthesize events and by their own efforts change the path of those events toward some innovation. The *Zeitgeist*, or "spirit of the times," model argues that events by themselves have a momentum that permits the right person at the right time to express an innovation. Accordingly, Martin Luther

(1483–1545), in nailing his theses condemning corruption in the Church to the church door at Wittenberg in 1517, may be viewed either as a formidable figure starting the Reformation or as the agent of Reformation forces already at work.

A variant of the *Zeitgeist* view for the history of science, proposed by Kuhn (1970), suggests that social and cultural forces develop paradigms, or models, of science at various stages and that scientific work is conducted within a given paradigm for a limited period until the paradigm is replaced. The change in paradigms is a byproduct of both the cultural needs of the age and the inability of the old paradigm to accommodate new scientific findings. Accordingly, Kuhn presents scientific progress as a cyclic process. Within a given scientific paradigm that is accepted by a consensus of scientists, an anomaly arises that cannot be explained or accommodated by the paradigm. A crisis is generated, and new theories compete to replace the inadequate paradigm. Finally, a single view gains the commitment and allegiance of a group of scientists who implement a scientific revolution, and a new paradigm is accepted. When an anomaly again arises, the cycle is repeated. Thus, Kuhn proposed a relativity in the understanding of theories, facts, and observations that is sensitive to the implicit assumptions of scientists.

Watson (1971) has proposed another manner of structuring the historical progress of science. Watson offered prescriptions, or dimensions for classifying psychological issues, by examining and describing the relationship between scientific findings and the prevailing cultural forces of a given age. Essentially, Watson's strategy evaluated a number of possible underlying assumptions and consequent implications of theoretical positions (details of Watson's prescriptive dimensions are given in Chapter 9). This approach is useful as an evaluative tool to compare the issues and implications for various theoretical positions within psychology.

Interpretations and explanations of historical events certainly help us bring order to the history of psychology. As we examine psychology's past and its contemporary state, we shall refer to the various interpretations of scientific history to understand the meaning of specific intellectual movements. However, this book may be best described as eclectic in orientation. As its author, I am not a historian, but rather a psychologist writing of the historical antecedents of my discipline in the clearest way I can, without any commitment or allegiance to a particular interpretation of historical events.

ORGANIZATION OF THE BOOK

This work is divided into two parts. The first deals with the evolution of competing models of psychology from the classic Greek philosophers to the emergence of empirical psychology in the 1870s. Although the study of psychology is our main concern, such a study must be placed within the broad, rich context of western European intellectual thought. In so doing, we implicitly recognize that psychology is an integral part of the tradition of Western civilization. The first part of this book, then, present's psychology's history—a history that is intimately linked to the milestones of

Western civilization. In particular, the close association of psychology's history with Western traditions flows logically from basic philosophical premises about the nature of the person, which date back to the ancient Greeks. However, in order to keep an accurate perspective on psychology, it is critical to recognize that important statements about human activity were made within the rich traditions of non-Western thought. Thus, the next section summarizes some of those movements before we proceed to the main themes of psychology within Western intellectual history.

The second part of the book, starting with Chapter 12, considers the major movements that developed as psychology became more distinct from philosophy, physiology, and physics. It is difficult to conceive of twentieth-century systems of psychology without an understanding and appreciation of the events preceding the last hundred years. As will become apparent, few of the critical issues that have emerged during the last hundred years of psychology are really novel. Emphases have shifted, new technologies for study have been developed, and new jargon has been invented, but essentially we are stimulated and perplexed by the same issues that confronted our ancestors in their wonder about themselves.

At the end of the book is a glossary of terms. In the study of the history of psychology, we confront terminology derived from a variety of disciplines, a reflection of the diversity of psychology's antecedents. Jargon describing concepts and issues from such disciplines as philosophy, physics, and physiology fit into the development of psychology. Accordingly, the glossary offers ready definitions of some of the terms needed to understand the evolution of psychological thought.

EASTERN TRADITIONS IN PSYCHOLOGY

As noted previously, psychology, as it emerged as a formal discipline of study in nineteenth-century Europe, was the product of an intellectual tradition that viewed human experience through a particular set of assumptions. The very conceptualization of psychology as we know it today was formed, nurtured, structured, and argued over during the 2,500 years of turbulent intellectual progress that have elapsed since the flowering of classical Greek thought. Psychology's reliance on Western intellectual thought must be appreciated, and this relationship justifies limiting the focus of this book to Western traditions.

Whereas the long intellectual tie between contemporary empirical psychology and Western thought is apparent, it is also important to recognize that non-Western philosophies have given considerable attention to the nature of the person and the internal world of individual reflection. So, before proceeding with our story, it is appropriate to pause briefly to review some of the alternative approaches to the subject matter of psychology, articulated through a variety of intellectual works in religion and especially in Eastern philosophies. These non-Western sources of psychology's past often brought new achievements to Western intellectual progress or resulted in the rediscovery of ancient writings preserved by Eastern scholars. For example, algebra, usually attributed to ancient Indian philosophers, was first used in the West by ancient Greeks of the fourth century before Christ (B.C.), but was lost during the Middle Ages. Western Europe recovered it as a result of contacts with Islamic culture during the Crusades. Arab scholars had preserved algebra, and through them its methodology and very name were reintroduced to the West (*al-jbr* means "to reunite separate or broken parts").

As we begin the study of psychology's past, starting with ancient Greek thought in Chapter 2, keep in mind the broader perspective—namely, that intellectual achievements were occurring simultaneously in other cultures and traditions. For the most part, these events were parallel developments with little interaction, but in some cases these advances enriched Western traditions.

The Crossroads: Persia and the Middle East

The Crusades, which are described within their historical context in Chapter 3, produced many benefits to Western intellectual progress, especially in providing contacts beyond the intellectual limits of western European thought of that period. Indeed, it was the scholarship of Muslim and Jewish teachers in Islamic territories that had preserved the essential body of ancient Greek writings and extended their interpretations in philosophy, science, and medicine. Islamic scholars were able to extend earlier intellectual achievements because of their contacts with Eastern civilizations, so that Eastern thought was transmitted from its origins to centers of intellectual achievement in the Arab world, and consequently to western Europe.

In much the same way as her Arab neighbors, Persia served as a conduit between East and West. Occupying roughly the territory of present-day Iran and the immediately surrounding area, the ancient Persians were an Indo-European tribe that came into contact with India to the east, Russia and the Slavic tribes to the north, and Arabia and the Middle East to the west. Led by great kings such as Cyrus (reigned 550–529 B.C.) and Darius (reigned 521–486 B.C.), ancient Persia grew in territory and power. However, when Alexander the Great (356–323 B.C.) defeated Darius III (reigned 336–330 B.C.) at Arbela, resulting in the latter's death, Persia became a province of Macedonia. Wheras Persia lost its empire, ancient Greece increased its contacts with the East—to the ultimate benefit of Greek intellectual life.

The central religious philosophy of ancient Persia was named after the priest and prophet Zarathustra (reigned ca. 628–551 B.C.), also known by the Greek name Zoroaster. Legend has it that he was born of the spirit of the supreme god, Ahura-Mazda, the Lord of Life. Zarathustra personified goodness, love, wisdom, and beauty, but was severely tempted by the devil to do evil. As a reward for his virtue, God gave him the *Avesta*, a book of knowledge and wisdom, which formed the basis of Zarathustran teaching. The *Avesta*, or what survives of it, is a collection of prayers, legends, poetry, and laws that describes the struggle between the god of good and the devil. Earthly existence is a transition in this conflict between good and evil, and it will last for 12,000 years. The virtues of purity and honesty will lead to everlasting life. Because they are targets of evil in life, the bodies of the dead must not be burned or buried, but rather left to birds of prey or thrown to the dogs and returned rapidly to nature. The supreme god, Ahura-Mazda, created and ruled the world and was assisted by lesser gods; Zarathustra taught that Ahura-Mazda had seven aspects for people to emulate or strive for: light, good mind or wisdom, right, dominion, well-being, piety, and immortality.

As part of this earthly conflict, individuals were engaged in a struggle between good and evil, and had the free will to choose between them. This psychology led to a code of ethics and values that stressed honesty and piety. The major sin in this code was unbelief, which was dealt with swiftly. The moral code was enforced by the priests, called *magi* (from the Persian word for "sorcerer") because of their reputation for wisdom, who were also practitioners of Persian medicine. As in pre-Renaissance Europe, religion and medicine were mixed in their service to the masses by the priestly class.

The legacy of Zarathustran philosophy and religion was far-reaching. The conflict between good and evil found expression in the works of the ancient Greek philosophers. The emphasis on one god was paralleled in Judaism, and there may have been other Zarathustran influences on Hebrew thought. Even the Christmas visit of the Magi and the birth of the boy-god have precedents in Zarathustran tradition. Occupying the bridge between the Hindu society of India and the Arabic and Greek societies of the Middle East, Persia had an influential and rich position and imposed an imprint on the mix of ideas.

India

As the birthplace of Buddha, the historical home of the Hindus and the metaphysical *Upanishads*, the target of repeated Muslim invasion, and the object of colonial exploitation by several European powers, India is a storehouse of deep intellectual variation. As a subcontinent filled with polyglot tribes, often clashing yet more often living in mutual tolerance, India's material and human resources have attracted outsiders throughout history. Western interest in India goes back a considerable time in recent history. Marco Polo visited India in the thirteenth century, and was followed 200 years later by a Portuguese navigator, Vasco da Gama. Columbus was seeking India when he discovered the Americas in 1492. In succession, the Dutch, French, and British established power bases and colonial economies in India.

Hindu Science and Philosophy. Much of the knowledge of ancient India comes from the *Vedas*, the *Book of Knowledge*. The *Vedas* are a collection of lessons, hymns, poetry, and prose that were compiled from oral recitations. The *Rig-veda* is perhaps most famous as a literary achievement, involving many hymns and poems praising various objects of worship, such as the sun, moon, wind, dawn, and fire. But the *Upanishads* are of interest to us because they represent the collected wisdom of Hindu scholars who thought about the person's relation to the world. An early expression of Hindu pantheistic philosophy, the *Upanishads* are a collection of over 1,000 discourses authored by various scholars. Written between 800 and 500 B.C., these teachings sought to describe individual relations to the universe. The *Upanishads* are important because several predominant themes in them reflect the unique character of Indian philosophy. Distrust of the intellect and sensory knowledge is a dominant theme, as is the search for self-control, unity, and universal knowledge. The process of attaining these goals involves shedding knowledge, participation, and even awareness of the particular and ephemeral. We are not body or mind or both; rather, we are an impersonal, neuter, and pervading reality. Within the lessons of the *Upanishads* are themes of special metaphysical knowledge that secure for us a release from the bonds of the particular and material. The *Upanishads* focus on methods of spiritual transcendence. Transmigration of a person's essence is viewed as punishment for evil living, and eventual release from successive reincarnations is the way in which we transcend these bonds. By eliminating individual desires through ascetic living, we can escape from our individualism and be reabsorbed into a whole unity of Being.

The goals expressed in the *Upanishads* lead to a psychology that is quite opposite to the basic philosophical tenets of Western psychology. Whereas the latter recognizes the individual asserting himself or herself as a process of successful development and adaptation—indeed, much of Western psychology actually describes and predicts ways to facilitate this individuation—the *Upanishads* propose the opposite. The mystical, impersonal, and unified themes of the *Upanishads* reveal a harmony that can be achieved by rejecting individual expression. These themes pervade Hindu and Buddhist thought and provide a striking contrast for understanding some of the basic differences between Indian and Western thought.

The Hindu philosophies have important implications for psychology. First, the individual is characteristically a part of a greater and more desirable unity. Individual growth, then, is away from individuality and toward an emergence into the bliss of universal knowledge. Second, the assertion of individuality is seen not as meaningful in itself, but rather as an activity to be minimized and avoided. Sensory and mental events are unreliable. Indeed, truth lies in transcending sensory and mental activities and voiding consciousness. Finally, the emphasis on humanism and the centrality of the individual self, expressed in some Western views of psychology, is out of synchrony with the major themes of Indian philosophy. According to the basic Hindu conceptualization, the integrity of the individual person is questionable, because the individual occupies an insignificant place relative to the entire, harmonious complexity that is the cosmos.

Buddhism. Although Buddhism spread to China, Japan, and Southeast Asia, it originated with the Indian philosopher and teacher Siddhartha Gautama (ca. 563–483 B.C.)—Buddha. Indeed, Buddhism served as a vehicle for exporting many Indian products besides philosophy. Buddhist missionaries introduced the decimal system to China, and the mathematical bases of Chinese astronomy came to China with Buddhism.

Like the Sophists of ancient Greece described in Chapter 2, Buddha traveled from town to town speaking to crowds of people who had heard of his reputation as "the enlightened one." His doctrine was assembled as threads (*sutras*) to jog a person's memory. Buddha taught a theology that bordered on the godless. He did not condemn the regular worship of the gods, but taught that some rituals were foolish. The pain and sorrow that pervade human experience overwhelmingly impressed Buddha. He found no order in the confusion of life, but rather some good and much evil, precluding any design by a knowing and personal deity. At best, Buddha taught a type of agnosticism, so that his religion became a prescription for virtuous living detailed by simple rules of deportment leading to a sense of subjective well-being.

Within this philosophy of religion, Buddha taught in a somewhat contradictory manner about the individual. In contrast to the Hindu traditions, but consistent with some expressions of modern Western psychology, Buddha dismissed the notion of a soul or mind as being merely a human invention needed to accommodate some unexplained aspects of experience. Sensory input is our only source of knowledge. The perceived unity of personality, according to Buddha, is caused by a succession of habits and memories. As individuals, we are not free to will our fate, but rather we are governed by the determinism of habit, heredity, and environmental events. The individual personality does not survive death.

Buddha's psychology sounds almost behavioristic and materialistic, similar to some twentieth-century expressions of psychology. Yet Buddha also accepted reincarnation and transmigration as unquestioned premises to his system. If there is no soul, what transmigrates? As far as we know, Buddha did not directly address this contradiction, but some resolution is provided in his belief in the goal of subjective well-being and the heritage of Hindu thought. A possible answer is that if we strive, through ascetic self-discipline and careful training, to attain the happiness of annihilating individual consciousness, then we begin to participate in the experience of the Spirit, which lies at our very essence. The spirit is that aspect of us that moves beyond individuality. Our separate beings are simply passing manifestations of little worth, and the study of psychological individualism is rather absurd, according to this view.

For psychology, as for other sciences, Indian achievement is not only significant, but also truly refreshing in the way that it conceptualizes human experience. The dominant theme of Hindu philosophy is to lose the individual, and this is the very antithesis of Western psychology. Even in Buddhism, where a psychological level is admitted, psychology is relegated to second-class status. Thus, Indian philosophy leaves little room for psychology in the Western sense of a discipline of scientific inquiry.

China

The Chinese have considered their country the "Middle Kingdom" between heaven and the rest of the earthly barbarians. Indeed, the emperor who first unified the nation, Shih Huang-ti (reigned 221–210 B.C.), started the Great Wall to keep out foreigners or barbarians. Within 10 years it extended 1,500 miles along China's borders. China's feudal age ended some 300 years before the birth of Christ, and literature, philosophy, and the arts flourished. Paper was manufactured as early as A.D. 100; books were commonly printed with block prints by the ninth century; A.D. 200 was the year of publication of the first Chinese encyclopedia. By 1041 the Chinese printer Pi Sheng made movable type of earthenware, and in 1611 the first known use of gunpowder in a war was recorded. When Marco Polo first arrived in China about 1270 to witness the absorption of yet another invading horde (this time those led by Kublai Khan), China's social and political system had operated on a national scale for almost 1,500 years. This brief list of ancient China's achievements provides a glimpse of the depth and wealth of Chinese civilization. Despite China's historical efforts to hoard and protect its achievements, Chinese culture became the dominant force in the Far East, spreading its influence throughout Asia. The West is a newcomer to culture and civilization when compared to China.

Early Philosophies. One of the earliest recorded works in Chinese literature is the metaphysical *Book of Changes*, the *I-Ching*. Written around 1120 B.C. and traditionally ascribed to Wen Wang, the book contains mystical trigrams that identify the laws and elements of nature. Each trigram consists of three lines. Some lines are continuous and represent the male principle of yang, indicating positive direction, activity, and productivity, and providing heavenly symbols of light, heat, and life. Other lines are broken and represent the female principle of yin, indicating negative direction and passivity, and providing earthly symbols of darkness, cold, and death. Wen Wang complicated the puzzling trigrams by doubling the strokes and increasing the yang and yin line combinations. Each arrangement signified some corresponding law. All history, wisdom, and reality lay in the combinations. Confucius placed it above all other writings. He is said to have wished for an additional 50 years to study the *I-Ching* further. This book is important because of the imprint that it left on subsequent Chinese philosophy. The "good life" taught in the *I-Ching* is a utopia, which is obtained through the keys to reality contained in the puzzles of the *I-Ching*. It underscores the uncertainty of theology and the relativity of morals. Thus, Chinese philosophy de-emphasized the search for absolute truth and universal principles and tended toward the practical.

Perhaps the greatest of the pre-Confucian philosophers was Lao-tze (604–531 B.C.), who wrote the *Book of the Ways and of Virtue*, *Tao-Te-Ching*, the most important work of Taoist philosophy. This system, literally meaning "the way" in the sense of a path to wise living, rejects intellectual enterprise in favor of a simple life that is close to nature. Lao-tze called for living in harmony with the laws and order of nature, and de-emphasized intellectual knowledge as a set of tricks or arguments designed to confuse people. The proper way of living is to find the laws of nature with which our lives must be harmonized. The person seeking Tao must begin the quest for wisdom with silence: "He who must speak about the way, does not know it." While denying the certitude of the intellect and stressing the relativity of knowledge, the Taoists did not offer an alternative, realistic prescription for the problems of living in society. A return to nature, if universally followed, would lead to the massive vulnerability of an entire people to forces that are part of nature—the aggression, poverty, and ignorance present in the "simple" life. Taoism and its idyllic versions throughout history have usually provoked a reaction, and in China the reaction came from one of the most influential philosophers in history—Confucius (551–479 B.C.).

Confucius. Legend has clouded the circumstances of Confucius' birth, suggesting that he was an illegitimate descendant of the legendary emperor Huang Ti (2697–2597 B.C.). At age 22, Confucius began his teaching, attracting groups of students, who lived with him. A lover of music, he taught his students only three subjects: poetry, history, and rules of propriety of deportment. His reputation for wisdom and honesty spread widely. He was made a government leader and held several posts. He became famous for reforms and the honesty of his administration. However, jealous factions succeeded in getting Confucius dismissed when he disapproved of his royal superior's licentious behavior and argued that a ruler must be a model of proper behavior for his subjects. For the next 13 years, Confucius and his students wandered the countryside as homeless pilgrims living off meager donations. Finally, following a change in leadership, Confucius was exonerated and given a pension to live out his last 5 years in peace, surrounded only by his students.

Confucius' major thoughts were collected in nine volumes. The first five books deal with the laws of propriety, a commentary on the *I-Ching*, a book of odes describing the principles of morality, a history of his own state, and a legendary history of China. These pedagogical works are interesting for their selection of lessons from history to demonstrate principles of virtue, wisdom, and perfection. The last four books, mainly assembled by his students after Confucius' death, contain his philosophical treatises.

Although he did not deny the existence of God, Confucius can probably be described as an agnostic. Confucius' moral teachings are based upon the individual's commitment to sincerity, honesty, and personal harmony. From the person's desire for goodness, the family structure can be nurtured. For Confucius, the family is the critical social unit supporting the individual as well as the broader, more complex society. Thus, social constellations are formed by loyalties based upon respect from people who are, in turn, pledged to conform to rules of proper conduct.

Confucianism is not a comprehensive philosophy. Rather, it consists of a series of practical teachings directed toward morals and politics. The ideal person is trustworthy, loyal, sincere, and intellectually curious, but reserved and thoughtful. Confucianism is a rather conservative outlook intended to preserve the unity of life, which will easily slip into chaos without such cautions. The history of China has been marked by cycles of chaos and order, and Confucianism seems to respond to these cycles by providing rules for people to live together successfully.

The teachings of Confucius defined the future course of Chinese political and intellectual life. As a practical philosophy applied to the everyday problems of individual morality and social interaction, Confucian philosophy led to a conservatism that has supported Chinese society through periods of severe havoc. The emphasis on the family, characterized by loyalty within prescribed relationships, provided the basic framework for political, educational, military, and economic institutions. As in the moral codes coming from Christianity, Buddhism, and Islam, psychology was absorbed in the teachings of moral deportment, and deviations from the rules were considered abnormal.

Later Philosophies. Following the death of Confucius, alternative philosophical systems were proposed, but in the end Confucianism triumphed. Examples of the various reactions to Confucius include Mo Ti (ca. 450 B.C.), known as a philosopher of universal love, who rejected Confucianism as impractical. Alternatively, he tried to develop a logical proof for the existence of spirits and ghosts. As the solution to social evil, Mo Ti advocated universal love, which would bring about a utopia, and his teachings became the basis for Chinese pacifism. In contrast to Mo Ti, the philosopher Yang Chu (ca. 390 B.C.) developed a theory based upon the denial of God and afterlife, leaving people helplessly subjected to natural fates. According to Yang Chu, in life the good suffer as well as the wicked, and the latter seem to have more fun. Complaining of the extreme positions of both Mo Ti and Yang Chu, Mencius (370–283 B.C.) presented a more moderate view and achieved a fame that was second only to that of Confucius. Mencius was interested in establishing a social order that allowed people to pursue the good life. In the practical vein of Confucius, he taught about benevolent leadership and individual goodness. These goals were to become social norms. Finally, another thinker, Chuang-tze (ca. 350 B.C.), came full circle to Lao-tze and the Tao by advocating a return to nature and a society without need for government. These reactions only underscored the primacy of the teachings of Confucius, who struck the proper chord of the applied and functional approach when he detailed his prescriptions for living.

Chinese history did not produce a scientific age like that of post-Renaissance Europe. Important scientific advances were made throughout Chinese civilization, yet science itself never became the dominant ideal for intellectual activity, as it did in nineteenth-century Europe. Rather, Chinese philosophy—especially Confucianism—seems to characterize better the major themes of Chinese thought and concerns. Issues of religion, morals, and politics were intermixed, and they influenced all intellectual concerns, including psychology. Superstition and skepticism, ancestor worship, social tolerance, goodness, and pantheism all provided dominant themes of Chinese thought and literature.

The place of psychology within this framework is certainly obscure. As a matter of practical consequence, psychology is limited to the extent of conformity or nonconformity with the moral code accepted by society. Fulfilling the prescribed codes of moral conduct became an important form of socialization. The codes themselves were imposed and accepted, with no further consideration given to individual expression or growth. On a more idealistic plane, psychological issues were integrated within the goals of such virtues as goodness and honesty. The themes of Chinese philosophy on the unity of the person as part of the family, society, the nation, and the cosmos all precluded the need for a psychology to study only one aspect of what the West considered a unified experience.

The remainder of the story in this book is told from a predominantly Western perspective in terms of psychology's emergence as an intellectual trend within the mosaic of Western civilization. The purpose of this brief and selected survey of the historical traditions of Asian psychology is to underscore the rich heritage from other civilizations for psychology as well as for any intellectual exercise seeking to understand human experience. Recognizing such recurring themes as unity, universal harmony, reflective knowledge, and virtuous living, we find psychology deeply embedded in the teachings of religion and moral philosophy. Thus, as we begin a more focused historical journey, we should be mindful of other rich traditions that readily accommodate the subject matter of psychology within alternative perspectives.

A NOTE ON RESOURCES

At the end of each chapter bibliographic materials are listed, usually under two categories: primary sources and studies. The primary sources include the writings of scholars discussed in the chapter. Citation references and dates are given for available publications. The studies list resource works or general commentaries on the period considered in the chapter. The research works cited reflect the exciting scholarly interest generated in the history of psychology in recent years. As a specialization, the study of the history and systems of psychology is a relatively recent development. Probably because of psychology's youth relative to other disciplines, the systematic study of its history was largely ignored before World War II. Several important and still interesting scholarly works, however, examined the history of psychology during the prewar period. The first was the erudite *History of Psychology* by G. S. Brett, published in three volumes between 1912 and 1921. Also in 1912, an anthology of excerpts of the psychological writings of scholars from Greek antiquity to the nineteenth century was published by B. Rand under the title *The Classical Psychologists*. In 1929, two Americans, W. B. Pillsbury and E. G. Boring, published books on the history of psychology. Of the two, Edwin Boring (1886–1968) became something of an institution and a spokesman for the history of psychology. His work *A History of Experimental Psychology*, published in 1929 and revised in 1950, became a classic reference for the study of the history of psychology.

Since World War II, the history and systems of psychology have evolved into a recognized field of specialized study. In 1966, the Graduate School at Loyola University Chicago awarded a PhD to Antos Rancurello, late professor of psychology at the University of Dayton, for the first discursive dissertation in psychology on a historical topic—a study of Franz Brentano. Subsequently, doctoral specialization in the history of psychology was offered in comprehensive programs at the University of New Hampshire and Carleton University. In 1966, the American Psychological Association established a division of the History of Psychology (Division 26); this was followed in 1969 by the formation of Cheiron: International Society for the History of the Behavioral and Social Sciences. The Archives of the History of American Psychology were started at the University of Akron in 1965. Most importantly, the *Journal of the History of the Behavioral Sciences* began publication in 1965 and continues to publish scholarly research of an interdisciplinary scope. All of these developments have stimulated research in the antecedents of modern psychology.

The bibliographic listings after each chapter are not intended to be exhaustive, but rather to represent the range of scholarship available to the reader who wishes to pursue the subject matter further. In addition, the bibliographic material following this first chapter presents some of the major classic and recent works in the history of psychology as general reference material.

BIBLIOGRAPHY

General Resources

BERRY, J., POORTINGA, Y., SEGALL, M., & DASEN, P. (1992). *Cross-cultural psychology: Research and applications*. Cambridge, UK: Cambridge University Press.

BORING, E. G. (1942). *Sensation and perception in the history of experimental psychology*. New York: Appleton-Century.

BORING, E. G. (1950). *A history of experimental psychology* (2nd ed.). Englewood Cliffs, NJ: Prentice Hall.

BORING, E. G., LANGFELD, H. S., WERNER, H., & YERKES, R. (Eds.) (1952). *A history of psychology in autobiography* (Vol. 4). Worcester, MA: Clark University Press.

BORING, E. G., & LINDZEY, G. (Eds.) (1967). *A history of psychology in autobiography* (Vol. 5). New York: Appleton-Century-Crofts.

COPLESTON, F. (1982). *Religion and the one: Philosophies East and West*. New York: Crossroad.

DENNIS, W. (1948). *Readings in the history of psychology*. New York: Appleton-Century-Crofts.

DIAMOND, S. (1974). *The roots of psychology*. New York: Basic Books.

DREVER, J. (1960). *Sourcebook in psychology*. New York: Philosophical Library.

DURANT, W. (1954). *Our Oriental heritage*. New York: Simon & Schuster.

GERGEN, K. J., GULERCE, A., LOCK, A., & MISRA, G. (1996). Psychological science in cultural context. *American Psychologist, 51*, 496–503.

HAYASHI, T. (1994). Indian mathematics. In I. Gratton-Guiness (Ed.), *Companion encyclopedia of the history and philosophy of mathematical sciences* (Vol. 1). London: Routledge, 118–130.

HEARNSHAW, L. S. (1987). *The shaping of modern psychology*. London: Routledge and Kegan Paul.

HEIDBREDER, E. (1963; orig. 1933). *Seven psychologies*. Englewood Cliffs, NJ: Prentice Hall.

HENLE, M., JAYNES, J., & SULLIVAN, J. (1973). *Historical conceptions of psychology*. New York: Springer.

HERRNSTEIN, R. J., & BORING, E. G. (1965). *A source book in the history of psychology*. Cambridge: Harvard University Press.

LINDZEY, G. (Ed.) (1974). *A history of psychology in autobiography* (Vol. 6). Englewood Cliffs, NJ: Prentice Hall.

MADSEN, K. B. (1988). *A history of psychology in metascientific perspective*. Amsterdam: Elsevier Science Publishing Co.

MARX, M. H., & CRONAN-HILLIX, W. A. (1987). *Systems and theories in psychology* (4th ed.). New York: McGraw-Hill.

MURCHISON, C. (Ed.) (1930–1936). *A history of psychology in autobiography* (Vols. 1, 2, 3). Worcester, MA: Clark University Press.

NAKAYAMA, S., & SIVIN, N. (Eds.) (1973). *Chinese science: Exploration of an ancient tradition*. Cambridge, MA: MIT Press.

NEEDHAM, J. (1970). *Clerks and craftsmen in China and the West*. Cambridge, UK: Cambridge University Press.

ORLEANS, L. A. (Ed.) (1980). *Science in contemporary China*. Stanford, CA: Stanford University Press.

PETERS, R. S. (Ed.) (1962). *Brett's history of psychology* (Rev. ed.). Cambridge, MA: MIT Press.

ROBACK, A. A. (1964; orig. 1952). *History of American psychology* (Rev. ed.). New York: Collier.

ROBINSON, D. N. (1981). *An intellectual history of psychology* (Rev. ed.). New York: Macmillan.

SAHAKIAN, W. S. (1968). *History of psychology: A source book in systematic psychology*. Itasca, IL: F. E. Peacock.

SINGER, C. J. (1959). *A short history of scientific ideas to 1900*. Oxford: Clarendon Press.

SPEARMAN, C. (1937). *Psychology down the ages* (2 vols.). New York: Macmillan.

WERTHEIMER, M. (1979). *A brief history of psychology* (Rev. ed.). New York: Holt, Rinehart, and Winston.

Approaches to the History of Psychology

BORING, E. G. (1955). Dual role of the *Zeitgeist* in scientific creativity. *Scientific Monthly, 80,* 101–106.

BROZEK, J. (1969). History of psychology: Diversity of approaches and uses. *Transactions of the New York Academy of Sciences, 31,* Serial II, 115–127.

BURGER, T. (1978). Droysen and the idea of Verstehen. *Journal of the History of the Behavioral Sciences, 14,* 6–19.

BUSS, A. R. (1977). In defense of a critical-presentist historiography: The fact-theory relationship and Marx's epistemology. *Journal of the History of the Behavioral Sciences, 13,* 252–260.

BUSS, A. R. (1978). The structure of psychological revolutions. *Journal of the History of the Behavioral Sciences, 14,* 57–64.

COAN, R. W. (1978). Toward a psychological interpretation of psychology. *Journal of the History of the Behavioral Sciences, 9,* 313–327.

FLANAGAN, O. J. (1981). Psychology, progress, and the problem of reflexology: A study in the epistemological foundations of psychology. *Journal of the History of the Behavioral Sciences, 17,* 375–386.

HELSON, H. (1972). What can we learn from the history of psychology? *Journal of the History of the Behavioral Sciences, 8,* 115–119.

HILGARD, E. R. (1982). Robert I. Watson and the founding of Division 26 of the American Psychological Association. *Journal of the History of the Behavioral Sciences, 18,* 308–311.

JAYNES, J. (1969). Edwin Garrigues Boring (1886–1968). *Journal of the History of the Behavioral Sciences, 5,* 99–112.

KANTOR, J. R. (1963, 1969). *The scientific evolution of psychology* (Vols. 1 & 2). Chicago: Principia Press.

KUHN, T. (1970). *The structure of scientific revolutions* (2nd ed.). Chicago: University of Chicago Press.

MACKENZIE, B. D., & MACKENZIE, S. L. (1974). The case for a revised systematic approach to the history of psychology. *Journal of the History of the Behavioral Sciences, 14,* 324–347.

MANICAS, P. T., & SECORD, P. F. (1983). Implications for psychology of the new philosophy of science. *American Psychologist, 38,* 399–413.

MAYR, E. (1994). The advance of science and scientific revolutions. *Journal of the History of the Behavioral Sciences, 30,* 328–334.

ROSS, B. (1982). Robert I. Watson and the founding of the *Journal of the History of the Behavioral Sciences. Journal of the History of the Behavioral Sciences, 18,* 312–316.

ROSS, D. (1969). The "Zeitgeist" and American psychology. *Journal of the History of the Behavioral Sciences, 5,* 256–262.

SHAPERE, D. (1976). Critique of the paradigm concept. In M. H. Marx & F. E. Goodson (Eds.), *Theories in contemporary psychology* (2nd ed.). New York: Macmillan.

STOCKING, G. W. (1965). On the limits of "presentism" and "historicism" in the historiography of the behavioral sciences. *Journal of the History of the Behavioral Sciences, 1,* 211–217.

TURNER, M. (1967). *Philosophy and the science of behavior.* New York: Appleton-Century-Crofts.

WATSON, R. I. (1971). Prescriptions as operative in the history of psychology. *Journal of the History of the Behavioral Sciences, 7,* 311–322.

WATSON, R. I. (1974). *Eminent contributors to psychology, Vol. I: A bibliography of primary references.* New York: Springer.

WATSON, R. I. (1976). *Eminent contributors to psychology,* Vol. II: *A bibliography of secondary references.* New York: Springer.

WETTERSEN, J. R. (1975). The historiography of scientific psychology. *Journal of the History of the Behavioral Sciences, 11,* 157–171.

2

Psychological Foundations in Ancient Greece

The common cliché holds that "history repeats itself." However, we may be closer to the truth if we assert that historical events are like snowflakes: Supposedly, no two snowflakes are exactly the same, although they may be similar. As we begin our sojourn through psychology's long past with the contribution of Greek thinkers, it may be appropriate to apply the analogy of snowflakes to historical events. We may be amazed at the similarities in the questions that human beings have asked about themselves—and at the similarities of their answers. However, we should also recognize that civilization has made some progress in the last 25 centuries; we will not have to close the book on psychology after simply reviewing Greek thought. Although both the formulations and the solutions of critical psychological issues in ancient and modern times are often strikingly similar, they are not identical.

Since the advent of human intelligence and understanding, people have thought about themselves with wonder. Why do we behave as we do? Why are we able to generate reasonable explanations of some actions but not of others? Why do we have moods? Why do we seem to know that we know? In the course of human experience, people have come up with answers for such questions, and usually their explanations have suggested some cause. For example, we run away because we are afraid. Or we cry because we are sad. The nature of these causal explanations has changed over time. The nineteenth-century French philosopher Auguste Comte characterized these causal explanations as a progression of intellectual stages. The most primitive level was labeled "theological," because people suggested that a god was the causal agent responsible for changes in themselves and in nature. Indeed, many ancient societies invented gods with tremendous power. The ancient Egyptians had a whole catalogue of gods ranging from the sun to house cats. Such spirits were used to explain human behavior, and people who wished to change themselves were best advised to pray or offer sacrifice to the relevant god. Moreover, changes in nature, such as volcanic eruptions or storms, were said to be a reflection of the displeasure of gods over some human activity. The theological stage confined people's explanations of themselves and their world to spiritual causes.

Comte's description of causal progression is discussed in a later chapter, but we should note here that he viewed the Greek thinkers as a transition between a theological stage and a later stage that focused on nature, or the environment, and the generalization of principles from natural laws. Prior to the flowering of Greek thought, the relationship between human beings and the environment was governed by a view that may be described as primitive animism; that is, early conceptualizations of life held that a spirit or ghostlike entity inhabits the body and makes the body alive and conscious. During sleep the ghost leaves temporarily, to return upon awakening, and at death the ghost permanently leaves the body. All psychological activities, including sensations, perceptions, thoughts, and emotions, are propelled by the ghost. A similar explanation was proposed for other aspects of nature that seemed to live or have movement, such as plants, animals, lightning, and rivers, so that the distinction between the animate and inanimate in nature was often blurred and ambiguous. Accordingly, a clear separation between the individual and the environment was not evident in the early study of human psychology.

EARLY EXPLANATIONS OF PSYCHOLOGICAL ACTIVITY

Many historians regard the birth of science in Western civilization as occurring when the Greeks became the first thinkers to shift the focus of causal explanations from god to nature, or to the environment. The early Greeks articulated their explanations of critical psychological issues along several categories, as diagrammed in Figure 2–1. Essentially, all five categories, or orientations, attempted to discover causal explanations of human activity by means of natural first principles, or at least analogies drawn from nature. The orientations differed in the emphasis they gave to various aspects of the environment, both internal and external to humans. Each orientation is presented in the following sections in rough chronology.

FIGURE 2–1 The major categories, or orientations, of early Greek explanations of human activity.

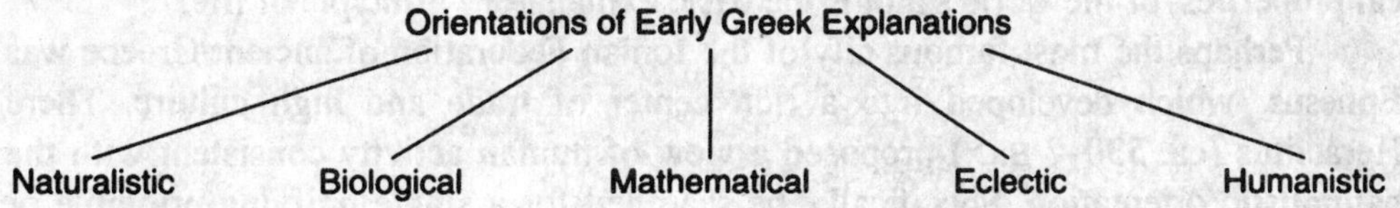

Naturalistic Orientation

All expressions of this interpretation looked to the physical environment, external to people, for causes of life-giving principles. The earliest, and perhaps clearest, expression of the naturalistic orientation is found in a group called the Ionian physicists, who lived in the sixth century B.C. The Ionian Federation of ancient Greece provided the setting for early advances in philosophy and science, which began predominantly in the city of Miletus.

These philosophers taught that life and physical matter are inseparable, so that people are intimately involved in the universe. Therefore, the determining physical principle from which all life flows had to be found in the universe.

Thales (ca. 640–546 B.C.) is widely recognized as an early sage of ancient Greece because of his introduction of mathematics and astronomy to Greek study. This pushed ancient Greek culture toward a commitment to science. According to Thales, water is the first element because it is intrinsic to all life. In reducing all of nature to water, Thales was stressing the unity of nature. Matter and life are inseparable because water is the origin of all nature as well as its final form. Thales expressed a monism that found the life-giving element water sufficient to explain all forms of nature, regardless of particular manifestations in time and place.

Another Ionian physicist, Anaximander (ca. 610–546 B.C.), advanced his teacher Thales's views of the universe by suggesting that the earth is a cylinder suspended in the center of the universe with the sun, moon, and stars revolving around it. Anaximander argued that it is the "boundless" space of the universe that contains the basic elements of nature. This boundless mass develops by its own amorphous forces the varied manifestations of nature. A student of Anaximander, Anaximenes (sixth century B.C.), speculated that the air around us, which he called *pneuma*, is this life-giving cause of nature. All three Ionian physicists represented a naturalistic orientation insofar as they searched for a first causal principle of life and found it in the physical world. Such a strategy was a radical departure from seeking explanations among the gods.

Another expression of the naturalistic orientation is derived from Democritus (ca. 460–362 B.C.), who traveled widely through the known world, supported by his father's generosity. For Democritus, our knowledge relies on our senses, which in turn receive "atoms" from objects in the world. Thus, the critical explanations of life are found in the atoms composing matter. Moreover, Democritus argued that the quantity of matter is always constant, leading to proposals for both the indestructibility of matter and its conservation. Atoms differ in size, weight, and configuration, but the relationships among atoms are completely governed by natural laws and not left to chance or spontaneity. Humans and animals consist of atoms that are the most sophisticated and mobile. Accordingly, Democritus saw in the materialism, or physical properties, of the world's atoms the basic explanatory principle of life.

Perhaps the most famous city of the Ionian Federation of ancient Greece was Ephesus, which developed into a rich center of trade and high culture. There Heraclitus (ca. 530–? B.C.) proposed a view of human activity consistent with the naturalistic orientation. Specifically, he searched for a single unifying principle or substance that could explain the nature of change and permanence in the world. His solution was fire, for both its physical properties and its symbolic value. Heraclitus felt that change is the most obvious fact of nature, and the physical properties of fire cause noticeable changes in other physical objects. Moreover, fire symbolizes the flux in nature. Thus in fire Heraclitus found a unifying substance in nature that serves as a basis for life.

The final representative of the naturalistic orientation, Parmenides of Elea (sixth century B.C.), attacked the problem of change using a rather different tactic than Heraclitus. Parmenides argued that changes in the world and all motion are superficial observations and distortions of our senses. Rather, the basic fact of nature is its permanence and immobility, which bring unity and form the basis of life. Accordingly, although Parmenides also based his solution to the question of the fundamental principle of life on matter, it was the unchanging character of matter that comprised the critical element.

Thus the naturalistic orientation viewed the environment as holding the key to the basis of life. Within this orientation, two clear trends are evident. First, there is an observational trend, represented by the Ionian physicists and Democritus, which proposed specific substances operating in our environment as the basis of life. Second, there is the view of Heraclitus and Parmenides, who hypothesized about the character of change and then deduced (to opposite conclusions) some implications about matter based on their hypotheses. Although the observational and the hypothetical deductive trends differ in their manner of dealing with the environment, both offered solutions to the character of life by examining the laws of nature and generalizing those laws to the causes of human activity.

Biological Orientation

As philosophers within the naturalistic orientation looked to the external environment in their search for the basis of life, philosophers with a biological orientation emphasized the internal state and physiology of humans as holding the clue to life.

Alcmaeon (fifth century B.C.) has been called the father of Greek medicine and is recorded as the first to use animal dissection and to discuss the optic nerve as well as the eustachian tubes. More germane to our purposes, he recognized the importance of the brain and clearly distinguished between sensory perceiving and thinking. He wrote that the causal determinants of human activity lie within the mechanisms of the body. The body seeks an equilibrium of its mechanisms, and this process explains the dynamics of human activity.

One of the more important advances in Greek philosophy and science was the separation of the practice of medicine from religion. This separation was personified in Hippocrates the physician (ca. 460–377 B.C.), who not only raised the level of medical investigation but also developed a code of ethics contained in the Hippocratic oath, followed by physicians to this day. Hippocrates, like Alcmaeon, emphasized the brain in psychological processes, and he approached the problems of medicine systematically, with what could be called a precursor of the scientific method. Relative to our concerns about psychological issues, Hippocrates contributed a theory of "humors" to account for the basis of human activity. He taught that the body contains four humors: blood, yellow bile, black bile, and phlegm. Borrowing the concept of equilibrium from his predecessors, Hippocrates argued that perfect health is a result of the proportionate mixture of these humors. The dominance of any of the humors results in characteristic indisposition. Interestingly, this theory outlasted Greek antiquity, even up to the nineteenth century, and our language still contains the phrase "bad humor" to describe someone who is not feeling well. Nevertheless, Hippocrates should be remembered for his positive efforts to free medicine from the superstitions that have historically plagued it.

The final representative of the biological orientation to consider is Empedocles (ca. 500–430 B.C.), a brilliant, eccentric, and eclectic physician whose interests and skills gained him fame as an orator, engineer, and poet. His psychology held that sensations are the product of particles from stimuli falling upon the "pores" of the sense organs. Thus, sensations have a time course, and their quality and intensity can be measured. He postulated that change develops from the conflicting forces of love and strife—that is, between attraction and repulsion. Moreover, human activity is intimately bound up in nature by means of an evolutionary process wherein change serves to differentiate aspects of the universe, followed by an amalgamation back to an indistinguishable mass. Thus, love and strife result in processes of development and decay. For human activity the focus of life is in the heart, which produces the dynamics of change.

The biological orientation tended to elevate the position of humans above the rest of nature by emphasizing the formulation of basic principles needed to account for human activity. In this sense, the biological orientation separated the uniqueness of human activity from the rest of natural relationships, in contrast to the naturalistic orientation, which emphasized human activity as a manifestation of the natural order. These early philosophers confined their explanations to primarily physiological means, and we will see how later developments made this solution inadequate.

Mathematical Orientation

Both the naturalistic and the biological approaches based their formulations of first principles firmly on the material of either the environment or the body. In contrast, the mathematical orientation attempted to extrapolate from the material level to a general principle for all life. By proposing a generalization not actually represented in the physical world but nevertheless used to explain physical reality, this orientation used the ordered beauty of mathematical structures to assert the unity of the world.

Perhaps the most famous mathematician of ancient Greece was Pythagoras (ca. 582–500 B.C.), who left a rich legacy to the modern world. After developing his mathematical system, familiar to us through Pythagorean theorems of geometry, Pythagoras examined the basis of life. He taught that we know the world through our sense impressions, but that this world is distorted and artificial. However, a second, more permanent reality exists in underlying relationships, essentially mathematical in nature, that are not available to the senses and must be discovered through intuitive reasoning. This second world of defined relationships explains all of reality by providing the essential unity of nature. Pythagoras further proposed the existence of an immortal entity as the life-giving principle. This life-giving element has functions of feeling, intuition, and reasoning, the first residing in the heart and the latter two in the brain. Both human and animal souls have feeling and intuition, but only humans have reasoning. Perhaps as the result of his exposure to Near Eastern mysticism in his wide travels, Pythagoras taught that at death the soul goes to Hades for cleansing and then returns to this life in a series of transmigrations that ends only at the completion of a life of definite goodness. Pythagoras founded a society of believers who continued adherence to his teaching for three centuries after his death. His influence as a mathematician and philosopher remains important to this day.

Although Pythagoras himself was by far the outstanding figure of this orientation, another person worth mentioning is Hippocrates the mathematician (ca. 500–450 B.C.). He wrote the first known book on geometry in 440 B.C., and Euclid was his most famous student. He is remembered as a systematist who reinforced the Pythagorean faith in the unity of numbers as the basis of life.

The mathematical orientation is interesting because it represents an approach to the problem of life's first principles that goes beyond the physical level. Although both the naturalistic and the biological orientations lent themselves to generalizations, they were firmly based in the physical world. The mathematical orientation tended to downgrade that world, and our knowledge of it, as untrustworthy. In its place it offered a different realm of mathematical relations, one we cannot know through our senses. However, by using our ability to reason we can arrive at some knowledge of this real but elusive world. Variations on this theme, stressing the unreliability of the senses and the need to extrapolate truth by our reasoning processes, will recur consistently throughout the history of psychology. Thus, the mathematical orientation gave us a de-emphasis on matter, or the material of the physical world, and an emphasis on a supposed overreaching form or structure of relationships.

Eclectic Orientation

Whereas the Pythagoreans built a system for explaining life based upon the ultimate nonphysical unity of mathematical relationships, a type of reaction occurred that was opposed even to the goal of trying to find any first principles. A group called the Sophists championed this approach, which we are calling eclectic because of its modest and practical directions. The Sophists of ancient Greece were learned men who went from place to place giving lectures and imparting wisdom to eager audiences able to afford it. In this sense, they constituted a mobile university of sorts by reaching larger groups than could be accommodated by the more traditional one-to-one arrangement of master and student. However, some Sophists became greedy and commercial in this enterprise, overcharging their constituents and causing the great philosopher Plato to ridicule them as pseudointellectuals. Plato's criticism has left the Sophists with a rather negative image that has masked some of the positive inheritance from this movement.

The best known of these wandering scholars, Protagoras (ca. 481–411 B.C.), admitted the value of sensory information as a guide to the pursuit of knowledge. However, he denied the value of making generalizations or extrapolating beyond the physical. The first principles of absolute generalization—that is, truth, goodness, and beauty—do not exist in themselves, and we only know of such concepts to the extent that they are embodied in people. This hypothesis has two far-reaching implications. First, the denial of first principles suggests that a search for the basis of life must be confined to the investigation of life as it operates in living beings. Such an operational attitude dictates that the study of living creatures is an end or goal in itself, and not simply a means to the end product of trying to find generalized, transcendent first principles. The second implication is that we must be constantly wary of assertions that generalize beyond what we observe. That is, we must be skeptical.

Another Sophist, Gorgias (ca. 485–380 B.C.), carried Protagoras's teaching further. His book *On Nature* stated the extreme position that nothing exists except what the senses perceive and that even if something did exist we could not know it or describe it to another person. Thus, Gorgias took Protagoras's assertion concerning the use of sense information from a *guide* to knowledge to the declaration that sense information is the *only* source of knowledge. Indeed, sense information and knowledge are synonymous descriptions of all we can know of life. This view was pursued further by Antiphon of Athens (ca. 480–411 B.C.), who elaborated on the value of sensory data and the limitations of knowledge.

The eclectic orientation was opposed to the pursuits of the naturalistic, biological, and mathematical strategies. According to the Sophists, a person's knowledge depends on that person's background of experience, thus precluding the possibility of objective truth. By denying first principles generalized from reality, they proposed a limited goal for seeking knowledge of life. Further, their reliance on sensory information stressed the importance of working on an operational level: If one wants to know about life, one should study life as it is presented to us by people living in the world. Coupled with the reinforcement of skepticism, this operational spirit resulted in a type of scientific method that cautioned against speculation beyond observable reality.

Humanistic Orientation

The choice of the description "humanistic" to label this orientation is meant to convey its goal of seeking out explanations of life by distinguishing people from the rest of life. In this sense, a humanistic approach places humanity on a higher plane than other life and emphasizes those characteristics that are considered to make humans unique, such as reason, language, and self-reflection.

The first person who explicitly held this orientation is Anaxagoras (488?–428 B.C.), who speculated on the origin and development of the world. He argued that the world was initially unordered chaos. Then a world-mind, or *nous*, brought order to the chaos and differentiated the world into four basic elements—fire, water, air, and earth. Like his Ionian predecessors, Anaxagoras taught that the world gradually evolved from these four elements. However, the addition of the knowing *nous* provides a new dimension. In postulating a mind to oversee the world's development, Anaxagoras attributed rationality and intentionality to this systematic agent of progress. Moreover, this nous permeates all life and forms a common basis that defines life itself. Anaxagoras attributed individual differences among people to biologically based variability. The essential nature of all people is commonly determined by the nous.

The great philosopher Socrates (470–399 B.C.) represents the full expression of the humanistic orientation and began a clear tradition that was developed further by Plato and Aristotle. Socrates derived inspiration from conflicting views of life. He held the conviction that a general conception of life is necessary. Moreover, it is the essential uniqueness of the individual that provides the key to understanding life. In opposition to the Sophists, he taught that without transcendent principles, morals would be debased and human progress would cease. Using what we now call the Socratic method, he first defined a critical issue at a general level, then ceaselessly questioned the adequacy of the definition, and finally moved logically to a clearer statement of the question to approach the resolution. Thus, he argued that the universality of knowledge allows a reasonable person to ascertain objective truth and make moral judgments. The philosophical substance of the teachings of Socrates is difficult to specify because he was not dogmatic and taught that his only certainty was his own ignorance. As a youth, Socrates studied the physical sciences, but he became increasingly skeptical, believing that resolving the facts and relations of the observable environment led only to new puzzles. He turned toward the individual, focusing at first on the psychological processes of sensation and perception. This led him to the conclusion that the acquisition of knowledge is the ultimate good. His turning from the physical level resulted in an emphasis on the role of the self and its relationship to reality. The uniqueness of the individual was expressed in his insistence on the immortality of the life-giving soul that defines a person's humanity. Socrates' teachings on politics and morals offended many Athenians, resulting in his forced suicide. However, he succeeded in establishing a clear direction for the pursuit of life's explanation. From Socrates we have a focus on people and their place in nature, a view that was articulated by his students and successors.

For Socrates and his successors, the study of human activity, whether through psychology or philosophy, must focus ultimately on ethics and politics. Moreover, logic must provide the method by which we gain knowledge of ourselves. Knowledge itself is inherently good because it leads to happiness, and ignorance is evil. Thus proper knowledge leads the individual to the proper action.

The outline of the five orientations provides us with a rich variety of strategies in the search for the basis of life. The naturalistic and biological views relied on physical explanations, whereas the Pythagoreans of the mathematical orientation asserted a basic unity to life from relationships that transcend physical expressions of life. Although the Sophists denied the possibility of this transcendence, their operational spirit and skepticism offered a methodological advance. However, it is

Socrates who culminated this development with a novel view, placing the humanity of people at the center of a system that holds general and absolute truth as a goal. This humanistic interpretation of life has profound implications for the study of people, and it is to the psychological views of Plato and Aristotle that we now turn to examine the elaboration of the concept of soul.

THE CROWNING OF GREEK PHILOSOPHY

Plato and Aristotle continued in the framework articulated by Socrates. Essentially, they tried to achieve a comprehensive framework of human knowledge designed to account for all of the following features found in human personality:

1. The intellectual abilities of unity, autonomy, consistency, and creativity
2. The behavioral manifestations of variability, contingency, and stereotypy
3. The purposeful or determined aspects of human activity

The teachings of Plato and Aristotle attained far-reaching influence throughout the ancient world. Through the military conquests of Alexander the Great, which are shown in Map 1, Greek philosophy and culture became part of many civilizations and formed an intellectual basis for subsequent philosophical developments.

Plato

Plato (427–347 B.C.) carried on the concept of his teacher, Socrates, by formulating the first clearly defined concept of immaterial existence. Plato's theory of Ideas, or Forms, held that the realm of immaterial, self-existent, and eternal entities comprises the perfect prototypes for all earthly, imperfect objects. The earthly objects are imperfect reflections of the perfect ideas or forms. Translating this theory to human activity, Plato asserted a psychophysical, mind–body dualism. In other words, human activity is composed of two entities: mind and body. Only the rational soul, or mind, can contemplate true knowledge, whereas the lesser part of the body is limited to the imperfect contributions of sensations.

Born into an old and established Athenian family and named Aristocles, Plato received his nickname from the Greek word *platon*, or "broad," which described his rugged athletic build. As a child and young man he excelled in mathematics, music, rhetoric, and poetry, and he fought in three battles, earning recognition for bravery. Around the age of 20 he came under the influence of Socrates, which led to a profound change in his life. Upon the death of his mentor, Plato traveled widely, studying mathematics and history at various centers of the ancient world. After his adventures he settled in Athens and opened his academy, which became the intellectual center of Greece.

The study of mathematics was central to Plato's teachings. Indeed, the portal of his academy contained the admonition, "Let no one without geometry enter here." Plato valued mathematics as the tool to develop logical thinking, and he worked on the systematization of mathematical knowledge. Moreover, he applied mathematics to the study of astronomy, leaving us a valuable methodological contribution.

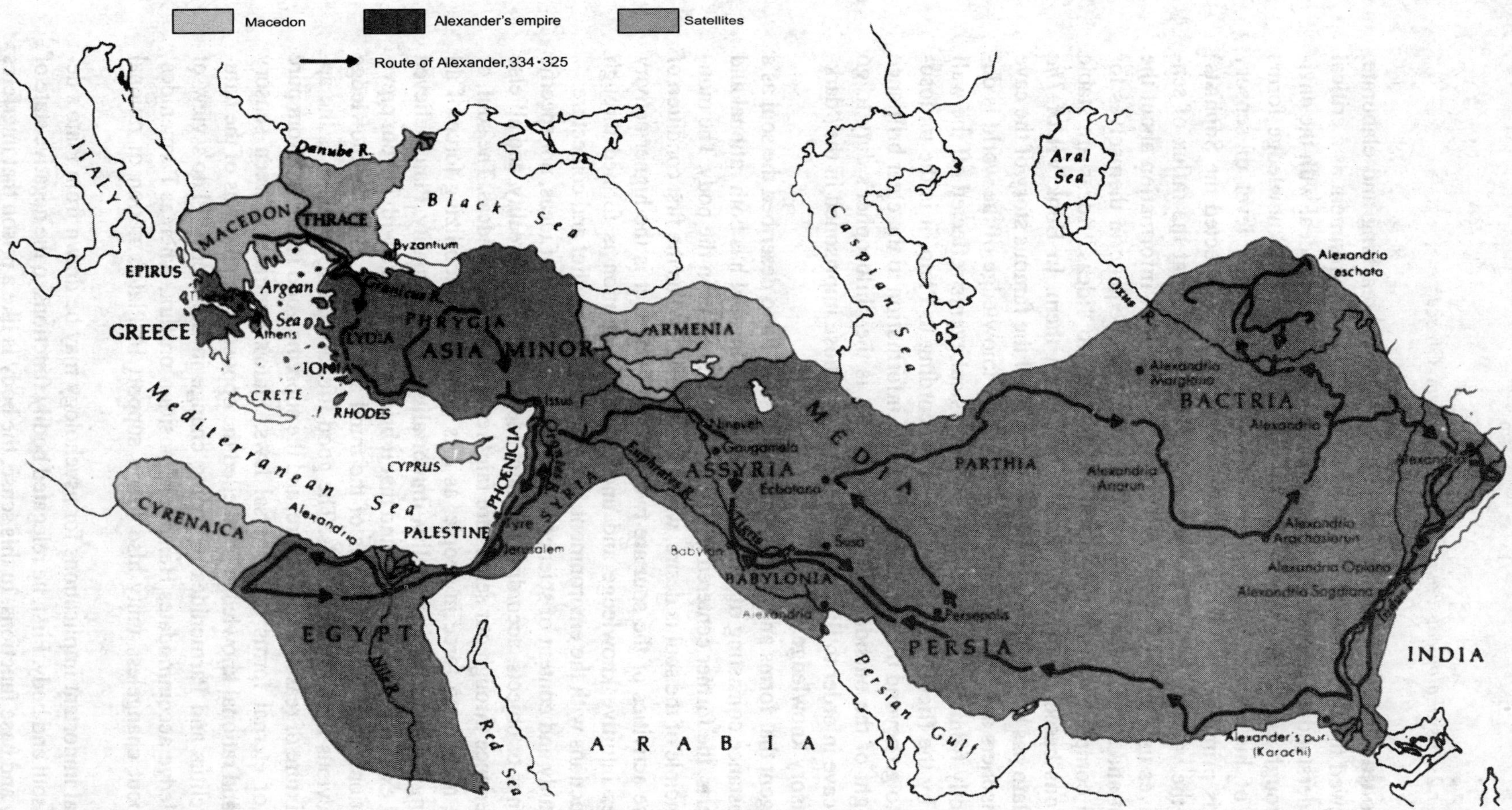

MAP 1. THE EMPIRE OF ALEXANDER THE GREAT AND HIS ROUTE OF CONQUEST. The shaded area shows the farthest extent of Alexander's conquests, from Macedonia to India. The major regions of ancient Greece—Thrace, Macedonia, Greece proper, and Ionia—are indicated, as are the important Greek cities. In addition, the Greek colonial settlements in Italy, Byzantium, and Egyptian Alexandria are shown, along with the ancient cultural centers of Babylon, Tyre, and Jerusalem.

Plato's teachings on psychological issues were far-reaching and elaborate. First, he viewed the interaction between people and their environment as a critical factor in understanding human activity. According to Plato, we deal with the environment through our senses, and this body-dependent type of knowledge forms one aspect of his mind–body dualism. However, this bodily level of sensory knowledge is primitive, distorted, and unreliable. Thus he rejected the Sophists' doctrine of the value of sense knowledge, arguing instead that the influx of sensory data gives us a percept, which he defined as a unit of information about the environment and subject to much flux. Percepts are inadequate in themselves for reliable and complete knowledge, but they give rise to "ideas." Ideas are stable generalizations based on percepts but not reliant on them. In Book VII of *The Republic*, Plato has his philosopher–hero Socrates tell the famous story of the cave in which prisoners are kept in darkness. Their only knowledge of the world is derived indirectly from the distorted images of physical events reflected off the wall of the cave by the flickering light of a fire. According to Plato, it is the philosopher's goal to go beyond the dark world of sense information to the clear brilliance of the sunlight of the outside world. Moreover, it is the philosopher's duty to go back to the cave in order to illuminate the minds of those imprisoned in the "darkness" of sensory knowledge.

The agent that forms and stores ideas is the soul. Plato described the soul as a spiritual substance consisting of reason and appetite. The soul has both rational and irrational parts, the former centered in the head and the latter in the body. The motivational principle of the soul is desire, which Plato described as the first condition of the soul. The activities of the soul are twofold: Pure intellect is the higher activity and provides intuitive knowledge and understanding; opinion is formed through bodily interactions with the environment, which give rise to belief and conjecture.

The study and content of science and philosophy consist of ideas, not specific concrete things or objects, according to Plato. Ideas are the sole reality, and all else that we experience through our senses are faint representations of ideas. The soul, or mind, is the mobilizing force in people, as it is part of the mobilizing force of all things, having the properties of vitality, immortality, and spirituality. Plato believed that the soul existed before the body and that it brings knowledge with it from previous incarnations, so that innate ideas of the mind are actually residual knowledge from the previous lives of individuals. The good life, according to Plato, is the appropriate mixture of reason and pleasure, and the supreme good is derived from pure knowledge of eternal forms of universal laws. Plato's contrast between sensory knowledge and rational knowledge reconciled the opposing conclusions of the naturalists Heraclitus and Parmenides regarding change in the world. Plato's view of sense knowledge accommodates Heraclitus' position on flux, whereas Parmenides' assertion about changeless unity also found support in Plato's notion of rational knowledge.

Several important implications for psychology may be drawn from Plato's description of soul and body. First, he relegated bodily functions to the negative state of unreliability and base functions. In this sense, the body is like a prison that interferes with the higher, more truly human functions of the soul. Second, Plato continued the tradition of Socrates with his view of the soul as containing all activities that separate humans from the rest of nature. Plato distinguished among a hierarchy of types of souls: nutritive, sensitive, and rational. At its highest level the processes of the human soul permit the formation of ideas in the intellect, leading to rational thought. Thus, the soul provides the order, symmetry, and beauty of human existence. Plato's conception of human beings presents a clear statement of mind–body dualism. At a physical level, there is motion in the world, eliciting sensations. Then, at an intellectual level, there is the formation of ideas that parallel, but go beyond, physical motion and allow abstractions from nature. Ideas do not rely on the physical level, and they become intellectually autonomous.

Plato applied his theory of the soul to politics and morals. Of interest to us is that these applications were marked by his basic distrust of human nature. Perhaps if people were pure souls, his predictions about government and society would have been more positive. However, he viewed the body as essentially evil, and believed that social structures must be built to protect people from themselves.

Aristotle

As a student of Plato for over 20 years, Aristotle (384–322 B.C.) fully appreciated Plato's mind–body dualism and his emphasis on the pure knowledge of the soul. Moreover, Aristotle brought to the study of Plato's teachings a recognition of the diversity and the dynamics of nature. Aristotle tried to understand the relationship between the abstract Idea, or Form, and the world of matter. His vast knowledge, especially of biology, facilitated his study, and the end product of Aristotle's search for knowledge was perhaps the most comprehensive and complete philosophy ever devised. Basic to Aristotle's view of life and the world was his belief that the world is ordered for some purpose or grand design and that all expressions of life are likewise propelled to develop according to some purpose.

Aristotle was born in Stagira, a small coastal settlement along the Aegean Sea in the region called Chalcidice bordering both Thrace and Macedonia. He journeyed to Athens, where he quickly established himself as a brilliant student of Plato. After Plato died, Aristotle went to Asia Minor and eventually served as the tutor of young Alexander the Great for 4 years. Probably with support from Alexander, Aristotle opened a school in Athens for the study of philosophy and rhetoric. Although he accepted the essential structure of Plato's system, Aristotle had vast knowledge of the physical world and attempted to incorporate that knowledge into the Platonic system. The end product of Aristotle's work was the categorization and systematization of all nature. In the process he dropped most of the pessimism of Plato's views on human nature.

Unfortunately, most of Aristotle's writings have come down to us in rather fragmentary form. He wrote approximately 27 dialogues, or books, but the original editions were destroyed in the repeated barbarian attacks and sacking of Rome, so that we have only dim reflections and notes on the original works and must rely on Arabic translations. The scope of Aristotle's treatises may be appreciated by categorizing his books under six general headings. The actual names of the books are those commonly titled in collected works or anthologies of Aristotle's writings:

1. Logic: *Categories, Interpretation, Prior Analytics, Posterior Analytics, Topics, Sophist Reasonings*
2. Science
   a. Natural Science: *Physics, Mechanics, Meteorology, On the Heavens*
   b. Biology: *History of Animals, Parts of Animals, Locomotion of Animals, Reproduction of Animals*
   c. Psychology: *De Anima (On the Soul), Little Essays on Nature*
3. *Metaphysics*
4. Esthetics: *Rhetoric, Poetics*
5. Ethics: *Nicomachean Ethics, Eudemian Ethics*
6. Politics: *Politics, The Constitution of Athens*

For our purposes in the history of psychology, it is appropriate to consider Aristotle's comprehensive system in terms of his views on logic and his books *Physics, Metaphysics,* and *De Anima.*

The core of Aristotle's methodological approach is found in his discourses on logic, which attempted to analyze the thought inherent in language. Aristotle's use of logic consisted of defining an object, constructing a proposition about the object, and then testing the proposition by an act of reasoning called a syllogism. This process may be seen in the following syllogism:

White reflects light.

Snow is white.

Therefore, snow reflects light.

The two processes in logic are deductions and inductions. Deductions begin with a general proposition and proceed to a particular truth; inductions start with a particular and conclude with a general statement. Aristotle's use of logic provided a systematic, common structure to his goal of accumulating all knowledge, and logic has provided an essential criterion for valid methodologies in science ever since. Specifically, the essential procedure in empirical science involves both deductive and inductive elements. The process of sampling a particular group or individual that is representative of a population involves a deduction from general characteristics of the population to specific expressions of those characteristics in individual or group samples. After describing samples, the process of inferring the descriptions back to the population from which the samples were drawn constitutes an inductive process. Finally, generalizing the conclusions about populations to all members of the population again involves deduction. Aristotle's specification of the rules of deduction and induction remains the guideline for strategies of empirical science.

Probably as the result of what he learned from his physician father, as well as from his own extensive travels, Aristotle had a wide-ranging appreciation of the natural world. His *Physics* defined the science of nature, and he provided an intricate system for cataloguing and categorizing the physical world. In so doing, he established general principles that govern and characterize the animate and inanimate parts of our environment. The structure of botanical and zoological classifications into genus and species have essentially been retained in the form taught by Aristotle. His views on the physical world evolved only after meticulous observation, and because of the clarity of his methodology, many scholars have attributed the foundation of science to Aristotle. Indeed, it is difficult to overemphasize the legacy of Aristotle's organization of scientific knowledge. He set the stage for all further developments in scientific inquiry by specifying the premises and assumptions that defined disciplinary study, and his legacy has remained largely functional up to the present. Although his specific observations on the physical sciences and biology contained many errors, Aristotle consistently tried to find the purpose or the design of nature. He examined the behavioral functions of animal biology in terms of such activities as movement, sensation, reproduction, and defense to determine how these behaviors fit into the survival and propagation of the individual and the species.

Metaphysics, meaning literally "after the physics," is the branch of philosophy that seeks the first principles of nature. Metaphysics may be divided into the study of the origins and development of the world (cosmology), the study of being (ontology), and the study of knowing (epistemology). Aristotle gave metaphysics its fullest expression and devoted considerable energy to this enterprise, which began with the search for the first principles and causes of life by the Ionian physicists. In his metaphysics, Aristotle distinguished among four types of causality:

1. Material cause—that out of which something is made. For example, the material cause of a table might be wood or plastic.
2. Formal cause—that which distinguishes a thing from all other things. The formal cause of a table is that it usually has four legs and a top positioned in a certain relationship.
3. Efficient cause—that by whose action something is done or made. The efficient cause of a table is the carpenter who constructed it.
4. Final cause—that on account of which something is done or made. The final cause of the table is the desire of someone to have a piece of furniture on which to place objects.

Using the four types of causality, Aristotle investigated the nature of being to find explanations of reality. He taught that all beings have two basic entities: primary matter and substantial form. The former is the basic material that composes all objects in the world; it is the essence of all things. The latter gives primary matter its existence. Thus, in the world there are no accidents of creation, no mutations. The direction of development is determined by the form or structure of each object governed by the urges of causality. For example, during gestation the embryo is propelled toward growth in specific ways determined by the form of the species. In Aristotle, then, we have the culmination of the Greek search for the first principles, because Aristotle's metaphysical principles explain the physical world around us.

In addition to explaining the physical world, Aristotle's metaphysical teachings construct a picture of the nonphysical, spiritual part of the universe—the soul. Aristotle's treatise on the soul, *De Anima*, contains the major pronouncements of his psychology, which defined the subject matter of psychology until the Renaissance study of science. Like Plato, Aristotle postulated a dualism of body and soul. The body receives information at a primitive sensory level through touch, taste, smell, hearing, and vision. The body gives existence to the essence of each person—the soul. However, because the soul is the life-giving element of all living existence, Aristotle proposed a hierarchical gradation of souls—vegetative, animal, and rational. The vegetative soul is shared commonly with all forms of life and is nutritive in the sense of providing for self-nourishment and growth; the animal soul is shared by all animals and allows for sensation and simple forms of intelligence; the rational soul is shared among all people and is immortal. All intellectual powers are contained in the rational soul, and in addition, the rational soul has a will, or volition. All movement originates in the soul, producing imagination, reason, and creativity. Moreover, self-reflection and the will result in purposive activity for humans, determining the specific direction of individual human activity.

Aristotle's detailed views on psychology focused on the relationship between body and soul. He stated that the emotions of anger, courage, and desire, as well as the sensations, are functions of the soul, but they can act only through the body. By asserting the critical importance of the biological foundations of life to a true understanding of psychology, Aristotle justified a physiological psychology. Moreover, he viewed ideas as formed through a mechanism of association. Specifically, sensations elicit motion in the soul, and motion grows in strength with increasing repetition. Accordingly, reliable repetitions of sensations establish internal patterns of events, and memory is the recall of series of these patterns. Aristotle distinguished between memory and recollection in a manner that parallels the contemporary distinction between short- and long-term memory. He also related the properties of physical events to the structure of human knowing by postulating 10 categories that allow their classification, comparison, location, and judgment. Aristotle's 10 categories are basically derived from the rational powers of the soul to classify our knowledge of ourselves and the environment. The categories may be summarized briefly as follows:

1. *Substance* is the universal category that essentially distinguishes an object to be what it is—for example, a man, woman, cat, flower, chemical, mineral.
2. *Quantity* is the category of order of the parts of a substance and may be discrete or continuous. Discrete quantities are numerical, such as 5, 20, or 40; continuous quantities may be parts of a surface or a solid, such as line, square, or circle.
3. *Quality* is an important psychological category because it portrays the abilities or functions of a substance. Aristotle discussed habit and disposition as qualities of the mind. A habit is a firmly established mental disposition that may be positive—such as justice, virtue, or scientific knowledge—or negative—such as erroneous knowledge or the vice of dishonesty. Quality in the human substance also refers to the capacity to operate or function—such as thinking,

willing, or hearing—and may also describe an incapacity—such as mental retardation, poor vision, or indecision. In addition, Aristotle used the category of quality for sense qualities to describe colors, flavors, odors, and sounds. Finally, he referred to the qualities of figure or shape, which may have degrees of completion or perfection.

4. *Relation* is the category that gives the reference of one thing to another—motherhood, superiority, equality, or greatness, for example.
5. *Activity* is the category of action coming from one agent or substance to another—running, jumping, or fighting, for example.
6. *Passivity* is the category of receiving action from something else or being acted on, such as being hit, being kicked, or receiving warmth.
7. *When* is a category that places a substance in time—now, last week, or in the twenty-second century.
8. *Where* is a reference to place—in school, in the room, here or there.
9. *Position* refers to the assumption of a specific posture, such as sitting, sprawled out, or standing.
10. *Dress* is a uniquely human category because it refers to attire or garb, such as wearing a suit, wearing makeup, or being armed.

Aristotle's 10 categories are listed to illustrate the detail of his comprehensive approach. The use of the categories is a psychological process, and Aristotle taught that the powers of the rational soul to know and to understand constitute the highest level of existence.

Although Aristotle is important because of his position as the culmination of Greek thought, the structure of his system and his conceptualization of human activity cannot be overstressed. The dominance of Aristotelian thought and methodology characterized the 1,500 years that followed. After dominating Greek and Roman thought, Aristotle's works were lost to western Europe, but were carefully preserved and nurtured by Islamic scholars, only to be rediscovered as western Europe shook off the ignorance of the first part of the Middle Ages, often called the Dark Ages because of the characteristic intellectual stagnation of the period. His system was the standard against which all other systems explaining human activity were compared. Only with the Renaissance did any serious challenge to Aristotle emerge, and even then, contrasting opinions were still dramatically influenced by his views. Aristotle crowned the development of classical Greek thought by his attempt to represent the world, in terms of physical, psychological, and moral knowledge, as a unitary system. He provided a philosophical synthesis that satisfied intellectual pursuit during his own time and survived until the seventeenth century.

The philosophy of ancient Greece leaves us with a rich myriad of views on the nature of life. The quest for the causes of life led to conflicting explanations of the provocative issue of first principles. As we shall see, few really new orientations have been added to the array offered by the Greeks. Rather, both the context and the methodology of each strategy were subsequently refined, and emphases were changed during the historical development of psychology. Therefore, psychology emerged from the Greek period with the basic issues and solutions fairly well defined. The classic Greek scholars successfully recognized the critical issues of psychology, and these scholars, especially Aristotle, tried to devise a systematic approach to investigate the issues. However, the emergence of science was slow, and alternative solutions to the nature of inquiry were offered before empirical science fully developed. The nonempirical, speculative approach to psychology constituted the major focus of psychological study until the use of empirical science emerged during the Renaissance.

CHAPTER SUMMARY

Ancient Greece provided the setting for the first detailed, recorded hypotheses about the causes of human activity in Western civilization. In the search for first principles of life, several systems of tentative explanations were offered. The naturalistic orientation, represented by the Ionian physicists Democritus, Heraclitus, and Parmenides, looked to some basic physical element in the world as this first principle. A biological orientation developed with Alcmaeon, Hippocrates, and Empedocles, which held that the physiology of the body contains the explanation of life. Pythagoras represented a mathematical orientation, postulating that the basis of life could be found in the essential coherence of mathematical relationships. The Sophists posited an eclectic orientation that denied the value of trying to seek out first principles. Rather, they advocated an operational attitude that relies on observations of life as it is lived. Finally, Anaxagoras and Socrates, rejecting the Sophists, proposed the existence of a soul that defines the humanity of people. This humanistic orientation developed the notion of the spiritual soul that possesses the unique human capabilities of the intellect and the will. The soul was elaborated as the central element in the interpretation of life offered by Plato and Aristotle. By the end of the Greek era the critical themes and issues of psychology as well as the methodological approaches were well identified and structured.

BIBLIOGRAPHY

Primary Sources

ARISTOTLE. (1941). *Basic works* (R. McKeon, Trans.). New York: Random House.

PLATO. (1956). *The works of Plato* (I. Edman, Ed.). New York: Modern Library.

RAND, B. (1912). *The classical psychologists*. New York: Houghton Mifflin.

Studies

BAUMRIN, J. M. (1976). Active power and causal flow in Aristotle's theory of vision. *Journal of the History of the Behavioral Sciences, 12*, 254–259.

JUHASZ, J. B. (1971). Greek theories of imagination. *Journal of the History of the Behavioral Sciences, 7*, 39–58.

LAVER, A. B. (1972). Precursors of psychology in ancient Egypt. *Journal of the History of the Behavioral Sciences, 8,* 181–195.

MANIOU-VAKALI, M. (1974). Some Aristotelian views on learning and memory. *Journal of the History of the Behavioral Sciences, 10,* 47–55.

ROYCE, J. E. (1970). Historical aspects of free choice. *Journal of the History of the Behavioral Sciences, 6,* 48–51.

SIMON, B. (1966). Models of mind and mental illness in ancient Greece: I. The Homeric model of mind. *Journal of the History of the Behavioral Sciences, 2,* 303–314.

SIMON, B. (1972). Models of mind and mental illness in ancient Greece: II. The Platonic model. *Journal of the History of the Behavioral Sciences, 8,* 389–404.

SIMON, B. (1973). Models of mind and mental illness in ancient Greece: II. The Platonic model, Section 2. *Journal of the History of the Behavioral Sciences, 9,* 3–17.

SMITH, N. W. (1971). Aristotle's dynamic approach to sensing and some current implications. *Journal of the History of the Behavioral Sciences, 7,* 375–377.

General Studies

BOURKE, V. J. (1964). *Will in Western thought.* New York: Sheed & Ward.

BURTT, E. A. (1955). *The metaphysical foundations of modern physical science.* New York: Doubleday.

COPLESTON, F. (1959) *A history of philosophy, Vol. 1, Parts I & II—Greece and Rome.* Garden City, NY: Image Books.

DURANT, W. (1939). *The life of Greece.* New York: Simon & Schuster.

KOREN, H. J. (1955). *An introduction to the science of metaphysics.* St. Louis: Herder.

MCKOEN, R. (1973). *Introduction to Aristotle.* Chicago: University of Chicago Press.

OESTERLE, J. A. (1963). *Logic: The art of defining and reasoning* (2nd ed.). Englewood Cliffs, NJ: Prentice Hall.

OWENS, J. (1959). *A history of ancient Western philosophy.* Englewood Cliffs, NJ: Prentice Hall.

ROBINSON, D. N. (1989). *Aristotle's psychology.* New York: Columbia University Press.

ROYCE, J. E. (1961). *Man and his nature.* New York: McGraw-Hill.

SAHAKIAN, W. S., & SAHAKIAN, M. L. (1977). *Plato.* Boston: Twayne.

SARTON, G. (1945–1948). *Introduction to the history of science.* Baltimore: Williams & Wilkins.

WATSON, R. I. (1971). *The great psychologists: From Aristotle to Freud* (3rd ed.). Philadelphia: J. B. Lippincott.

3

From Rome through the Middle Ages

Rome existed as a republic for 500 years under a constitution that vested authority in a senate of wise men. The republic survived wars and internal dissent until the rise of Julius Caesar (100–44 B.C.). The republic ended with Caesar and his successors and was replaced by the empire, perhaps the most remarkable political institution in the history of Western civilization. At the height of its influence, the Roman Empire covered the entire Western world, from the Near East to the British Isles. The Roman civilization absorbed the cultural influences of the ancient societies of Mesopotamia, Egypt, Israel, and Greece. Moreover, the Romans assimilated new peoples into the mainstream of Western civilization. From the East, the Armenians and Assyrians were brought under Roman rule; in the West, the Romans conquered vast areas of North Africa, Spain, France, and Britain (Map 2). Along the frontiers of the empire, Roman culture touched German, Slavic, Nordic, and Celtic tribes. From the time of Augustus (63 B.C.–A.D. 14) until the barbarians began sacking the Western Empire around the year 400, the entire Mediterranean world enjoyed relative peace and orderly administration under the *Pax Romana*. Indeed, the eastern part of the empire lasted until 1453, when Constantinople (present-day Istanbul) was finally

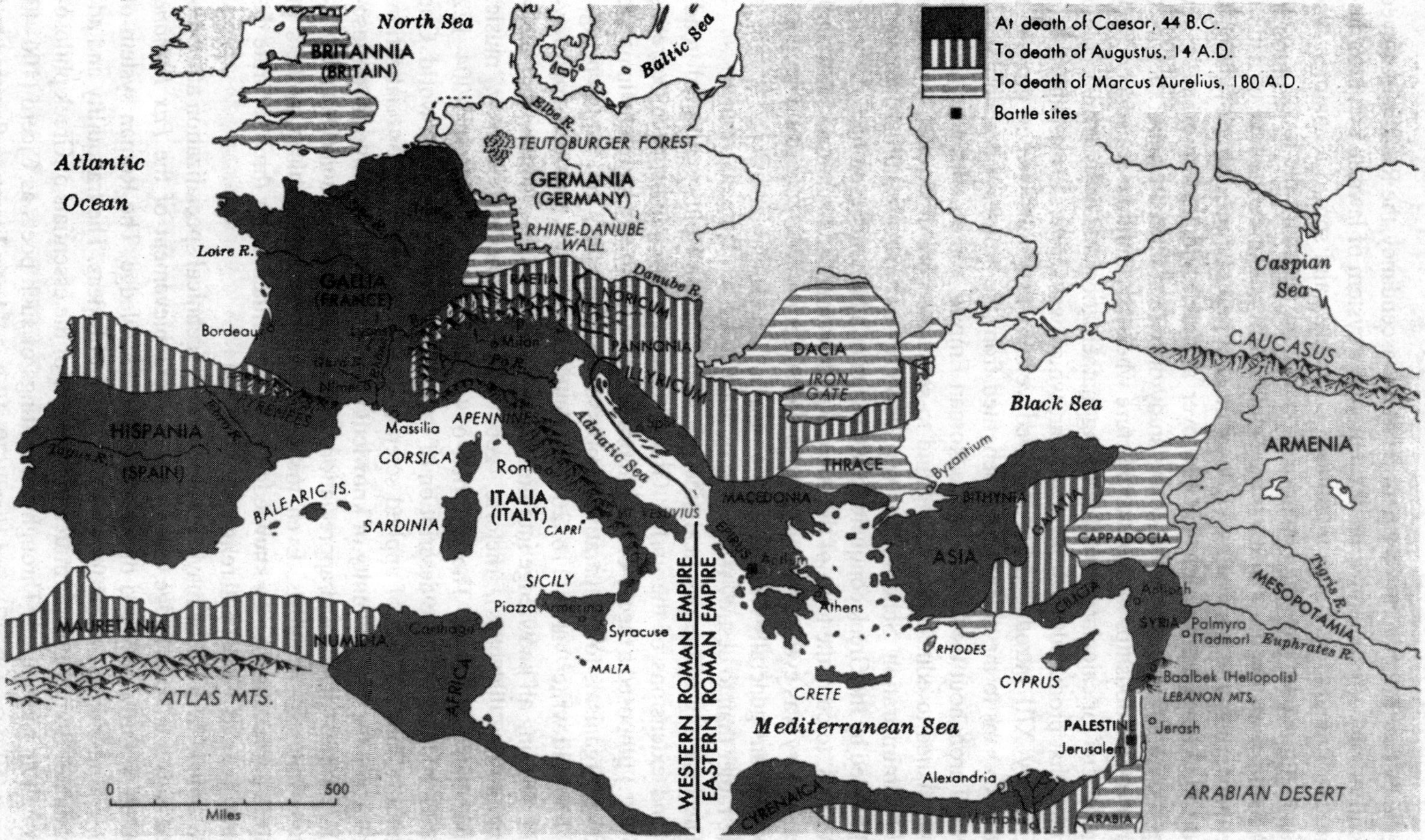

MAP 2. THE GROWTH OF THE ROMAN WORLD. Important battle sites are shown, as are the major provinces and cities of Roman commerce.

conquered by the Turks. During the period of their ascendancy, the Romans achieved successful rule by effective government. Through a system of laws and civil administration, the Romans were able to develop commerce and spread a common language and culture over diverse populations.

As administrators and builders, the Romans did not share the love of natural science that formed the basis of the philosophical systems of their Greek forebears. The Romans valued application and use over abstract studies. For example, the Romans did not dramatically advance the study of pure mathematics, but they used mathematical relationships in their architecture when they built the aqueducts. They used the abacus for calculations and teaching mathematics, and devised an accounting of time that produced the Julian calendar, which was universally accepted until Pope Gregory XIII introduced an improved version in 1582. Science prospered under the Romans to the extent that it benefited from the progress of technological advances. Throughout the history of the Roman Empire, centers of higher learning were established to educate the young and to serve the goal of supporting Roman rule and administration. Scholars and scribes were sent to Alexandria, the city in Egypt founded by the Greek conqueror and the site of a Hellenistic cultural revival in Roman times, to copy the texts of the ancient philosophers and scientists. Although the great library was eventually destroyed and even set on fire by Julius Caesar himself, the Romans generally recognized the value of Greek scholarship and sought to preserve rather than to destroy it.

The fostering of the practical side of science by the Romans resulted in advances and extensions of the earlier Greek advances. The philosopher Lucretius (99–55 B.C.) proposed a theory of natural order that recognized a hierarchy in nature from lower organisms to comparatively sophisticated mammals and human beings. The scholar and writer Varro (116–26 B.C.) developed an early version of an encyclopedia, dividing all knowledge into nine disciplinary studies: grammar, logical argumentation (or dialectics), rhetoric, geometry, arithmetic, astronomy, music, medicine, and architecture. The historian of Greek origin Polybius (ca. 204–122 B.C.) attempted a systematic description of the geography of the known world. One implication of the practice of applied science was the tendency for specialization. The Greek emphasis on the unity of knowledge had produced the universal philosophers. In contrast, the Roman appreciation of technical knowledge and detailed applications required specialists. Even the great teaching and scholarly centers of Alexandria agreed that human knowledge is best examined under three separate departments: science, ethics, and religion.

Although the Romans may have emphasized technical specialization at the expense of universal knowledge, their remarkable achievement of the *Pax Romana* contributed to the widespread dissemination of knowledge. The Roman system of government provided a means for the rapid spread of ideas. The tranquility and administration of Roman rule permitted the transfer of the essentials of Greek philosophy throughout the empire. Through the writings of such poets as Cicero (106–43 B.C.), Livy (59 B.C.–A.D. 17), and Virgil (70–19 B.C.), Latin literature flowered and successfully adapted the Greek heritage for a wider audience. Moreover, the empire provided the setting for the emergence of new institutions, most notably Christianity. Before considering the impact of early Christianity on the formulation of psychological thought, it is appropriate to examine briefly some advances in philosophy that the Romans developed as extensions of the Greek concept of the soul.

ROMAN PHILOSOPHY

The Stoic and Epicurean philosophies of Rome contributed to the development of psychology in ways that paralleled the fate of the natural sciences in Rome. Both philosophies were limited in scope and were expressed mainly in Roman religious practices. They did not follow the Greek attempts to devise a comprehensive system of human knowledge, for which the role of psychology was central. Rather, the Roman philosophies were specialized and limited to rather general attitudes toward life. The psychological implications of these views, in turn, were limited to guidelines of deportment and moral values. Likewise, the revival of Plato's teachings known as Neoplatonism enjoyed major influence in Roman intellectual spheres just as Christianity was expanding to include significant numbers of followers within the empire.

Stoicism

The Stoic period of Rome (roughly 500–200 B.C.) was characterized by a system of beliefs contained in the religion of ancient Rome, which greatly affected the moral and social values of Romans. The Stoics derived their views from the teachings of the Greek philosopher Zeno (ca. 336–264 B.C.), who believed in two basic types of matter—passive and active; that is, matter that is acted upon and matter that acts. The human soul's ability to act through intellectual capacities leads to the conclusion that human reason is intimately bound up in the universe of matter. Human freedom was simply described as the capability of cooperating with the causality of the universe. This latter view of freedom held the key to Stoic belief. It is the universe that determines life. Fate, derived from the laws of nature or whims of the gods, was the critical thesis of Stoicism. The Romans developed an elaborate religion to accommodate and cooperate with fate. Thus, in retreating from Aristotle's notion of the soul, the Stoics shifted the emphasis from inner determinism to universal determinism governed by the forces of fate. Within this perspective humans were once again viewed as a part of the environmental order.

Stoicism led to a personal resignation of the individual to the dictates of fate. In practice, this attitude advocated an abdication of personal responsibility and a surrender of individual initiative. Although the pessimistic overtones of Stoicism precluded individual behavior from degenerating to frivolity, Stoicism as a philosophy accepted the view that the individual is a reactive, not an active, organism. This theme of contrasting active and passive assumptions about the essential nature of human existence recurs consistently throughout the development of psychology. The Stoic solution left the person as part of the environment and subject to the governing pressures of environmental determinants.

Epicureanism

A somewhat later development (approximately 50 B.C.–A.D. 100) was the philosophy of the Epicureans. In dramatic contrast to the conservative Stoics, Roman followers of the Greek philosopher Epicurus (ca. 342–270 B.C.) held the sole principle that the end or goal of life is happiness. This value was reflected in the festivals and games of imperial Rome as well as in the religion that eventually asserted the deification of the emperor. The Epicureans denied the spiritual and immortal soul of the Stoics, suggesting instead that the soul is a material part of the body. The soul has knowledgeable functions of sensation and anticipation and an activity function of passion. However, the soul operates through the mechanical physiology of the body. The senses assumed a critical importance for Epicurean psychology, as thought processes are established through atoms of the environment striking the atoms of the soul. The concepts of reason and freedom, although acknowledged, exist only as individual expressions not connected to any universal, metaphysical principles. Rather, the guiding determination of human activity is the seeking of pleasure and avoidance of pain. Thus, we can see some similarities between the Roman Epicureans and Greek Sophists. The Epicureans reduced the concept of the soul to an emphasis on sensation. Moreover, this parsimonious explanation of life affirmed the view that the mechanisms of bodily functions were central to understanding life. The social and moral implications of this view are a rather mundane, self-seeking direction of individual behavior.

Both Stoicism and Epicureanism remained influential after the zenith of their respective teachings. Indeed, as Greek and Roman philosophical systems were absorbed into Christian theology, shades of these views were retained during the first centuries of the establishment of Christian belief and doctrine.

Neoplatonism

The last great pagan philosopher, Plotinus (ca. 203–270), was of Egyptian origin but spent most of his life in Rome, where he revived interest in the classic Greek philosophers, especially Plato. Plotinus argued that matter exists only as a formless potential to acquire form. Every form that matter assumes is made possible by the energy and direction of the soul. Nature itself is the total energy and universal soul, articulated into varying forms of life. Every form of life has a soul that determines the direction of growth. In human beings the vital principle within the soul molds individual progress toward maturity. The soul provides our knowledge of the environment through the generation of ideas, derived from sensations, perceptions, and thoughts. Ideas themselves transcend matter and provide the uniquely human experience of communication with the universal soul of nature. Reason is our ability to use ideas. It provides the highest form of life, allowing the individual ultimately to be conscious or aware of the creative direction of the soul.

Plotinus taught that the body is both the agent and the prison of the soul. The soul is capable of the highest form of activity—reason—which depends on sensory information but transcends the sensory level by the creative use of ideas. God is universal unity, reason, and soul. The human soul desires to seek God, but this attraction is the only certainty we have about God. Thus, life is a process whereby the soul seeks dominance over the body by rejecting the material world and finding universal truth in nature and God.

The importance of Roman Neoplatonism lies in its reception by Christianity. Greek philosophy entered Christianity in its Neoplatonist expression, so that Plato's teaching on body and soul were "Christianized" and dominated early Christian views on psychology. Christianity, in turn, dominated western Europe on the demise of the Roman order.

CHRISTIANITY

The life of Jesus, as interpreted by his followers, has offered an example and an invitation that have dramatically altered the lives of people. Beyond the religious significance of his claim to be the Messiah of the Hebrew prophesies, the story of Jesus has had a tremendous impact on the evolving importance of the soul in psychology's history. Specifically, his birth and life of poverty and his admonition to avoid worldly goods placed great emphasis on the spiritual. Moreover, his promise of love and salvation filled ordinary people with hope of deliverance from earthly problems of loneliness, poverty, and hunger. His death and bodily resurrection turned the natural order of the universe upside down, reinforcing the prominence of the spiritual life. The story of Jesus offered a message of universal appeal. The political tranquility of the *Pax Romana* provided an opportunity for that message to reach the millions of people under Roman rule.

It is difficult to distinguish the actual teachings of Jesus from the interpretations of his followers, who were intent on extending the message of Jesus beyond his Jewish context. However, like his immediate predecessor, John the Baptist, Jesus was preaching a renewal of religious commitment. Further, Jesus declared himself to be the fulfillment of the Hebrew prophesies of the Messiah. Some of his listeners perceived that the Messiah should offer the Jewish faithful a political deliverance from the occupation of Rome. However, Jesus made clear that he was not challenging Roman or any other secular authority. His kingdom was not of this world; rather, his domain consisted of the peace and love of the spiritual life in God. Whereas his teachings were consistent with the Jewish tradition, they were also amenable to the Greek concept of the body–soul dualism. Indeed, Jesus' message supported a dualistic view by enhancing the dignity and ultimate value of the spiritual, immaterial existence of the soul. Moreover, Jesus preached that human beings are distinct from the rest of nature because God has favored people by offering them the chance for immortality and salvation.

The apostles and their immediate followers took advantage of the communications network provided by Rome. If Christian teachings were to encompass more than a cult of Judaism, it would be necessary to appeal to cultural traditions beyond the Hebrew structure of the Torah. Such a movement was especially compelling in light of the loss of the Jewish foundation of Christianity after the destruction of Jerusalem in A.D. 70 by the Roman general Titus and subsequent dispersion of Jews from Palestine. Accordingly, during the first few centuries of Christianity, missionaries wandered clandestinely throughout the empire, and the center of Christian thought evolved to Rome, although important leaders of Christianity were also found elsewhere, principally in Antioch and Alexandria.

The Early Leaders

Saint Paul (ca. 10–64), the fervent missionary to the non-Jewish world, may be called the first Christian theologian. As a young Jew in his native Tarsus, he was forbidden to study classic Greek and Roman literature, but he picked up enough Greek to communicate in that language. In addition, his contact with Greeks and Romans of his time, as well as his possession of Roman citizenship, afforded a basis for his missionary activities. There is a definite Stoic influence in his writings on the strict morality of Christian society. Paul championed the separation of the new religion from Judaism and successfully fought against enforcing the practice of circumcision, a strict and basic requirement of Judaism. More importantly, his teaching identified the message of Jesus with the culture founded on Greek philosophy. Like the Stoics and Neoplatonists, Paul viewed the physical body as evil and inadequate and preached about the spiritual wisdom and perfection acquired through Jesus. He taught that Jesus is more than the Messiah fulfilling the Jewish prophecies. Rather, Jesus is God who came to the world to redeem all people, who had been condemned by the evil of original sin. As such, Jesus was the universal savior. By sacrificing himself, Jesus allowed all people to participate in the glory of perfect wisdom and knowledge. Thus, Paul radically transformed early Christianity by preaching the hopeful message of Jesus in a form that could be understood by the vast majority of the Roman Empire.

It was in the flourishing intellectual center of Alexandria that the relationship between Christianity and Greek philosophy became firmly established. Two Christian teachers, Clement (ca. 150–220) and Origen (185–254), both reconciled the Hebrew origins of Christianity with pagan Greek philosophy. The prolific Origen managed a Greek translation of the Hebrew Old Testament and provided comments and interpretations amenable to Greek understanding. The net result of his efforts was the assertion that the God of the Hebrews is the first cause or principle of life. The Hebrew doctrine of monotheism and the Greek tradition of polytheism were resolved by the concept of the Trinity. Employing the Aristotelian distinction between essence and existence, God was perceived as pure essence, capable of three expressions of existence: as the creative Father, the redeeming Son, and the Spirit that gives knowledge. Accordingly, the Trinity readily accommodates the essential tenet of Christianity that God sent his Son as the embodiment of supreme reason who organized and saved the world. Similarly, the view of the individual was Christianized within a basically dualistic context. Each person is composed of an essence, the soul, which takes on an existence through the body. The immortal soul passes through stages to eventual attachment with the body. After death, the soul continues in stages until it is eventually united in the perfect wisdom of God. All life and the sequence of the soul's development fall within the grand design of God. Thus, the Alexandrian teachers succeeded in giving Christianity a Greek foundation by incorporating the influences of Plato and Aristotle, and they added to that foundation the determinism of the Stoics.

The early theologians provided a basis for the greater appeal of Christianity, but Christianity had to cope with negative pressures from both within and outside. Internally, dissension among Christians spread through a variety of heresies. Perhaps the most important deviation from orthodox Christianity came from the Gnostics, whose mystical writings questioned basic beliefs on the resurrection and divinity of Jesus. Early Church councils settled such disputes, and gradually Church doctrine acquired form. Externally, Christians suffered waves of persecutions that did not end until the Emperor Constantine issued his Edict of Milan in 313, granting religious toleration throughout the empire.

A final problem resolved by early Christianity concerned the related issues of the gradual disintegration of the Western empire and authority from within the Church. Both issues paved the way for the emergence of papal supremacy, which had powerful implications for the European intellectual climate in succeeding centuries. The early Church had recognized the bishop of Rome as first among equals with respect to other bishops. Christianity in Rome took on many forms of Roman pagan worship in terms of liturgical dress and ritual. Indeed, the bishop of Rome assumed the title of Pontifex Maximus used by the pagan high priest. With the weakness of a series of emperors and eventual movement of the center of the empire to the East, the people of Rome began to call upon their bishop to assume the responsibilities of civil government. The evolution of papal authority was gradual and did not reach a culmination within the church until the split between Eastern and Western Christianity in 1054. Nevertheless, the centralization of authority and its identification with the pope had a tremendous impact, as we discuss later.

The Church Fathers

With the cessation of state persecution of Christianity by the Edict of Milan and the concurrent deterioration of civil authority in the Western empire, Western society began a restructuring of its values along Christian directions. A popular theology arose that contained many of the rituals of the earlier cults. The use of incense, candles, and processions as well as the veneration of saints were all adapted to Christian liturgy and served a need that was understood by the masses. As the cities declined and society assumed an increasingly agrarian character, the liturgical year was adapted to the agricultural cycle. Official Church policy tolerated some excesses that resulted in the rather primitive practice of Christianity because such customs and rites reinforced the moral teachings of the Church. In other words, the Church became the source of order and organization for both individual and social behavior. In the vacuum created by the breakdown of civil government, the Church assumed a position of the sole institution of social structure, but the Church was presiding over a decaying society with an eroding intellectual level. Accordingly, the customs and traditions of the practice of Christianity were used to preserve some semblance of moral order in the people.

Church Defenders. A group of churchmen of the fourth and fifth centuries left Christianity with the basic formulations that prevail today. Saint Jerome (340–420) chastised the people and clergy of Rome for their worldliness, and then retired to the Palestinian desert to live a crusty existence writing nasty letters to other church leaders. However, he used his classical education in the monumental task of translating the Bible into Latin, the universally understood language. Saint Ambrose (340–397), as bishop of Milan, defended the basic doctrine of the Church and served as a model of charity to the poor. Saint Anthony (ca. 251–356) in Egypt and Saint Basil (330–379) in Palestine founded the monastic movement in the Eastern empire, which stressed the value of the hermit's solitude to achieve human perfection. When monasticism spread to the western areas of the empire, it gradually acquired more community organization and became an important movement to preserve learning in feudal Europe.

The teachings of the Church scholars were integrated with the scriptural sources of Christianity in a series of Church councils that standardized Christian teachings. The first Council of Nicea (325) produced a common creed accepted by all Christians, and deviation from the Nicene Creed was considered heretical. The bishops were charged with insuring that the practice of religion conformed to the defined doctrines, and increasingly the bishop of Rome assumed a precedence over the other bishops. Emperor Valentinian III issued an edict declaring that Pope Leo I (ca. 400–461) and his successors, as bishops of Rome, had authority over all Christian churches. Although the bishops of Constantinople, Alexandria, Jerusalem, and Antioch protested his edict, the papacy was increasingly recognized as the primary source of authority within Christian society.

Saint Augustine. The writings of Saint Augustine (354–430) are critical to the history of psychology because of his reliance on Platonic thought. After receiving a sound background in classic Greek philosophy, he wandered from his native North Africa to Italy, taking various teaching posts. He led a rather Epicurean existence, reflected in his supposed injunction, "Lord, make me pure, but not right now!" While in Milan, he became infatuated with Neoplatonism and the writings of Plotinus. Finally, at age 33, he experienced a revelation from Christ and was baptized by Saint Ambrose. He returned to North Africa, founded a monastic group, and lived in poverty. In 396 he was elected bishop of the city of Hippo and remained there preaching and writing for the last 34 years of his life.

Two of Augustine's works are important to the historical evolution of psychology. His *Confessions*, written about 400, is perhaps the most famous autobiography in history. With keen introspection and masterly detail, he described how one person found peace through faith in God and resolved the conflict between the passions and reason. For Augustine, the mind is the receptor for divine wisdom and shares in the glory of God. Through it we can acquire a type of knowledge that is unknowable through the bodily senses. Moreover, this interior sense of the soul or mind allows us a level of consciousness that transcends, yet completely explains, physical reality. Thus, Augustine played down the rationality of the mind, which is dependent on unreliable sensory information. Rather, he proposed a more psychological view of the mind insofar as consciousness, or the self of the individual, endowed with the grace of divine wisdom, determines the direction of activity. According to Augustine, only by removing the faulty impressions of sensory knowledge can we reach this level of consciousness.

He wrote *City of God* in installments from 413 to 426 as a response to the outcry over the barbarian Alaric's sacking of Rome. Specifically, many people argued that this shocking event was the fault of Christianity, which had undermined the glories and power of imperial Rome. Augustine countered by asserting that Rome fell to invasion because of the inherent decay in the pagan society, which antedated the Christian era. Borrowing from Plato's notion of an ideal republic and Christian teachings on good and evil, Augustine suggested that humanity could be divided into two cities, or societies. The earthly city is concerned with worldliness and dominated by the evils of materialism. The city of God is everlasting with God and is identified for us by the Church. This city is spiritual and embodies goodness. Historically, people may vacillate between the cities and only at the last judgment will membership in each city be finally separated into those condemned to the sin and evil of Hell, and those who win happiness and perfection in God.

For our purposes in the study of the history of psychology, Augustine may be remembered for two great accomplishments. First, he completed the "Christianization" of Greek philosophy by affirming the Platonic relationship between body and soul. By relegating sensory information to a primitive level and positing a transcendent consciousness, Augustine taught the ideal of the mind reflecting upon itself as the key to ultimate beauty and love in God. This view dominated Christian thought until the end of the Middle Ages, so that all intellectual endeavors that studied life, including psychology, were done in a Platonic context. Second, he established a justification for a special relationship between the Church and the State. Augustine related the Church to the city of God. Worldly government would always be faulty and inferior to Church rule. Augustine was far more influential in the West than in the East. Because the Eastern empire was stronger there, the Church was subordinate to the State. However, in the West, with the deteriorating civil government of Rome, Augustine's arguments justified the Church's filling the void in civil as well as spiritual administration.

THE DARK AGES

A series of barbarian threats culminated in the sacking of Rome in 410, marking the first time in 800 years that the city fell to an enemy. From then until 476, when the succession of non-Roman Western emperors stopped and the Western empire came to an end, repeated invasions took their toll. Rome was reduced from a city of 1,500,000 to 300,000 people. The new tribes settled into various parts of the Western empire: The Germans moved into Italy, the Visigoths into Spain, the Franks changed Gaul to France, and the Angles and Saxons took Britain (Map 3). These tribes could not sustain the system of commercial centers in great cities, which had been managed from Rome. As a result, western Europe became rural. The strength of law given by Rome declined and was replaced by violence and individual aggression.

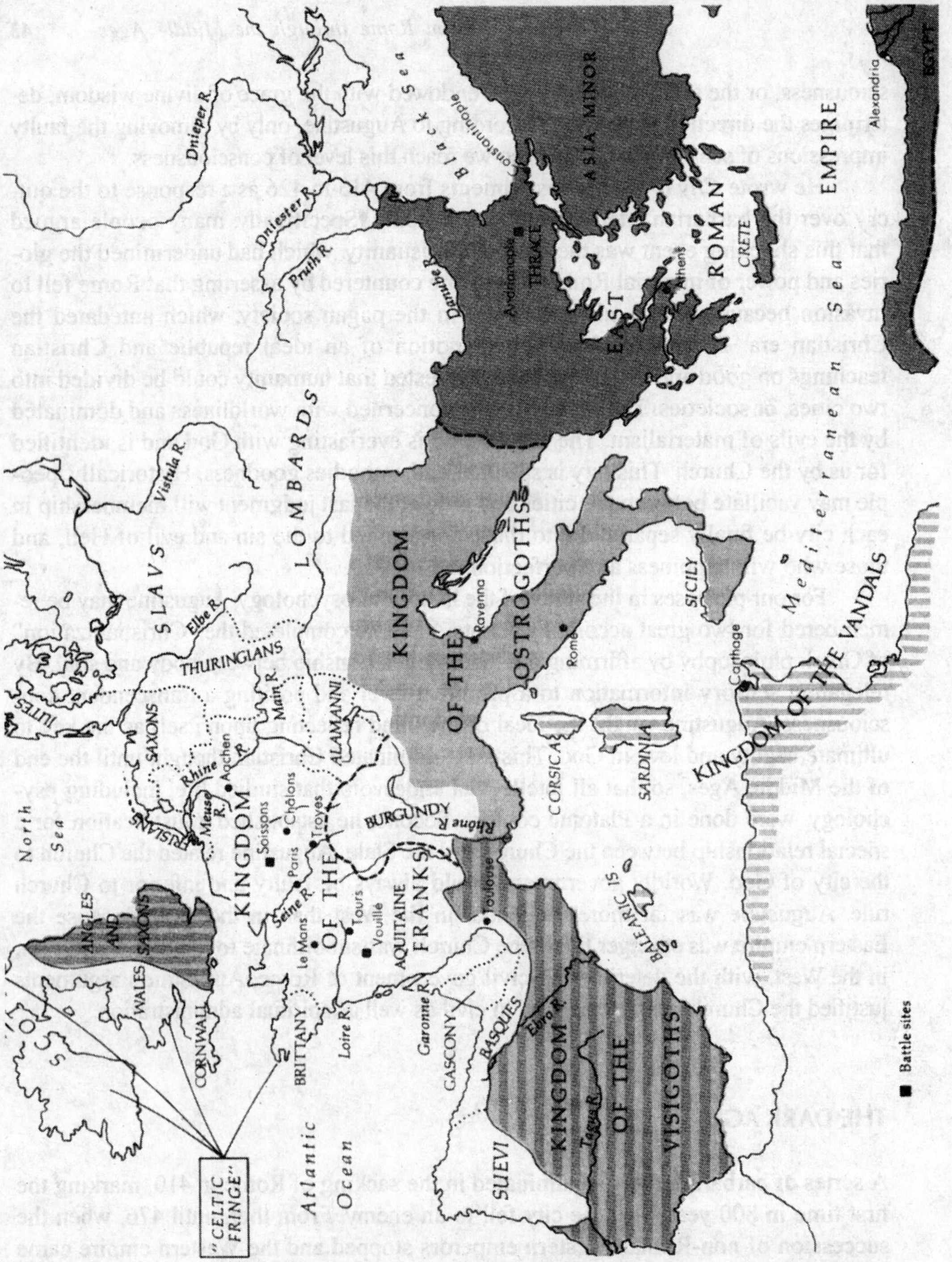

MAP 3. THE FRAGMENTED WEST AND THE EASTERN ROMAN, OR BYZANTINE, EMPIRE (ca. A.D. 500). In the west the kingdoms of the Angles and Saxons, the Franks, the Visigoths, the Suevi, the Vandals, the Ostrogoths, and Burgundy are shown. In addition, the less organized settlements of the Celts, the Frisians, the Jutes, the Thuringens, the Lombards, and the Slavs are indicated. In the east, the area governed by the emperor from Constantinople is darkly shaded, and the major cities of the Byzantine Empire are shown.

The Eastern Empire

The seat of civil government in the Roman Empire moved to the city of Constantinople, and the Byzantine Empire, with its own culture, gradually emerged while civilization in the West deteriorated. Under the leadership of Emperor Justinian (483–565), the Eastern empire flourished and a new code of law bearing the emperor's name clearly differentiated the culture and society of the East from the chaotic situation in the West. Great universities became centers of excellence at Constantinople, Alexandria, Athens, and Antioch, specializing, respectively, in literature, medicine, philosophy, and rhetoric.

Steadily, the Byzantine Empire acquired a character of its own. The Latin language gave way to Greek, and Christianity took on a Greek flavor in both ceremony and theology. The social chaos of western Europe and the increasing menace of Islamic tribes to the south made contact with the West difficult. The empire began to decline and became insulated and corrupt. However, the Byzantines supported a system of colonies in the Balkans and in the present-day Ukraine, introducing the Greek alphabet, culture, and religion. In 989 Vladimir (972–1015), grand duke of Kiev, became a Christian and brought his Ukrainian and Russian nations under the influence of the Byzantine culture. After the Byzantine Empire was finally obliterated with the fall of Constantinople to the Turks in 1453, Russia, then known as the duchy of Moscow, became the remaining repository of the Byzantine culture.

Islamic Civilization

The birth of Mohammed (570–632) in the poor desert region of Arabia marked one of the most extraordinary phenomena of the medieval period. Within a century, Mohammed's followers had conquered most of the Byzantine territories in Asia, all of Persia, Egypt, and North Africa, and were preparing to invade Spain. In 610 Mohammed experienced his first vision of the angel Gabriel, who informed him that he had been chosen as the messenger of God, or Allah, and began to reveal the sacred writings that eventually formed the holy book of Islam, the Koran. Mohammed gathered avid followers among the nomadic tribes of Arabia, and he soon conquered the holy cities of Mecca and Medina. By the time of his death Mohammed had established the essential doctrines of Islam, and his successors extended the theocratic state to an ever-expanding empire.

As the Muslim invaders occupied Christian territories once under the rule of the Byzantine Empire, they encountered the cultural heritage of Greek scholarship in philosophy and science. The Islamic intellectuals appreciated the Greek culture and borrowed freely in their formulations. Most importantly, they preserved the writings of the ancients at a time when scholarly works were being destroyed by the barbarian aggression in the West. During the rule of the Abbasid caliphs (750–1258), centered at Baghdad, the works of most classical Greek scholars as well as more recent commentaries were translated into Syrian. Islamic scholars, who in turn contributed to the development of arithmetic and algebra, also studied the mathematical treatises of the Greeks. Hospitals were established throughout the Islamic world, with the most famous in Damascus, which provided the setting for the education of physicians. Islamic physicians developed anesthesia and surgical procedures and published books on pharmacology.

One of the more famous scholars of Islam during the medieval period in the West was Abu ibn Sina, known as Avicenna (980–1037). He was a renowned physician who published a synopsis of medical treatments in his *Canon of Medicine*. As a philosopher, Avicenna was well acquainted with the writings of Aristotle. Avicenna's philosophy antedated by almost two centuries the revival in the West of interest in Aristotle, which was known as Scholasticism. Essentially, Avicenna accepted the metaphysics and psychology of Aristotle and attempted to reconcile them with his faith in Islam. He viewed the essence of the human soul as the extension of God's essence, and believed that through the rational powers of the soul we can share in the perfect knowledge of God. He treated the acquisition of sensory knowledge in some detail, concluding that the characteristic mind–body dualism of human beings reflects an interaction between sensory and rational knowledge. His synthesis of Aristotelian thought and Islamic faith was a remarkable tribute to Islamic scholarship.

Islam as a crusading, religious movement threatened the existence of Christianity. Its successes in the eastern Mediterranean virtually wiped out Christianity in that area (Map 4), and the possibility of Islamic forces overrunning western Europe was not completely eliminated until the seventeenth century. Nevertheless, in the attempt to preserve intellectual life from destruction, the Western Church scholars of later centuries were aided by the efforts of Islamic scholars who maintained the libraries of their vast empire, so that the classic authors of antiquity could be reintroduced to western Europe.

The Feudal West

The situation in western Europe after the capital of the empire moved from Rome to Constantinople steadily deteriorated. Plagued by wars, famine, and disease, Western social structures regressed, as did the general intellectual level, so that even the most elite of social classes were largely ignorant and illiterate. Because the invaders were eventually converted to Christianity, one institution that survived the devastation was the Church. The Church, left as the sole international institution in western Europe, sought to preserve a semblance of order and culture.

The Church institution most responsible for preserving the remnants of intellectual life in the West was monasticism. The founder of the monastic movement in the West was Saint Benedict (480–543), who in 529 opened the greatest monastery of the medieval period at Monte Casino in central Italy. In contrast to Eastern monasticism, which stressed the solitary existence of the hermit, Saint Benedict defined monasticism in terms of a group of men living in absolute poverty, chastity, and obedience. The famous Rule of Saint Benedict governed monastic life throughout western Europe, and variants of the rule are still followed today by contemplative religious orders of monks and nuns. A succession of monasteries composed of clerics living under the commune law of Saint Benedict spread throughout Europe and North

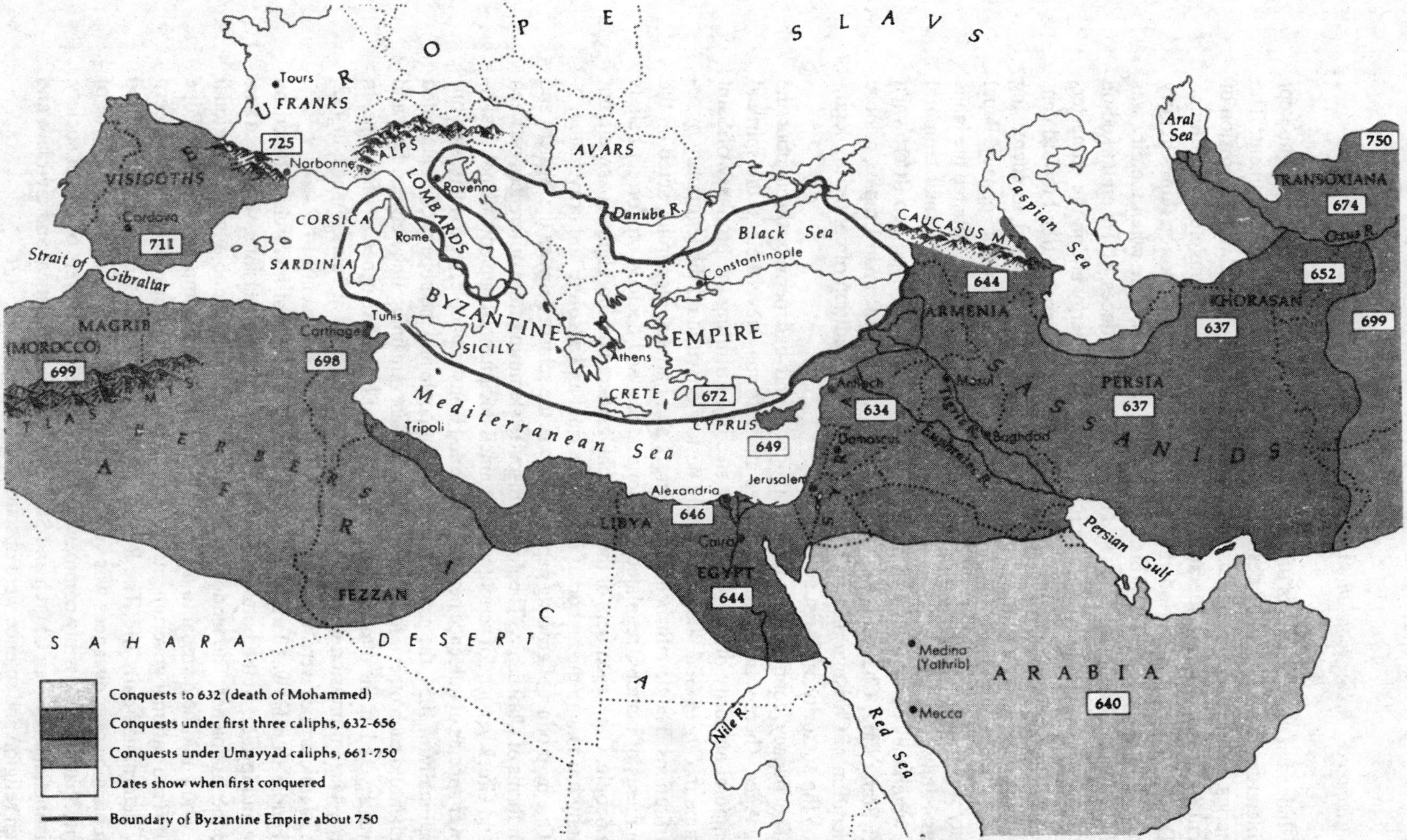

MAP 4. THE EXTENT OF THE MUSLIM CONQUESTS FROM THE TIME OF MOHAMMED (632) UNTIL THE OMMAYYAD CALIPHS OF BAGHDAD (750).

Africa. Indeed, it was the monks of faraway, but relatively tranquil, Ireland who kept Latin literature from the total destruction so widespread on the Continent. The monasteries, although not distinguished as intellectual centers, slowed the erosion of intellectual life and saved art, literature, and philosophy from complete eradication.

During the medieval period, the papacy gained enormous power. Pope Gregory the Great (540–604) was elected bishop of Rome in 590, a time when the city was decimated by a bubonic plague that had carried off his predecessor. He embarked on a reform movement that tightened discipline among the clergy and in the monasteries and improved the civil administration of the city. As a result a trend toward centralization within the Western Church was established, which led to increasing standardization in church practice and heightened the authority of the papacy. Papal authority was further enhanced in 756 when the Frank king, Pepin, donated the lands of central Italy to the Pope. These papal estates made the popes formal temporal rulers, which they continued to be until 1870. Finally, in 800, Pope Leo III crowned another Frank king, Charlemagne, as emperor of the Holy Roman Empire, a loose confederation of Christian princes, thus beginning the tradition of conferring legitimacy on the authority of Christian rulers.

The papacy continued to have both good and bad periods throughout the Middle Ages. There were cycles of abuses and reforms. Nevertheless, the spiritual and temporal authority of the popes grew steadily. Fulfilling the prophesy of Saint Augustine in *City of God*, the popes could bestow or withhold the legitimacy of social institutions. They confirmed the emperors and national monarchs, acted on the appointments of bishops, regulated the monasteries, and decided on correct beliefs for the people. Other members of the hierarchy gathered temporal power as well, but ultimately it was with the bishop of Rome that final and absolute authority rested.

One by-product of growing papal power was the schism between the Western and Eastern forms of Christianity. Theological disputes stemming from differences between Latin and Greek versions of scriptural and council documents, coupled with political rivalry over spheres of influence between Rome and Constantinople, led to increasing bitterness. Finally, in 1054, the patriarch of Constantinople and the pope excommunicated each other and each other's followers, so that the last remaining link between East and West was severed. In the meantime, the Holy Wars of Islam threatened all of Europe. In the West, Muslim invasion through Spain was finally stopped at Tours, France, in 732, but the Islamic armies were not completely driven out of Spain until 1492.

Feudal Europe was largely a loose collection of social hierarchies based on service and loyalty. At the bottom of the hierarchy were the peasants, who owed service to the landowner. The landowner, in turn, owed allegiance to a local or regional noble, who could be a vassal to a king, to the Holy Roman Emperor, or to the Pope himself. True national governments were yet to emerge, and issues of local concern mostly determined daily life. The Roman system of roads had been neglected, and communication over distances was very difficult. The papacy was the only source of authority that could possibly command the obedience of all levels of feudal society. The distinctions between Church and State, between ecclesiastical and civil law, and between religion and science were obscured.

By the year 1000, intellectual life in western Europe was isolated and losing ground. Most of the classic writings were lost, and the Church censored others because of their pagan authors. The cultural life of Europe was found largely in religious expressions of art and music. However, also by the year 1000, Europe was almost completely Christian. From Ireland in the west to Poland and Lithuania in the east, from Scandinavia in the north to the Mediterranean, people shared a common religion and a common allegiance to the papacy. Whereas feudal disputes among emerging nations would continue, sometimes with great severity, the era of greatest devastation was over, and intellectual activity slowly reemerged.

THE CRUSADES

In a sense, the Crusades represented the zenith of power for Christianity. They were a series of eight military or quasi-military campaigns from 1095 to 1291 to secure the Holy Lands of the Near East from Muslim control. Although unsuccessful in the long run, the Crusades reflected a fervent expression of Christianity. However, on another level, the Crusades may be viewed as the beginning of the reawakening of western Europe. They brought contact and commerce with other civilizations. Moreover, the Crusades produced stimulation from the intellectual life of Islam, where scholarship had fared much better than in Europe. Islamic scholars had preserved the Greek masters; mathematics, architecture, and medicine had flowered under Islamic rule. These new ideas were brought back to Europe, along with more complete copies of the ancient writers. The Crusades began to shake Europe out of its feudal provincialism. The political life of Europe and the rise of national states were facilitated by the Crusades, and this movement was at the expense of the papacy.

The Crusades were a product of a homogeneous Christianity that permeated all aspects of western European life. At the same time, however, the Crusades were symptomatic of great changes that were about to occur in western Europe. First, the papacy was powerful enough to decry the reports of persecution of Christians by the new Turkish rulers of Palestine, who had replaced the relatively tolerant Egyptian Fatimids. The First Crusade backed up papal indignation with impressive military force, but the popes lost control of the later campaigns, so that the net impact of the Crusades upon papal power and prestige was negative. Second, the Crusades filled a vacuum created by the weakness of the Byzantine Empire, which was no longer powerful enough to serve as an effective buffer between the Turks of the Middle East and the Christians of western Europe. After reaching its zenith of power and culture under Emperor Justinian, who contributed his name to the Byzantine revision and codification of Roman law, the Byzantine Empire sank into internal discord and strife, and its governing effectiveness diminished. Finally, the city-states of the Italian peninsula, such as Genoa and Venice, were developing as mercantile centers and needed extended markets. Thus, the Crusades served as a catalyst to move western Europe out of feudalism and intellectual lethargy.

As military enterprises and religious movements, the Crusades were failures. However, the Crusades did succeed in propelling western Europe into a more mature period of consolidation and organization. First, because the Crusades required the raising of large armies on an international scale, they fostered the restructuring of rivalries from a local level to a national identity. Second, because the Crusades opened the possibility for vast commercial markets, they facilitated the development of mercantile economics. Finally, the Crusades brought back the classical scholarship of antiquity. Fortunately, western Europe was ready to discard feudalism and begin the rebirth of intellectual life.

At the beginning of this chapter, we observed that the Romans inherited the Greek systems of philosophy and science and devised details of applications. However, with the fall of Roman rule in the West, scholarly pursuit, including the study of psychology, was halted and regressed. The theocratic character of feudal society mixed religion with psychology and science, so that psychology was reduced to the practice of Christianity. This loss of psychology to religion occurred on two levels. Psychology became part of the moral doctrines on behavior taught by the Church, and psychology became immersed in the mythology of Christian practice. On the first level, psychological explanations of any activity had to conform to the tenets of Christianity. For example, the directives of propagation within the exclusive bounds of marriage governed individual sexual activity, and any deviation was simply defined as wrong and abnormal. On the level of the practice of Christianity, psychology was confused with the superstitions of widely believed mythology. Mental illness and social deviancy were considered evil curses or demonic possession. The recognized cure for such maladies did not involve understanding or study, but rather prayer or exposure to relics. Medieval Europe was indeed an age of faith, and science, including psychology, was dormant.

CHAPTER SUMMARY

Roman culture adopted classic Greek philosophy but developed unique Roman perspectives, as illustrated by the Stoics and Epicureans. The Stoics held a conservative view of humanity, determined by the fates of nature. Human adjustment consisted of cooperating with universal designs. Conversely, for the Epicureans happiness consisted simply of the seeking of pleasure and avoidance of pain. The teachings of Plato were revived by Plotinus, and dominated Roman philosophy during the early years of Christianity. Both the missionary zeal of the Christian apostles and the tranquil efficiency of Roman administration contributed to the rapid spread of Christianity. The teachings of Jesus and interpretations of the Christian message evolved from a Hebrew basis to a foundation in Greek philosophy. In addition to the early fathers of the Church, Augustine successfully put a Platonic imprint on Christian theology. With the fall of the Western empire, intellectual life in Europe came to a virtual halt, and only the monastic movement preserved a rough semblance of Greek and Roman civilization. The papacy assumed a leading role not only in

spiritual direction but also in civil administration, culminating in the call for the crusades. By the time of the crusades, however, Europe was relatively peaceful and intellectual life began to stir. Exposure to the cultural inheritance of Islam revived European interest in the masterpieces of ancient civilization, and a great intellectual awakening was about to shake Europe out of the intellectual nadir of feudalism.

BIBLIOGRAPHY

AUGUSTINE. (1948). *Basic writings of St. Augustine* (W. Oates, Ed.). New York: Random House.

AUGUSTINE. (1958). *City of God* (G. G. Walsh, D. B. Zema, G. Monahan, & D. H. Honan, Eds. and Trans.). New York: Image Books.

AUGUSTINE. (1955). *Confessions* (J. K. Ryan, Ed. and Trans.). New York: Image Books.

COPLESTON, F. (1961). *A history of philosophy, Vol. II, Medieval philosophy, Part I—Augustine to Bonaventure*. Garden City, NY: Image Books.

DURANT, W. (1944). *Caesar and Christ*. New York: Simon & Schuster.

DURANT, W. (1950). *The age of faith*. New York: Simon & Schuster.

MORA, G. (1978). Mind-body concepts in the middle ages: Part I. The classical background and the merging with the Judeo-Christian tradition in the early Middle Ages. *Journal of the History of the Behavioral Sciences, 14,* 344–361.

OATES, W. (Ed.). (1940). *The Stoic and Epicurean philosophers*. New York: Random House.

PAGELS, E. (1979). *The Gnostic Gospels*. New York: Random House.

WINTER, H. J. J. (1952). *Eastern science*. London: Murray.

4

The Reawakening of Intellectual Life

During the period between 1000 and 1300, the map of Europe began to take its present form. The recovery in Europe led to the emergence of nation-states that consolidated civil administration under the political leadership of a monarch. In England the political power of the king relative to the rights of the landed barons and church authorities started to be defined through the Magna Carta, forced on the king in 1215. Similarly, the peoples of France, Spain, Portugal, and Denmark acquired national identities and cultures under the centralized order of powerful aristocracies. In the East, the monarchy of Poland merged that country initially with Hungary and later with Lithuania to form a powerful confederation that defended the Latin Christianity of western Europe from the Eastern Orthodoxy of Russia. Only in Germany, where direct and effective unification was stymied by the combined power of the landed barons and the papacy, and in Italy, which was directly under papal rule, were political consolidations under single monarchs delayed.

THE PAPACY AND CHURCH POWER

The papacy triumphed over this age. As indicated in Map 5, most of western Europe was under papal control. The institutional Church occupied a privileged role in society.

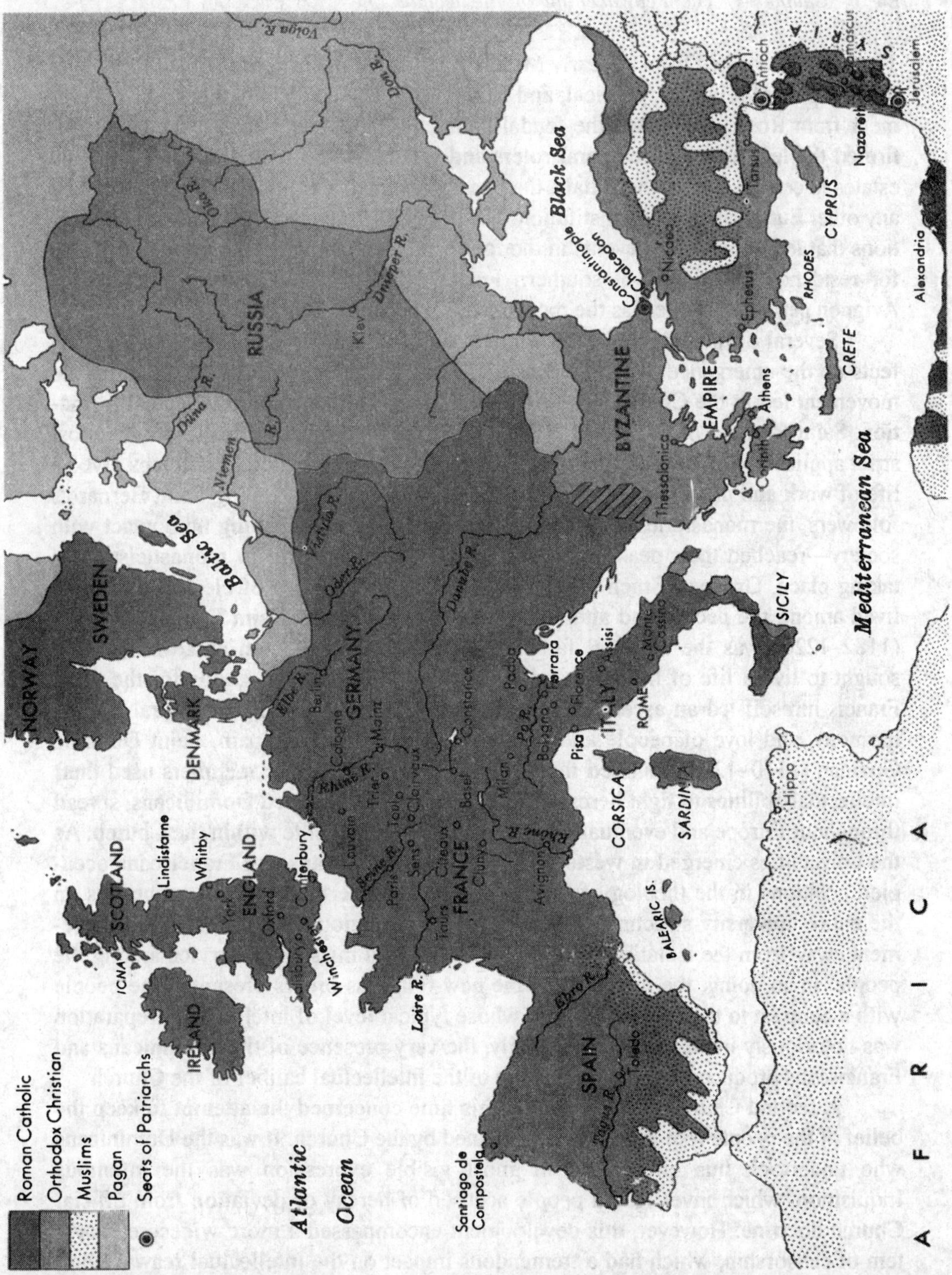

MAP 5. WESTERN CHRISTENDOM (ca. 1100). Areas of major cities and university locations in western Europe are shown, as are the remnants of the Byzantine Empire and the advancing Ottoman Turks, who finally captured Constantinople in 1453.

The papacy emerged from the early Middle Ages as the major source of authority over every aspect of religious, political, and cultural life. The essentially theocratic government from Rome dominated the feudal hierarchy. Moreover, because the pope confirmed the legitimacy of temporal rulers and was himself a temporal ruler over vast estates in central and northern Italy, the papacy assumed a political role unmatched by any other European political institution. The power of the papacy had political implications that led to disastrous events in the fourteenth century, when the popes left Rome for residency at Avignon in southern France (1309–1377). However, prior to the Avignon period, the pope was the most powerful person in Europe.

Several developments in the Church are of interest to us because of their effects on the emergence of intellectual life. First, a dramatic reform in the monastic movement led to the Church's having a more direct influence on intellectual formation. Saint Bernard of Clairvaux (1091–1153) founded monasteries under the most strict application of Benedictine rule. This order of clerics, called Cistercians, lived a life of work and prayer, avoiding all intellectual enterprises. Among Saint Bernard's followers, the monasteries—founded as self-sufficient units having no contact with society—reached their peak. However, in Italy a new reform in monasticism was taking place. Groups of men and women following the rules of clerical discipline lived among the people and attempted to serve their needs. Saint Francis of Assisi (1182–1226) was the spiritual father of several orders of men and women who sought to live a life of humility and to sacrifice their material goods for the poor. Francis himself led an ascetic life of absolute poverty, rejoicing in natural beauty, harmony, and love of people and the world. Meanwhile, in Spain, Saint Dominic Guzman (1170–1221) founded the Order of Preachers, whose members used their intellectual abilities to fight heresy. These men, popularly called Dominicans, spread throughout Europe and eventually formed an intellectual elite within the Church. As the universities emerged in western Europe, the Dominicans and Franciscans occupied positions in the theology faculties and in turn exerted powerful influences on the entire university structure. Together, the new religious orders fostered a movement away from the isolation of the monasteries to a mission of service among the people. In so doing, the members of the new religious orders presented the people with a contrast to their parish priests, whose typical level of intellectual preparation was notoriously inadequate. Accordingly, the very presence of the Dominicans and Franciscans produced a general uplifting of the intellectual caliber of the Church.

A second Church development at this time concerned the attempt to keep the belief of the people free from error, as defined by the Church. It was the Dominicans who undertook this effort, and its most visible expression was the infamous Inquisition, which investigated people accused of heresy or deviation from official Church doctrine. However, this development encompassed a more widespread system of censorship, which had a tremendous impact on the intellectual reawakening of Europe. Specifically, all intellectual activity, from the writing of books to teaching, had to be scrutinized for possible errors. In those institutions most directly controlled by the Church, such as the universities, this censorship was sometimes heavy-handed, stifling imaginative inquiry. The system produced an Index of Books, forbidden for the faithful to read, and the censors had power to condemn individuals to either death or imprisonment. The censorship system was applied unevenly from country to country, depending on the extent of papal power and of cooperation from civil authorities. However, the system was generally effective, and it gave the Church direct control of intellectual inquiry. Out of fear—or perhaps simply to avoid potential problems—many scholars were forced to work in secret or at least outside those institutions under Church control.

The Inquisition lasted several centuries in a checkered history of cyclic intensity. In addition to finding heretics, the Inquisition extended its authority to investigating demonic possession, witches, and others with deviant behavior. Unfortunately, many persons who would today be judged mentally ill, retarded, or socially incompetent were caught up in the Inquisition and suffered tortures and death because of their nonconformity. As late as 1487, the Dominican friars Jacob Sprenger and Heinrich Kraemer published *Malleus Maleficarum* (*The Hammer of Witches*), an encyclopedia of demonology and witchcraft with suggested remedies and tortures. This sordid chapter in the treatment of the mentally ill extended to the seventeenth and eighteenth centuries, as evidenced by such colorful characters as Cotton Mather and by the Salem witch trials of colonial America.

The cruel and abusive nature of the Inquisition was a means of social control in the name of Christian orthodoxy. The Inquisition was an invention of a society that subordinated all human activity to the doctrines of the Church. In turn, the Church defended its authority as based on God's will, and the people accepted Church authority on faith. Accordingly, psychology was identified with Christianity, and an understanding of individual behavior and mental activity required an appreciation of the person's desire to achieve eternal salvation. In this age of unquestioned Church supremacy, there was simply no room for other considerations. Those who questioned the authority of the Church or deviated from the social teachings of the Church, for whatever reason, were considered abnormal. Indeed, any deviation from Church teaching was contrary to what was perceived as the natural order, and therefore some abnormal and powerful agent, such as the devil, was assumed to be responsible. Thus the policing function of the Inquisition, although a sad episode in Western history, was not an unreasonable reflection of this age.

THE UNIVERSITIES

One of the characteristics of the breakdown in communication and commerce during the early Middle Ages was the loss of Latin as a universally understood language. What began as dialects of Latin after the fall of Rome emerged by 1000 as the distinct Romance languages of French, Spanish, and Portuguese. Italian evolved more slowly, but by 1300 Dante chose to write *The Divine Comedy* in the Italian dialect of Tuscany. Similarly, the Old German of the northern European tribes evolved into a family of language forms, including a forerunner of English. The French influence contributed to the development of the hybrid of modern English after the

Norman invasion of England in 1066, and the Middle English of Chaucer's fourteenth-century *Canterbury Tales* still reflected a heavy French influence. By 1300, varieties of local languages were established in spoken communication at the expense of Latin. The painstaking process of hand-copying books was still predominantly done in Latin, but that, too, was soon to change.

Education in medieval times was largely limited to moral instruction that took place in schools related to cathedrals, monasteries, and convents. The Church supported fundamental education, and the Fourth Lateran Council of Rome (1215) instructed bishops to establish in every cathedral teaching positions in grammar, philosophy, and church law. However, the cathedral schools soon proved insufficient to cope with growing numbers of clerics and nonclerics interested in learning and scholarship. A revival of interest in Roman law led to the founding of Europe's oldest university in Bologna in 1088. Soon the two schools of law—church and civil—were described as a *universitas scholarium* in Bologna, and the university was recognized by a papal decree as a place where one could engage in *studium generale*. Under the auspices of wealthy princes and Church leaders, universities shortly spread throughout Italy: Modena (1175), Vicenza (1204), Padua (1222), Naples (1224), Siena (1246), Rome (1303), Pisa (1343), Florence (1349), and Ferrara (1391). Although strong in law and medicine, these centers of learning provided Italy with the push toward the renaissance of intellectual activity in all areas of inquiry that flowered in the fourteenth and fifteenth centuries.

The University of Paris, perhaps the greatest center of philosophy and theology during medieval times, was started in 1160 and enrolled as many as 5,000 to 7,000 students. The curriculum began with the study of the seven "arts" of grammar, logic, rhetoric, arithmetic, geometry, music, and astronomy. Students then went on to study philosophy and, finally, theology. By the fourteenth century, the University of Paris consisted of 40 colleges, or residencies, of which the Sorbonne remains the most famous. As France emerged from the medieval period to the Renaissance, the sense of national identity grew faster than in other parts of Europe. Moreover, France became the most populous country in Europe, which gave the French monarchs large armies to support their political ambitions. As the leading university in the leading European country, the University of Paris acquired great prestige throughout Europe, and the opinions of the Paris scholars were considered definitive statements throughout the Continent. In particular, the faculty of theology at the University of Paris, consisting mostly of Dominican priests, embodied the culmination of intellectual activity at that time. Noblemen, kings, emperors, and even the pope deferred to the theological interpretations emanating from Paris.

Elsewhere in continental Europe the spread of universities as centers of learning continued. In Portugal, the king founded a university in Lisbon in 1290, which was later moved to the old Roman town of Coimbra, where it still prospers. Great university centers were founded in the Spanish-speaking cities of Salamanca (1227), Valladolid (1250), and Seville (1254). In German-speaking nations, universities were founded in Vienna (1365), Heidelberg (1386), and Cologne (1388). The oldest university of central Europe is in Prague, founded in 1348 by King Charles IV, who had studied at the University of Paris. The Jagellonian University, founded in the Polish capital of Krakow in 1364, gained a reputation in humanistic studies and in astronomy. It was here that Copernicus received his education.

In England, groups of students gathered at Oxford as early as 1167, and by 1190 a university had developed. As a result of disturbances in the town of Oxford in 1209, when the townsfolk killed several students, some scholars and students journeyed to Cambridge, and by 1281 another university was functioning there. Four faculties of arts, Church law, medicine, and theology were organized in both English universities. By 1300 Oxford rivaled the University of Paris for prestige and scholarly productivity.

The organization of the universities was a critical stage in the reawakening of European intellectual thought. Not since the Greek and Roman academies had centers of scholarship existed in western Europe. However, we must not forget that the Church, with its system of censorship, pervaded medieval societies, including the universities. Theology was recognized as the most sophisticated discipline, and the theology faculty dominated other faculties of the universities. The medieval universities had their shortcomings. Internally, the dominance of theology often restricted independent study that relied on methods of reason rather than faith. As institutions of the Church, the universities required conformity to Church discipline, so that intellectual pursuit was qualified. Both the Church and the national monarchs exerted external pressure through their control of financial support. Often those pressures were political and clearly violated the integrity and independence of the universities. Nevertheless, the early universities filled an invaluable role in the rebirth of intellectual life in western European culture. With their libraries and learned teachers, the universities attracted people to the common pursuit of intellectual activity.

SOME EMINENT THINKERS

The accomplishments of a few of the outstanding university scholars of awakening Europe attest to the significant level of learning in the medieval university. Moreover, these scholars reflected the steady questioning of Church authority by those who demonstrated the benefits of pursuing knowledge through means other than those based on faith. This movement allowed for the eventual emergence of science and the triumph of reason over faith in scholarly inquiry.

Pierre Abélard

Pierre Abélard (1079–1142) was a brilliant philosopher and one of the pioneers of the University of Paris. Born in Brittany, he eventually wandered to Paris, where he studied under the Platonic philosopher William of Champeaux (1070–1121) at the cathedral of Notre Dame. At that time a philosophical controversy raged concerning the metaphysical problem of universal beings. Plato had argued, from the presence of constant change in physical appearances, that the universal is more lasting, permanent, and therefore real, compared to particular events based upon sensory information. In other words, people change, but not humanity. Conversely, Aristotle argued that the universal is a mental idea to represent a classification for particular physical manifestations; that is, we use the concept of humanity to classify people as separate from other animals. The Church had some vested interest in the dispute because it considered itself to be a spiritual universal, greater than the sum of individual believers; that is, the Church was not simply a mental abstraction. William of Champeaux took the extreme Platonic view that universals are the only reality and individuals are only incidental manifestations of that universal. Abélard used his excellent rhetorical and logical skills to show rationally the absurdity of William's extreme position—that is, the relegation of individual persons simply to instances of a universal defies our observed order of reality in nature. Soon Abélard was teaching at the Notre Dame School as a canon of the cathedral.

However, before proceeding with his brilliant teachings, Abélard's career was interrupted by one of the most famous romantic tragedies of medieval Europe. He fell in love with the intelligent and beautiful Héloïse, niece of the head canon of the cathedral. As nature took its course, Héloïse became pregnant, prompting Abélard secretly to take her to his sister's home in Brittany, where she gave birth to their son. Héloïse refused to marry Abélard because such a formal union would preclude his ordination as a priest and a promising career as a high Church official. Rather, she preferred to remain his mistress. However, after leaving the child in Brittany, they were secretly married in Paris, although they continued to live separately. After continued fights with her uncle, Héloïse again fled—this time to a convent. Thinking that Abélard had forced Héloïse to become a nun to cover his own transgression, the uncle and some aides pounced on Abélard and, to paraphrase Abélard, severed that part of his anatomy that had done the harm. Thus, Abélard became a monk and Héloïse became a nun, limiting their future intercourse to a romantic and somewhat spicy written correspondence.

Abélard then resumed with intensity his study and teaching. He tried to place Christian thought on a rational plane and deal with the critical relationship between faith and reason. His method, based upon the ancient teaching technique of Socrates and adopted by later writers, was to exhaust all sides of an issue by outlining in question-and-answer form the logical consequences of philosophical and theological assumptions. He taught that if truth is given from God, then both faith and reason will, through parallel directions, reach the same conclusion. Abélard's efforts were an important contribution to the development of science because he relied on arguments that appealed to reason.

Abélard's writings were confined to logical discourses on the manner of knowing God and nature, and the weight of his conclusions gave legitimacy to the place of reason in the pursuit of knowledge. He did not discard faith as a source of knowledge, but his success in securing at least the acceptance of reason as coequal with faith in intellectual inquiry was a major accomplishment. Both his views and his caustic manner of presentation led Abélard to trouble with his superiors, and in 1140 he was condemned to cease teaching and writing by Pope Innocent II. Soon after, he died a lonely and bitter man. However, he introduced to medieval philosophy a systematic method that rested firmly on reason, independent of theology. Others followed who explored the implications of Abélard's teachings to their fullest.

Roger Bacon

Called by many scholars the greatest medieval scientist, Roger Bacon (ca. 1214–1292) was born in Somerset in southwestern England and spent most of his student and teaching years at Oxford University. Exposed to ancient Hebrew teaching by Jews in England, he went to the University of Paris and was further educated in ancient and modern languages. However, he was not impressed with the metaphysical and logical methods in the approach to natural philosophy or science while in Paris. After joining the Franciscan order, Bacon returned to England to teach natural philosophy at Oxford. Bacon emphasized the importance of study through systematic observation and a reliance on mathematics, his first love, to describe careful observations.

Bacon wrote extensively on philosophical and moral issues, but he was clearly at his best on scientific matters. His substantial contributions to science are rather minor and limited to some discourses on optics and on the reform of the Julian calendar. Nevertheless, his position in the emergence of scientific activity is well deserved. First, he revived interest in the ancient authors, especially mathematicians such as Euclid. Second, and more important, he stressed that empirical demonstration, based upon the deliberate observation of the physical world, will gain more than logical arguments. In other words, the validation of truth by sensory agreement among observers, aided by mathematics, holds the key to science. Thus, empiricism was reintroduced to science. Bacon's emphasis on an inductive approach in science was in contrast to the prevailing interpretation of Aristotle, which stressed definition and classification in the study of the physical world. Medieval knowledge of Aristotle's writings had emphasized logical reasoning as the primary method of demonstration—at the expense of Aristotle's views on the importance of observation. There is an underlying truth to the anecdote describing medieval monks standing in the monastery courtyard, arguing endlessly over the number of teeth of the horse, until finally, a brash young novice walked over to a horse, opened its mouth, and counted the number of teeth. Bacon's major accomplishment reinforced Aristotle's traditional teachings on the importance of observation. Thus, it was established that knowledge can be gained by both logical deduction, based on reason, and inductive empiricism, based on careful and controlled observations through the senses.

Albertus Magnus

Albertus Magnus (ca. 1193–1280) was a Dominican scholar who worked in schools and monasteries in Germany and also spent two periods of time at the University of Paris. He was one of the first western European scholars to review completely all of the known works of Aristotle, a daring accomplishment for a

Christian intellectual at a time when Aristotle's writings were considered heretical by the Church. Albertus Magnus wrote extensively on Aristotelian logic as the basis of correct reasoning and proposed six principles of logical inquiry. The metaphysical treatises of Albertus included an evaluation of the interpretation of Aristotle's views offered by the Muslim scholar Averroës (1126–1198), who taught in Spain and Morocco about the relationship between Aristotle's philosophy and the tenets of Islam. In addition, Albertus produced books on ethics, politics, and theology. Interestingly, Albertus' writings on psychology were rather comprehensive, dealing with such topics as sensation, intelligence, and memory. Although Albertus did not add much original work on psychology, his reliance on Aristotle as an authority was a significant step in itself. Moreover, he took Aristotle's mind–body dualism and related the potential of the soul to the Christian ethic of seeking eternal salvation. The end result of this merger, contained in his treatise *De Potentiis Animae* (*On the Powers of the Soul*), was a proposal for a dynamic psychology of human striving for goodness and intellectual fulfillment in the knowledge of God. Albertus' views on psychology were innovative because he elevated human rational powers as a source of salvation in addition to faith.

Albertus Magnus was a brilliant and prolific scholar who successfully ignored the intimidating censorship of the Church and drew inspiration from non-Christian resources of scholarship. He championed exact observation and his detailed studies of plant life contributed to the science of botany. He visited areas in Europe or corresponded with other observers to classify and describe varieties of fauna. As a naturalist, he reinforced Bacon's teachings on the importance and efficacy of careful empirical observation. Moreover, as one of the first Christian scholars to rely on the work of the pagan Aristotle, Albertus provided a refreshing source of intellectual stimulation that was to have a tremendous impact on the awakening of scholarly pursuit.

SCHOLASTICISM

One of Abélard's students, Peter Lombard, used the question-and-answer method in an influential book, *Sententiarum Libri IV* (*Four Volumes of Opinion*), written around 1150. His book attempted to reconcile the Bible with human reason. This work advanced the use of reasoning in addition to faith as a source in the pursuit of knowledge and became a classic of Christian theology. At the same time, Aristotle's works, translated from the Arabic, reached wider audiences in the universities, especially the University of Paris. Albertus Magnus had set the stage by liberally using Aristotle's teachings to explain nature and human psychology. However, it was a Christian age, and faith was dominant. Somehow, Aristotle's teachings on metaphysics and on the soul had to be systematically reconciled with Christian theology. This task was accomplished by Thomas Aquinas. The results are known as Scholasticism, which opened the door for the life of the mind by admitting human reason, along with faith, as a tool for seeking the truth. Indeed, it may be argued that the very admittance of reason spelled the end of the dominance of faith as the source of human knowledge.

Thomas Aquinas

Saint Thomas Aquinas (1225–1274) was born of a German father and a Sicilian mother descended from Norman invaders, in his father's castle near the town of Aquino, which lies within sight of the great Benedictine abbey of Monte Casino, where he received his early schooling. He grew to such robust proportions and exhibited such quiet studiousness that he was dubbed the "dumb ox of Sicily." No appellation could be further from the truth. In 1882, Pope Leo XIII commissioned a group of Dominican priests to assemble a compendium of Aquinas' writings. After well over a century of continuous effort and many changes in personnel, the group is still working and is far from finished. Aquinas joined the Dominicans in 1244, and by the following year he was studying with Albertus Magnus in Paris. He spent most of his remaining years teaching in Paris and occasionally in Italy. During all of this time he consistently defended reason against those arguing for faith alone as the source of truth. To prove his thesis he attempted to reconcile Aristotelian philosophy with Christian thought, a goal similar to Augustine's reconciliation of Plato with Christianity 800 years earlier. In 1272, the ruler of Naples, Charles of Anjou, asked Aquinas to reorganize the university there. Shortly after beginning this effort, Aquinas reportedly received a vision showing the completeness of divine knowledge, causing his own comprehensive work to have only minimum value, and he ceased writing. He was known throughout Europe as a learned man, yet he had great humility and gentleness, and he continually asserted his own inadequacies. The scope and depth of Aquinas' scholarship mark him as one of the outstanding intellectuals of Western culture.

His greatest work, the *Summa Theologica* (*A Summary of Theology*), represents the embodiment of Christian thought in its most detailed and comprehensive presentation. Underlying his system is Aristotelian logic. Thomas used the logic of the question-and-answer method to arrive at the essential truth of God. He took

THOMAS AQUINAS (1225–1274). (Courtesy, New York Public Library.)

Aristotle's metaphysical principles of matter and form and described a dynamic relationship between the body and the soul, Christianizing the system in the process. A diagrammatic presentation of Aquinas' relationship of body and soul is shown in Figure 4–1, which expresses the basic features of Aristotelian dualism. In this description, the person is defined in terms of essence and existence. The essence of a person is the universal that classifies the nature of all people. It is composed of the physical world, from which the body is derived, and the soul, which is immortal and has primary functions of intellect and will. From the potential to the principle of actualization, a person's existence defines her or his individuality. Thus, the person consists of necessary bodily and spiritual constituents, whose dynamic interaction results in the sharing of humanity, expressed individually.

According to Aquinas, the human person is not simply a physical machine propelled by external stimuli or environmental pressures. Neither is the person a soul imprisoned within a body, as Plato and Augustine taught. Rather, the person is a dynamic entity, motivated internally by the soul. The human soul possesses five faculties, or powers:

1. Vegetative: concerns the functions of physical growth and reproduction. To accomplish growth, the organism seeks food and nourishment.
2. Sensitive: concerns the ability of the soul to accept information about the external world through the five senses.
3. Appetitive: concerns the desires and goals of the organism and the ability to will.
4. Locomotive: describes the capacity to initiate motion toward desirable goals or away from repulsive environmental objects.
5. Intellectual: relates to the power of thinking, or cognition.

Aquinas' psychology, then, contained two key elements of human learning. First, there is an environmental dependency, in that our knowledge is based on input through the senses. However, the sensory input does not enter an empty, passive intellect. Rather, sensory knowing is acted on by the second element, *sensus communis*, or the center of common sense, which actively organizes, mediates, and coordinates the sensory input. Thus, Aquinas argued for two types of knowledge. Sensory knowledge, shared with other animals, provides information about physical reality, whereas human reason provides abstractions of universal principles. The

FIGURE 4–1 A diagrammatic representation of Aquinas' concept of the dualistic relationship between body and soul, constituting the nature of the human person.

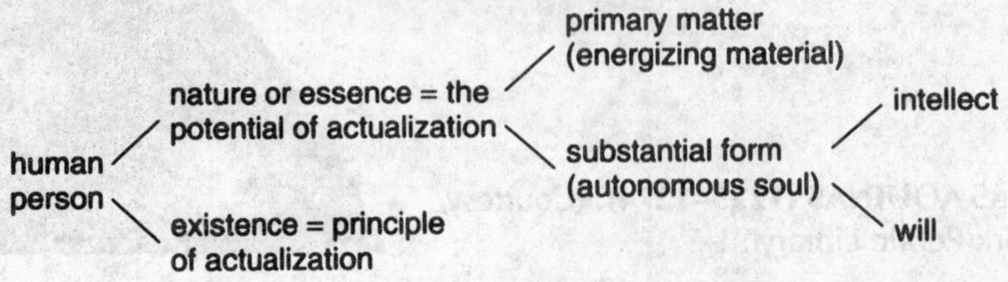

soul, then, accomplishes the highest and most powerful form of human activity through the rational intellect. It is the capacity to reason that makes humanity unique and also joins humanity to God. Moreover, although freedom motivates the will, human freedom lies in the intellect. Freedom increases with increased rationality, wisdom, and knowledge. The pursuit of wisdom is the highest calling of people, and the act of understanding characterizes the proper human state.

The motivational factor in Aquinas' psychology is the will, comprising the critical force of growth and movement. The intellect is subordinated to the will because the will determines the direction of the intellect. The proper end, or goal, of the will is to seek goodness. Although God is the ultimate good, according to Aquinas, the will seeks intermediary goodness in the form of earthly beauty, harmony, and organized proportion, which are pleasing to the soul. The autonomous soul, composed of intellect and will, is a unified, active entity that is dependent on sensory input while serving as the supreme arbitrator of sensory knowledge.

Aquinas made a significant step toward the emergence of modern science when he spelled out how we can know ourselves, our environment, and even God. He dismissed the ideas of those philosophers influenced by Plato, such as Augustine, who stressed the unreliability of sensory information. Rather, knowledge is a natural product of the bodily senses. Limited as it is to the constraints of natural laws, sensory knowledge is nevertheless trustworthy. Further, by our own powers of reason, we may derive supersensory knowledge, such as the existence of an impersonal First Cause or a Prime Mover, through analogy based on sensory knowledge.

It is beyond our scope in this book to present a complete overview of all of the Thomistic contributions to Scholasticism; we must be content with some conclusions. Thomas Aquinas and the Scholastic movement represent a transition to the emergence of science. Prior to the Scholastics, Aristotle was viewed as suspect, if not outright heretical. After Aquinas, Aristotle's teachings were mandatory in Christian universities. Using that criterion, Aquinas was successful in reconciling Aristotle with Christianity. More important, he completed the intellectual justification that elevated reason to a level with faith as a source of truth and knowledge. This contribution was critical to the emergence of science. By accepting Aquinas' defense of reason, the Church also accepted a new set of rules for evaluating intellectual activity. In a very real sense, Aquinas' strategy of studying theology inadvertently left the Church vulnerable to scrutiny under a new standard, which is reason. Before the triumph of Scholasticism, the Church held its authority on faith, based on Scripture and the revelation of Christian tradition. After the work of Aquinas was accepted, the Church was required to respond to rational arguments.

The Papacy and Authority

At this point, it is important to summarize briefly the position of the Church and the papacy in the late medieval period. Superficially, the power and authority of the Church were at their zenith, present in every facet of life. However, two forces were operating that eventually undermined the power of the Church and led to the questioning of papal authority. The first was Scholasticism itself, which destroyed the effectiveness of Church dictates based on faith alone. The medieval scholars, culminating in the work of Aquinas, had secured the place of reason and demonstration in the seeking of truth and knowledge. After the admission of Scholastic thought to Church teaching, the papacy could not successfully revert to the demands of obedience justified by faith alone. The acceptance of rational argument imposed a serious restriction on the basis of traditional papal authority. Subsequent intellectual progress in science made the foundation of papal authority on faith increasingly untenable, and papal authority steadily eroded during the following 500 years. Finally, in 1870, the First Vatican Council declared that, as part of Roman Catholic doctrine, the pope is infallible when teaching on matters of faith and morals. Although contemporary political events contributed to the reactionary clerical forces that devised the 1870 statement (see Chapter 11), it is important to recognize that papal infallibility is a formal policy based on faith. Interestingly, the very need for the Church to make such a pronouncement in 1870 stemmed from a process that began in the thirteenth and fourteenth centuries: Admitting reason as a source of truth precluded exclusive reliance on faith.

The second force eroding papal authority was political. The rising nation-states of western Europe competed with the papacy and other Church institutions for revenues. When temporal authority was centralized in the monarchy in France, money was needed for government and military operations. After exhausting available resources, the monarchy began taxing the wealth of the Church in France. This issue led to a church–state confrontation. Philip IV (Philip the Fair) of France seized Pope Boniface VIII in retaliation for the pope's dictum against Philip's taxation of Church property. Boniface died shortly after his abduction, and Philip succeeded in getting a French bishop elected pope as Clement V. Fearing the reprisals of pro-Boniface forces in Rome, Clement settled in 1309 at Avignon, a Church-owned town in France, near the Italian border. There the next six popes, all French, remained until 1377. The period of the Avignon papacy was marked by the depths of corruption. To support the sumptuous papal court, anything could be bought, from bishoprics to dispensations from Church law to indulgences. Moreover, the return to Rome after the Avignon papacy was accompanied by a schism within the Church, caused by rival popes, with each claimant having the support of competing political factions. England, Flanders, most of the German states, Poland, Hungary, Bohemia, and Portugal followed the Roman pope, Urban VI; France, Naples, Spain, and Scotland declared for the Avignon pope, Clement VII. In an attempt to resolve the conflict, a Church council held at Pisa elected a compromise pope. However, when the Roman and Avignon popes refused to yield their claims, there were three rival popes. The confusion was finally settled by the Council of Constance (1414–1418), and after 39 years of schism the Church was once again unified under a single pope in Rome—Martin V—who was accepted by all Western countries. However, during the schism the papacy had become a political ploy, and political corruption was rampant. Although the Church was reunited, the papacy never completely recovered its prestige and authority.

THE ITALIAN RENAISSANCE

From the end of the fourteenth century to the beginning of the sixteenth century, a remarkable event of profound cultural significance occurred in Italy (Map 6). This rebirth, or Renaissance, of European culture was characterized by a turn toward humanism in art, literature, and music. It marked a change in emphasis from the dominance of traditional Christian themes to a glorification of humanity, often reverting to styles of painting and sculpture reminiscent of Epicurean Rome.

Beginning in Florence under the beneficial patronage of the Medici family, the rulers of that city, a new wave of artistic accomplishment spread throughout Italy. Florence itself was beautified by the architectural achievements of Brunelleschi and Verrocchio, and civic buildings, churches, and palaces were filled with the sculpture of Ghiberti and Donatello. Paintings of humanistic and religious subjects by Fra Angelico, Ghirlandaio, and Botticelli attracted students from all over Europe. Under the enlightened rule of Lorenzo de Medici, the Magnificent (1449–1492), Florence became the focus of the Italian Renaissance in art, music, and literature.

The genius Leonardo da Vinci (1452–1519) typified the "Renaissance person" by the depth of his abilities as painter, inventor, and scientist. Beginning in Milan, where his Last Supper may still be admired, the Renaissance traveled with da Vinci through Florence and finally to Rome. It was da Vinci, the scientist and engineer, who translated the humanistic spirit to anatomical drawings and machine designs. His genius captured the imagination of Europe, elevating the physical reality of the body in the world to new visions of capabilities. Equally competent as an inventor, da Vinci's curiosity led him to develop such diverse inventions as a machine gun, a machine for cutting screws, and an adjustable monkey wrench. His fascination with physics drew him to studies of motion and weight, and he devised sophisticated experiments in magnetism and acoustics. In his study of anatomy, he made systematic comparisons between the structure and mechanics of human and animal limbs. Da Vinci's genius has been called one of the most remarkable of all time. His imagination and compulsion for perfection blended well with the spirit of the Renaissance, and he demonstrated the vast potential for the free exercise of human intelligence, which subsequently served as a model and an inspiration.

Throughout Italy, in Mantua, Ferrara, Naples, and Venice, the entire peninsula was infused with the excitement of the Renaissance. In Rome, supported by the patronage of powerful popes such as Julius II (reigned 1503–1513) and Leo X (reigned 1513–1521), the classical art of the ancient world was recovered and imported to adorn buildings and avenues. To insure a return to the glories of old, new construction was undertaken, most notably Saint Peter's Basilica and the palaces of the Vatican. The immense dome of Michelangelo Buonarroti (1475–1564) crowned the physical glory of the capital of Christendom, and his *Pietà* remains to adorn its entrance. Moreover, in his *Last Judgment*, painted over the walls and ceiling of the Sistine Chapel in a palace adjoining the basilica, the full magnificence of the Renaissance shines forth. These 200 years produced an awesome statement confirming that Europe had indeed emerged from the Middle

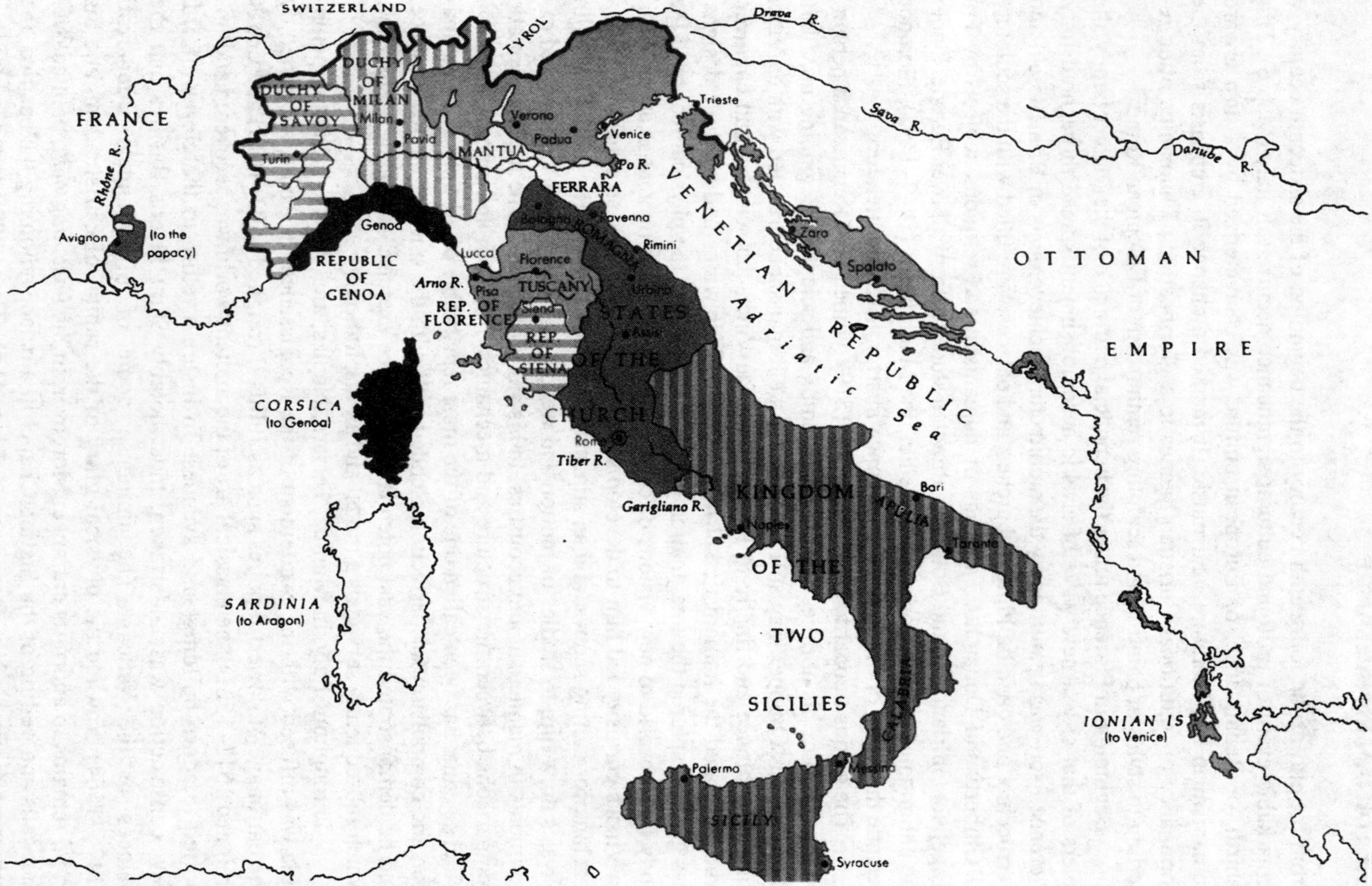

MAP 6. RENAISSANCE ITALY ABOUT 1494.

Ages and that a new age of enlightenment had arrived. Italy still capitalizes on the glory of this era of cultural achievement as tourists annually flock to gaze in wonder at the creations of men.

The Italian Renaissance spread throughout Europe. As far away as Krakow in Poland, there are splendid tombs in the Wawel Cathedral, designed and executed by Italian artisans, reflecting specific characteristics of the Florentine Renaissance. The spread of Renaissance ideas was facilitated by improvements in printing. By the early fifteenth century, Venetian printers were producing books of superb quality. In the German city of Mainz, Johann Gutenberg (1400–1468) perfected movable type, a process used for centuries by the Chinese, ending the age when access to written resources was limited to a select few. The spread of reborn cultural influences served to complement the rational spirit developed by Scholasticism. Together, these movements set the stage for a new strategy in the pursuit of human knowledge and values, a strategy that involved the emergence of science. However, the authority of the Church remained a challenge that had yet to be finally resolved.

THE CHALLENGE TO AUTHORITY

The combined developments of the elevation of reason by the Scholastics, the political divisions of the papal schisms, and, most notably, the Renaissance were radically changing European society. These forces had a negative effect on the authority of the Church and, in particular, on the power of the papacy. The direct challenge to Church authority occurred in two parallel spheres—from within the Church itself and from external intellectual progress. The former consisted of the Protestant Reformation, which questioned the structure of the Church and threatened the very survival of the Roman institution. The latter hurled a direct challenge against Church teachings, pitting knowledge based on reason against knowledge based on faith. Both developments seriously undermined the strength of Church authority in European society.

Within the Church: The Reformation

The causes of the Protestant revolt against papal authority, leading to doctrinal disputes, have been extensively argued by historians. Certainly we are not able to add much to their analyses, but a few observations are appropriate. First, the political scene contributed greatly to strife between the papacy and national states, especially in Germany and England. The popes, who so generously supported the Italian Rcnaissance, were true temporal leaders, controlling vast stretches of estates in central Italy and maintaining large armies. This temporal power, coupled with the medieval prerogatives of the Church in all Christian nations, created a perception of the papacy as threatening to the emerging national identity and the monarchical consolidation proceeding quite steadily in individual European nations. Accordingly, the political climate encouraged any dissension within Christendom that would work to the detriment of papal authority.

A second contribution to the Reformation concerned the revived intellectual atmosphere of Europe. One by-product of the rise of the universities, the elevation of reason by the Scholastics, and the cultural innovations of the Renaissance was the creation of an intelligentsia that had access to, and understanding of, ancient pagan and early Christian documents. These scholars focused on the abuses of Church authority, such as the selling of indulgences to underwrite the construction of Saint Peter's Basilica. The errors of such practices—departing as they did from early Christian teachings of antimaterialism—became more evident to early reformers. This conclusion occurred to Martin Luther and formed the basis of his indictment of the Church. Luther certainly did not seek an alternative in the study of reasoning processes; rather he reverted to a greater reliance on faith. Nevertheless, it was the intellectual spirit and freedom of the times that permitted him to follow his convictions.

A third force supporting the Reformation was the interest in humanism, so beautifully articulated in the artistic productivity of the Italian Renaissance. Of interest to us is the humanistic attitude in philosophy expressed most clearly by Desiderius Erasmus (1469–1536). He was born in Rotterdam, and his early inclination toward the scholarly life led him to take vows as an Augustinian priest in 1492. As a scholar and teacher, his career took him all over western Europe and put him in touch with the major figures of his time. He studied at Paris, Oxford, and Louvain University in Belgium. He was a great friend of Thomas More, a future chancellor of England, and he tutored the sons of Henry VII while they visited Bologna. His satirical *The Praise of Folly* mocked the hypocrisy of contemporary moral life, and he lectured the monarchs of Europe in his *Education of a Christian Prince*. Although his greatest contribution may lie in his revision and translation of the New Testament from Greek and Latin, it was his notes compiled while pursuing this work that proved to be a popular resource of his views as a humanist. Despite some errors contained in the work, Erasmus presented a critical study of the basis of faith from Scripture in light of rational scholarly scrutiny. As such, Erasmus revealed the context of the writers and the pitfalls of subsequent interpretations. Although the Church reforms of the Council of Trent (1545–1563) condemned this translation and critique, Erasmus' work showed that even the most sacred resource of faith could be better understood in terms of the human context of its authors.

The leaders of the Reformation on the Continent, Martin Luther (1483–1546), John Calvin (1509–1564), and Ulrich Zwingli (1484–1531), were motivated by a sincere desire to correct the abuses of Church authority. In the process, they and their followers questioned the tenets of Roman teachings. From that point, European Christianity became fragmented, and society was divided along religious lines. Similarly, in England, political dissension between King Henry VIII and Pope Clement VII led to England's separation from Roman direction, although it still retained the basic doctrine of Catholicism. The Church attempted to minimize losses through internal reform, and the Council of Trent (1545–1563) succeeded in restoring discipline in the clergy and an evangelical spirit in the Church itself, which was represented by the newly founded Society of Jesus, or Jesuits, in 1540. Nevertheless, the Reformation marked the definitive end to the medieval society that fell under the authority of the pope.

Within Science: The Copernican Revolution

Aside from the internal dissension of the Church, the times were ripe for a direct challenge to the authority of the Church through the acceptable strategy of reasoned arguments. Since ancient times, the problem of motion in the universe had baffled scholars. The resolution that prevailed up to the Renaissance was the Ptolemaic, or geocentric, explanation, which placed the earth at the center of the universe. Such a view fit well into certain religious and theological considerations, including Christianity, because an earth-centered universe placed humanity in a unique position as the special creation of God.

Nicholas Copernicus (1473–1543) was born in the merchant town of Torún in northwestern Poland and sent to Jagellonian University in Krakow to study for a clerical career. Not satisfied with Scholasticism, he began to pursue mathematics and astronomy. Some of his notes and primitive instruments are still preserved in the museum of that university. He eventually journeyed to Bologna, where he studied law and medicine, and around 1512 he settled as a cathedral canon in northern Poland near the Baltic Sea. There, he practiced medicine and advised the Polish monarchy on matters of currency reform. All this while, he was studying the question of planetary motion, using data reported by the ancients. His findings led him to conclude

NICHOLAS COPERNICUS (1473–1543). (Courtesy, New York Public Library.)

that a heliocentric, or sun-centered, universe provided a simpler, more parsimonious explanation of planetary motion. He tested his ideas in correspondence with various scholars throughout Europe. His major findings, contained in *De Revolutionibus Orbium Coelestium* (*On the Revolutions of Celestial Planets*), were not published until the year of his death.

It should be remembered that Copernicus did not present convincing new evidence. Indeed, it was left to later scholars with better instrumentation, such as Kepler, Galileo, and Newton, to provide empirical observations to support the heliocentric theory. Rather, Copernicus used the basic tools of logic, through mathematical demonstration, to provide his simpler explanation of planetary motions—tools sanctioned by the Church after the success of Scholasticism.

Although Copernicus was encouraged by certain Church authorities, such as Leo X, and he dedicated *De Revolutionibus* to Pope Paul III, Church leaders soon recognized the danger inherent in his work. *De Revolutionibus* was placed on the Index of Forbidden Books in 1616. The theological implication of this work meant that humanity was a part of a small planet in a vast universe. This realization had a far-reaching effect on Church history, requiring a reappraisal of the position of humanity in relation to the rest of the universe and to God. Indeed, the conclusion of Copernicus was truly a revolution in thinking that continued through the rise of modern science, culminating in the writings of Darwin, who firmly placed humanity in the natural order, subject to the same constraints and determinants as other species of life. Copernicus challenged the authority of the Church, which had supported the Ptolemaic theory. The opposition between faith in the Ptolemaic view and rational justification of the heliocentric theory was ultimately resolved in favor of the latter. Thus, the Copernican revolution may be interpreted as the beginning of the drift away from an emphasis on God and the hereafter to an examination of humanity existing in the environment of the present.

CHAPTER SUMMARY

The five centuries from 1000 to 1500 saw the consolidation and then the fractionation of Christianity and the decline of the papacy. The authority of the Church faced serious challenges. On a political level, the rising nation-states of Europe successfully competed with the papacy and undermined both the temporal and the spiritual powers of the Church. On an intellectual level, the teachings of Pierre Abélard, Roger Bacon, and Albertus Magnus led to a revival of interest in the ancient writers with their emphasis on rational thought to secure human knowledge. It was Thomas Aquinas who reconciled Aristotle's rationalism and Christian theology, which resulted in the Church's accepting both reason and faith as sources of human knowledge. On a cultural level, the Italian Renaissance lifted Europe into a new era of humanism that glorified humanity and shifted attention to the present needs and desires of people. Erasmus translated this humanistic attitude into scholarly pursuits that revealed the frailties and needs of the human authors of Scripture. All of these forces eroded the authority of the Church, leading to dramatic confrontation, both from inside and outside the Church. The Protestant Reformation took advantage of the rift between Christian monarchs and the papacy, successfully fragmenting the unity of Western Christendom. However, it was Copernicus who used the strategy and tools of reasoned arguments to arrive at his heliocentric theory of planetary motion. This bold assertion successfully demonstrated a truth arrived at through reason that differed from the conclusion supported by the authority of the Church. As a result, reason triumphed over faith, and the age of science began.

BIBLIOGRAPHY

Primary Sources

ERASMUS. (1941). *The praise of folly*. Princeton, NJ: University Press.

MCKEON, R. (Ed.) (1929). *Selections from medieval philosophers*. New York: Scribner's.

AQUINAS, T. (1945). *Summa Theologica* (A. Pegis, Trans.). In *Basic writings of Thomas Aquinas*. New York: Random House.

Studies

CROMBIE, A. G. (1959). *Augustine to Galileo*. New York: Anchor.

DIETHELM, O. (1970). The medical teaching of demonology in the 17th and 18th centuries. *Journal of the History of the Behavioral Sciences, 6,* 3–15.

DURANT, W. (1953). *The Renaissance*. New York: Simon & Schuster.

DURANT, W. (1957). *The Reformation*. New York: Simon & Schuster.

JACKSON, W. T. H. (1962). *The literature of the Middle Ages*. New York: Columbia University Press.

KIRSCH, I. (1978). Demonology and the use of science: An example of the misperception of historical data. *Journal of the History of the Behavioral Sciences, 14,* 149–157.

KUHN, T. S. (1959). *The Copernican Revolution: Planetary astronomy in the development of Western thought*. New York: Modern Library.

TUCHMAN, B. W. (1978). *A distant mirror: The calamitous 14th century*. New York: Ballantine Books.

5

The Emergence of Modern Science

Psychology was recognized as an independent science by the end of the nineteenth century. During the preceding two centuries, models were developed of what psychology should study and how such study should be conducted. Specifically, during the seventeenth and eighteenth centuries, competing models of psychology vied with each other for dominance. We deal with this very important period, under the structure of national movements advancing philosophy and science, in Chapters 6 through 9. The present chapter sets the intellectual background, especially in the physical sciences, for the articulation of models of psychological inquiry. In approaching the intellectual background of science, we first consider particular themes and issues to pursue, then backtrack to follow through on another theme. This approach is necessary to deal with the volume of material, but it is also somewhat artificial; therefore, we must remember the simultaneity of events, despite their successive presentations.

The sixteenth and seventeenth centuries witnessed the successful demonstration of the value of empirical science. The empirical disciplines triumphed at the expense of speculative approaches, particularly metaphysics. Recalling Comte's hypothesis from Chapter 2 concerning the stages of intellectual progress, we may consider the sixteenth and seventeenth centuries as the transitional phase of the post-Renaissance development of empiricism. Somewhat paradoxically, it may be argued that the downfall of Aristotelian metaphysics, resulting from the rise of empiricism, was initiated by the Scholastic reliance on reason as a source of knowledge, which in turn was based on Aristotelian teachings. In other words, the Scholastic elevation of reason as a source of knowledge made possible the efficacy of observation, which is the basis of empiricism. Accordingly, Aristotelian philosophy, supported by the Scholastic affirmation, was a comprehensive system that accommodated both metaphysical and empirical approaches.

ADVANCES IN SCIENCE

Post-Copernican advances in science and mathematics were crucial to the eventual success of science. With the downfall of Church authority based upon faith, the age of reason began; the human intellect was valued and used to generate knowledge. Accordingly, a trend started that witnessed the triumph of science. In a very real sense, science, based on reason, was viewed as a replacement for religious doctrine, based on faith. Science and scientific methods were valued as the best approach to any area of investigation. This trend culminated in the nineteenth century, when physics was seen as the queen of the sciences, and the more closely any discipline emulated physics, the greater the value placed on that disciplinary inquiry. Thus, whereas Niccolo Machiavelli (1469–1527) in *The Prince* used common sense and compelling logic to analyze the principles of effective leadership, Marx and Engels, writing in the nineteenth century, were able to look to their "age of the proletariat" as a utopia of economic and political life derived scientifically. Machiavelli is viewed as an artist; Marx and Engels are called political scientists. A similar movement occurred within psychological inquiry, as we shall see, so that one of the models of psychology in the nineteenth century was almost bound to emulate physics. At any rate, we shall now review some of the major people and events that formed the foundation of assumptions and methods common to all approaches that we call scientific.

Francis Bacon

One of the most influential, colorful, and brilliant men of Elizabethan England, Francis Bacon (1561–1626), was born in London to the Lord Keeper of the Great Seal for Queen Elizabeth I. Prone to the solitary life of a scholar but finding that he was left with little wealth at his father's death, Bacon had to rely on his own resources for support. With his birth and lineage, as well as his superb education in law, literature, and diplomacy, Bacon yearned for a secure political appointment, but that eluded him until James I, who ascended to the throne in 1603, appointed Bacon Solicitor General in 1613. His political career accelerated, and by 1618 he was Chancellor of England. However, Bacon was caught in the crossfire of a movement to discredit the king, and his position was threatened in 1621, when he was impeached for corruption and abuse of his offices. Pardoned from prison and a heavy fine, Bacon retired a wealthy man to pursue his interests in philosophy and science.

FRANCIS BACON (1561–1626).
(Courtesy, New York Public Library.)

In his writings, Bacon's essential goal was to reorganize the approach to scientific study. Although Aristotle and the Scholastics recognized both deductive and inductive logical reasoning, Bacon noted that the deductive was stressed at the expense of the inductive. In other words, the traditional approach to science created a rigid mental set that limited investigation of the person in the environment to a relatively sterile procedure filled with a priori assumptions. From the assumed nature of humanity (that is, the relationship between body and soul), particulars of human life or the physical world were deduced. Bacon believed that the validity of this approach was limited to the extent that the underlying assumptions were correct or relevant.

In his work *Novum Organum* (*A New Instrument;* 1620), Bacon called for better situations to study the world directly—more laboratories, botanical gardens, libraries, and museums. By eliminating any preconceived notions of the world, the scientist could then study people and the environment using detailed and controlled observation. On the basis of such observations, best expressed quantitatively, cautious generalizations could be made. Thus, Bacon emphasized an approach to science that stressed practical observation as the primary setting for scientific inquiry.

Bacon held that the method of science must be predominantly inductive, proceeding from particular to general. Moreover, he qualified this position by building in several critical elements for scientific inquiry. First, the scientist's study of particulars must be done through observation. Sense validation of quantitative observations becomes an important source of agreement among scientists. In other words, if one scientist describes a particular event by observing its measurement, a second scientist could repeat the observation and support the first scientist. Presumably, if enough scientists agree about a certain observation, this agreement itself represents a compelling argument for the validity of a finding. A second implication of Bacon's method was that scientists must rid their inquiry of any influences not derived from observation. Thus, the scientist must be skeptical and not accept formulations that cannot be tested through observation. Rather, the scientist must take a critical view of the world and proceed carefully with the study of observables. Bacon, then, presented a strong statement of empiricism as the basis of science. Scientists must experience particular events through their ability to observe them via sensory processes. Moreover, Bacon indicated that observations have no value if made casually and carelessly. Rather, he advocated the goal of controlled observations. Thus, Bacon's empiricism was expressed in terms of a systematic inductive method.

As a scientist, Bacon was concerned about the process of discovery and the demonstration of discovery in the generation of new knowledge. Empirical science for Bacon was a new, refreshing approach to the ageless puzzles of the universe. Bacon's views on scientific methods required a dependence on sense information about environmental events. This perspective in British science became a prevailing theme and formed the basis of the subsequent British empirical tradition of psychology, which is presented in Chapter 7.

Galileo Galilei

Advances in astronomy and mathematics provided the quantitative foundation for Bacon's methodological innovation. It was Galileo Galilei (1564–1642) who asserted that science is necessarily synonymous with measurement. Born in Pisa, Galileo received a sound education in the classic languages and mathematics from his Florentine father. While at the University of Pisa, he was introduced to the mathematics of Euclid, which opened a new world for him. At age 25, he was appointed to the professorship of mathematics at Pisa. In 1592, he left for a teaching position at the University of Padua, where he set up a laboratory for experiments in physics. His findings on the velocity of moving objects were later confirmed and elaborated by Newton.

In 1609, Galileo built his first telescope, then constantly improved the magnification. He made accurate observations of stellar constellations, the moon's surface, and sunspots. His colleagues at Padua would not accept Galileo's discoveries, and he left the university for a secure stipend provided by the grand duke of Florence. In his writings and lectures, Galileo asserted that only the heliocentric explanation could account for his astronomical data, and he accepted the Copernican system as fact. This assertion brought him to the attention of the Jesuits, the newly founded society of militant scholars determined to defend papal authority. Convicted by the Inquisition, Galileo was forced to recant publicly his belief in the Copernican system, and he was allowed to retire to Florence. However, he continued to study and made significant contributions to mechanics and astronomy.

Although Galileo had the misfortune to work in a place under the jurisdiction of the Inquisition, his trial and conviction enhanced his popularity in Protestant northern Europe. His works were widely read and acclaimed, and he secured acceptance of the

Copernican teachings. Galileo's synthesis of science and mathematics extended beyond the limits of Copernican astronomy. Galileo's studies led him to the view of a mechanical world inhabited by mechanical people. His telescope was essentially a mechanical extension of the senses. This interpretation of human activity had enormous implications for psychology. First, it implied that human activity itself is ultimately subject to mechanical laws. Second, the emphasis on mathematical relationships in the universe suggested that it is profitable to examine external, environmental forces as the source of human activity, rather than limiting examination to internally generated sources, such as the Scholastic interpretation of the will as the motivational principle of human existence. Galileo distinguished the world in terms of primary qualities, which are unchangeable and quantitative, and secondary qualities, which are fluctuating, unstable characteristics knowable through the senses. Primary qualities, such as motion, position, and extension, are subject to mathematical relations and descriptions, whereas secondary qualities, such as colors, sounds, and tastes, are elusive and reside in the consciousness of the perceiving person. Ultimately, he thought scientific discovery might allow secondary qualities to be expressed in terms of the mathematical relationships of primary qualities. Galileo's work drew a sharp dichotomy between science and religion, reinforcing separate and rival interpretations of life.

Johann Kepler

Johann Kepler (1571–1630) was another scientist whose experiments supported the Copernican view. Born in Germany, Kepler spent his most productive years in Prague. He had a brilliant mind and constantly wondered about the universe, generating a myriad of hypotheses. His major findings included the proof of elliptical planetary orbits, rather than the circular orbits proposed by Copernicus. His discovery that planets closer to the sun have a more rapid orbit than those further away anticipated Newton's findings of gravitation and magnetism. Kepler consistently wrote of the harmony and order of the universe that impressed him so much. Moreover, he provided the detailed mathematical proof required by the Copernican system, thus winning its further acceptance.

Like Galileo, Kepler believed in the fundamental mathematical basis of the universe. His empirically derived mathematical laws of planetary motion convinced him that the mathematical basis of the physical world must have parallel expression in other levels of reality, such as the psychological world of the person. Kepler also studied vision and made important contributions to our understanding of binocular vision and visual accommodation. He proposed critical hypotheses that were tested eventually in the nineteenth-century psychophysics movement that served as an immediate precursor of modern psychology. Finally, Kepler provided strong evidence to support the distinction between primary and secondary qualities in the world—the former being absolute, immutable, and objective; the latter, relative, fluctuating, and subjective. This distinction was a source of controversy in the subsequent development of models of psychological inquiry.

Isaac Newton

A mathematical genius who formulated the basics of modern physics, Isaac Newton (1642–1727) represents the climax of the scientific development started by Copernicus. Born in the midlands of England, Newton entered Trinity College of Cambridge University in 1661 and began his studies of mathematics, astronomy, and optics. In 1669 he was appointed professor of mathematics at Cambridge and remained in that post for 34 years. Following in the spirit of Francis Bacon, Newton devised a methodological strategy that attempted to remain true to the level of observations without going beyond what they could directly support. Accordingly, he approached a problem by thinking of every possible solution, and then tested the mathematical and experimental implications of each hypothesis.

Newton is credited with devising calculus, independently of the German philosopher Leibniz, whom we consider later. As a physicist, Newton used his mathematical tools to investigate light. In 1666, viewing light projected through a prism, he discovered that white light is actually a compound of the color spectrum. However, Newton's most significant contributions are contained in his *Principia Mathematica (Principles of Mathematics)* (1687), a classic in the history of science. His views of the mechanics of the world were spelled out in the three laws of motion:

1. Every object remains in a state of rest or steady motion unless acted on by external forces.
2. The change of motion is proportional to the external impressing force and is made in the direction of the straight line in which that force is impressed.
3. To every action there is an equal and opposite reaction.

Newton then developed the principle of gravitation and applied it to the planetary system, greatly elaborating on Kepler's work and proposing a mechanical model of the universe. Thus the Copernican system was completed.

ISAAC NEWTON (1642–1727).
(Courtesy, New York Public Library.)

Newton's mechanical conceptualization of the universe embodied the revolutionary conclusion of complete determinism. He provided evidence and derived formulations that described the orderly nature of matter. His physics of matter offered a framework for examining its transformation, leading to the study of gases and the chemical elements. The conservation of matter formed a basis for the investigation of the relationships of mass and weight and eventually led to the development of molecular theory and the study of the transformation of forces. The success of Newton's research supported the physical basis of biology and ultimately reinforced views focusing on the mechanical laws that govern living organisms.

Newton's method, firmly based on observation, embodied three rules of reasoning to guide empirical investigations:

1. Causal explanations of observed events are confined to the observed events and nothing more.
2. The same causes are responsible for the same observations.
3. The guiding logic of empirical investigation is inductive, which provides explanations that may be accepted until new observations call for modifications of explanations or new hypotheses.

Accordingly, Newton advocated a close adherence to observation and careful induction. Casual generalizations and all speculations are to be avoided. The first stage of scientific inquiry is skepticism, and subsequent stages are guided by observations.

Newton's views were not universally accepted. Many religious leaders attacked his predominantly mechanical view of the universe, which left little room for God. Indeed, their perceptions were valid, and Newtonian science provided new criteria for evaluating the products of intellectual scholarship. Newton did, however, enjoy wide prestige, and in his old age he was praised as the greatest living scientist. From our perspective in the history of psychology, Newton's work had a powerful implication. By careful observation and sophisticated quantification, Newton examined the most immense problem in all of the physical world—namely, the relationships among heavenly bodies—and he showed that they all follow the same rules. If the universe is so orderly, many scholars thought that surely mental activities must be governed by some system of laws.

Other Scientists

The sixteenth and seventeenth centuries were exciting times, in that exploration of the world revealed almost unlimited horizons of adventure and discovery. The Portuguese and Spanish dominated the sixteenth century, but the English gained supremacy of the sea in the seventeenth century and began their empire of expansion beyond their small island. In 1600 a physician to Queen Elizabeth I, William Gilbert, published a work describing the magnetic compass, a device well known to Arab scholars, which facilitated British navigation. With discoveries of previously unknown societies and unknown species of plants and animals, European science received an impetus for expansion.

In medicine and physiology, great advances were made in the understanding of bodily processes. William Harvey, a physician, studied his patients, animals, and cadavers and, in 1628, published a work explaining circulation of the blood. In 1662, Robert Boyle published his finding that the pressure of any gas varies inversely with its volume. With a colleague, Robert Hooke, Boyle related his law to bodily heat and posited a reasonable explanation of respiration. By 1690 Anton van Leeuwenhoek had developed the microscope, and a new world was opened for investigation. In Bologna, Marcello Malpighi, while investigating the lungs of frogs in 1661, discovered how the blood passes from arteries to veins; he called these tiny fibers capillaries. By the end of the seventeenth century, the results of careful empirical investigation were finally beginning to erode much of the superstition surrounding the mechanisms of the human body.

Not all advances of science were automatically accompanied by a complete reliance on reason and rejection of faith. For example, the brilliant French scholar and writer Blaise Pascal (1623–1662) studied the effects of atmospheric pressure on a column of mercury and devised the first barometer. His mathematical studies led him to the development of probability distributions, most notably "Pascal's triangle" of probabilities, and the formulation of the binomial theorem. However, during the period of his most productive investigations, Pascal became involved with religious issues. In particular, Pascal and his sister Jacqueline were followers of the Jansenist movement of seventeenth-century French Catholicism. The Jansenists advocated a complete commitment to faith and were eventually condemned by the Vatican for holding essentially Lutheran and Calvinistic beliefs. Despite Pascal's scientific achievements, his religious convictions led him to downplay science as an inaccurate effort because it is based on reason and the senses, which Pascal asserted were faulty. Rather, Pascal accepted the existence of the mysteries surrounding God and individuals and believed that only religion can deal with this level of knowledge. Pascal was an exception to the personal belief in science held by most scientists. Nevertheless, his life demonstrates that an acceptance of reason did not necessarily lead to a rejection of faith. Most scholars tried to accommodate both perspectives, and the triumph of reason and science was a gradual process.

The theories and scientific strategies proposed by Bacon, Kepler, Galileo, and Newton, along with specific scientific discoveries of practical usefulness, clearly demonstrated the value of empirical science. Indeed, the emergence of the natural sciences of biology, chemistry, and physics, all grounded in observational methods relating scientific findings to mathematics, provided a tested model for successful inquiry. In addition, the definition of the planetary system was characterized by the order of matter and motion. A mechanical model that described the physical basis of matter provided an attractive direction for psychological study as psychology made the transition from speculative to empirical inquiry.

THE LEARNED SOCIETIES

Scientific progress in the seventeenth century developed an organizational structure of its own in the form of societies of men dedicated to the advancement of disciplinary study. The form and power of these societies varied from country to country. In southern Europe they tended to be secret so as to avoid conflict with Church authorities. Elsewhere, they were recognized and sometimes supported by the government. The societies all shared two traits. First, they attempted to be independent associations devoted to advancing scientific knowledge and divorced from official Church or governmental control. Second, they were formed to compensate for the lag in scientific progress within the universities. As mentioned earlier, the universities came under the control of the government and the Church, and science could not truly flower under such bureaucratic control. Moreover, the theology faculties within the universities were still dominant and slow to yield to scientific inquiry. This is not to say that science was absent from the universities. As we have seen, many eminent scientists were in fact university professors. However, academic freedom is a twentieth-century concept, and whether opposition arose from the hostility of the theologians or from the petty jealousies that have always plagued academia, scientific study was a new enterprise that ran against university conservatism. Accordingly, the learned societies filled an important role not fully met by the universities.

The first societies in Italy were secret, permitting scientific communication while protecting scientists. The Accademia Secretorum Natural was founded in Naples in 1560, and the Accademia dei Lencei began in Rome in 1603. Galileo's experimental work provided the inspiration for the Accademia del Cimento, founded in Florence in 1657. In northern Europe, scientific societies were started in Berlin (1700), in Uppsala, Sweden (1710), and in St. Petersburg (1724). The Académie des Sciences in Paris was founded by Jean Baptiste Colbert (1618–1683), the financial wizard under Louis XIV, and received a royal charter in 1666. However, the Académie was closely tied to the king, and in turn to the Church, so that some of its activity was at times inhibited. It was disbanded by the Revolution in 1793 but managed to revive and continues to this day, still with vast government support. Perhaps the strongest learned society was the Royal Society, officially called the "Royal Society of London for Improving Natural Knowledge" when chartered by Charles II in 1662. The Royal Society was always privately endowed and managed to be independent. As time went on, membership in the Royal Society became an honor bestowed on distinguished scientists, and projects sponsored or certified by the Society were valued.

The tradition of learned societies has continued to this day. In some nations, such as Russia and some eastern European countries, they became official agents of governments that determine scientific policy; in others, such as the United States, they are essentially private foundations with limited government support. During the last century, many disciplinary societies (e.g., the American Psychological Association, the American Psychological Society, and the Psychonomic Society) as well as interdisciplinary societies (such as the Society for Neuroscience and the

Society of the Sigma XI) have appeared. Such organizations serve vital functions as advocates of scientific study and as conduits of communication among scientists. At their inception in the seventeenth century, the learned societies played a critical role in advancing science. They were organizations of scientists set up to direct, sponsor, certify, and evaluate scientific enterprise under the criteria established by a community of scholars. For the most part, the scientific societies succeeded, remaining aloof from the political and religious pressures of the State and the Church.

ADVANCES IN PHILOSOPHY

The scientific progress of the sixteenth and seventeenth centuries provided methodological advancements in the approaches to scientific issues and established the importance of quantification. In addition, a coherent body of knowledge of the physical world began to emerge from the empirical efforts of this period. Despite these advances, psychology was far from ready to pursue the scientific study of human activity. The major obstacle remained the problem of defining the nature of the person. How the scientist pursues the study of psychology depends completely on how human activity is initially viewed. Should psychology study mental activity? Behavior? Consciousness? Furthermore, precisely how should these terms be defined so that various empirical approaches can be applied? These questions are basically philosophical, and the answers are necessarily based on preconceptions about the nature of the individual. Two parallel trends emerged during these centuries that would eventually lead to the formulation of psychology. The first, consistent with the steady reliance on empiricism, was methodological; it developed as the natural and physical sciences made dramatic advances in the accumulation of findings. The second intellectual trend was more philosophically oriented; it consisted of arguments exploring the relationship between body and mind (or soul) and the functions of each constituent in what we may call human activity.

Baruch Spinoza

The philosophical system of Baruch Spinoza (1632–1677) offered an alternative to the theistic morality of the Scholastic philosophers by commenting on the individual, society, and government from a naturalistic perspective. Born in Amsterdam of emigré Portuguese Jews and educated in the Jewish tradition at a synagogue school, Spinoza supported himself by grinding and polishing lenses for eyeglasses and microscopes after his father died in 1654. His Hebrew first name, for "blessed," was translated into the rough equivalent in Latin, so that he is generally known as Benedict Spinoza. He secured a position as a tutor in a progressive Latin school, gaining exposure to the Scholastic philosophers. In 1670, Spinoza published *Tractatus theologico-politicus* (*Treatise on theology and politics).* In this work, he set forth his conception of God as not being the personalistic patriarch who, according to traditional Judeo-Christian teachings, guides the world. Rather, he postulated that God is the underlying principle that brings unity to matter and mind; God is synonymous with nature. By this definition of God as nature, Spinoza identified himself as a pantheist. Spinoza taught that, despite the absence of evaluative and determining judgments by a personal God, people must still seek to live an ethical existence by striving for virtue based on natural laws. Nature itself has both the power of motion, seen in the movement of all objects in nature, and the powers of generation, growth, and feeling in all living organisms. Thus, Spinoza sought to reconcile the conflict between science and religion by redefining the deity in terms of the universe, revealed initially by Copernicus.

Spinoza viewed the mind and the body as different aspects of the same substance. The mind is the internal manifestation and the body the external manifestation of the individual's unity. Spinoza, then, was one of the first post-Renaissance philosophers to offer an alternative to the Aristotelian notion of mind–body dualism. Rather, Spinoza's stress on the integrity and unity of human existence interpreted the mind and body as different ways that scholars have found to describe different functions of common human experience. From a psychological perspective, Spinoza described the mental functions of feelings, memories, and sensations as mechanical processes mediated by the physical senses and originating through the stimulation of physical objects. This conclusion about the relationship among the physical environmental stimuli, sensory processes, and mental activity places all three elements of experience on a single continuum, underscoring the unity achieved by input from all three sources. Higher mental processes of perception and reason, as well as what Spinoza called intuitive knowledge, are derived not from the external world but from the mind acting on itself. The mind, then, is not an entity or agent, but rather an abstraction: The mind and the activities of the mind are identical. Spinoza wrote that the essential state of the person is to act. Action ultimately motivated by self-preservation is guided by desire. According to Spinoza, the wise person can resolve the conflicts of desires, but for most of us conflicting desires give rise to emotions. Absolute freedom does not exist for the individual, and people are governed by the desires that eventually secure self-preservation.

Spinoza's notion of self-preservation, because it contains the major motivational elements of human activity, is critical to his psychological views. Survival, for Spinoza, was a biological predisposition, a hypothesis that anticipated the evidence of Darwin in the nineteenth century. The individual struggle for survival was seen as the source of all motives and desires, although a person may not be always conscious or aware of the ongoing struggle. Echoing an Epicurean theme, Spinoza asserted that all desires ultimately involve the seeking of pleasure and avoidance of pain. Desires give rise to emotions that, in turn, have both physiological and mental aspects, again reiterating the emphasis on the unity of experience. Indeed, Spinoza's description of the physiological and mental relationship during emotional states was remarkably similar to the theory of emotions offered by William James and Carl Lange in the nineteenth century (see Chapter 12). Ultimately, however, reason must prevail over the emotions if we are to achieve the relative freedom to act.

These highlights of Spinoza's views allow us to draw a few conclusions. First, Spinoza offered a dynamic, action-oriented conception of the mind–body relationship. Mind and body are the same, and personal harmony may be achieved in the mediation of conflicting desires through the highest intellectual powers of reason. Second, Spinoza's system was deterministic, derived not from the providence of God but through natural laws. In his emphasis on natural laws, Spinoza offered a philosophical view of determinism paralleling the advances in science that were to culminate in the mechanical determinism of Newtonian physics. Thus, although he did not deny the existence of God, Spinoza relegated God to a role far removed from human activity and placed humanity securely in the natural world under the same constraints of natural laws as other forms of life. Third, according to Spinoza, the particular dynamics of human activity nevertheless make humanity unique in its intellectual abilities. Specifically, Spinoza acknowledged the central role of rational activity in modulating emotional states. Harkening back to Plato's negative evaluation of the emotions, Spinoza argued that the emotions are a necessary part of human experience, arising from the desires of self-preservation. However, reason must control the emotions, and rational acts constitute the uniquely human ability of individuals to direct their own lives properly in accordance with natural laws.

Spinoza's views were unpopular and distorted in England, and his influence in France was minimal, as Descartes dominated French philosophy. However, as we shall see in Chapter 8, Spinoza's teachings were attractive to German philosophers, who accepted his views and developed the concept of the essential dynamic action of the mind.

René Descartes

We are considering René Descartes (1596–1650) last, and out of chronological order, because his views represented a stepping-off point for philosophical developments up to the nineteenth century. His philosophy was the first comprehensive system since the contribution of the Scholastics, and he is labeled the first modern philosopher. He was born in La Haye, in central France, the son of a prosperous lawyer, who left him an annual income for life. After receiving an early education from the Jesuits, he went on to earn degrees in civil and church law at the University of Poitiers. His insatiable interest in mathematics led him to pursue the examination of philosophical issues by the methods of mathematical reasoning. From 1628 on, he lived in Holland, with only occasional visits to France. Although he affirmed his Christian belief throughout his life, he was nevertheless controversial, and perhaps felt that a quiet life of scholarship in Holland, away from the intellectual control of France, would give him greater personal freedom.

His *Discours de la méthode,* published in 1637, described the evolution of his thinking. Starting with complete doubt and skepticism, Descartes went on to the first principle of certainty and validity: "I think, therefore I am" *(Cogito, ergo sum; Je pense, donc je suis)*. This famous statement asserts Descartes' affirmation of the realization of experience. He is stating that the only sure fact about which we have

RENÉ DESCARTES (1596–1650).
(Courtesy, New York Public Library.)

absolute certainty is our own experience and our awareness of the knowledge of ourselves. By defining the self in terms of the subjective knowledge of the experienced idea as the first principle, Descartes radically departed from previous views, which had always begun with the external world and then concluded to the mind as necessary for knowing this external world. Reversing the traditional views, Descartes asserted that our knowledge of ourselves is the most certain principle, and the reality of the external world may be questionable.

However, to deal with the external world, Descartes used the concept of God—that is, because we know the idea of perfection, some entity must possess complete perfection, and this entity is God. Rejecting Platonism, Descartes asserted that the perfect God would not create people with unreliable senses; therefore, sense information is an accurate depiction of the environment, ordered again by the perfection that is God. The critical factor in Descartes' thinking is his reliance on self-awareness of our ideas, which then permit us to know God and eventually our external surroundings. Thus, for Descartes, the ideas of the self, God, and the dimensions of space, time, and motion are all innate to the soul, or mind; that is, they are not derived from experience but from the essential rationality of the mind.

Descartes developed his views on the relationships between mind and body and among the individual, the environment, and God in his subsequent works: *Meditationes de prima philosophia* (*The meditations on first philosophy;* 1641), *Principia philosophiae* (*Principles of philosophy;* 1644); *Traité des passions de l'âme* (*Treatise on the passions of the soul;* 1650), and *Traité de l'homme* (*Treatise on man;* 1662). Descartes' system recognized the advances in the natural sciences, which conceptualized the physical world as governed by mechanical laws. With the exception of God and the human rational soul, all reality is physical and can be explained through mechanical relations. Descartes believed that as science progressed and revealed the intricacies of the activities of life, the operations of human existence would fall within the same guiding principles as all life, and only the human capacity to reason would lie beyond these mechanical principles. Accordingly, Descartes' system held two levels of activity in the universe: the physical world of matter following the order of mechanical laws, and the spiritual world represented solely in human reasoning.

The famous Cartesian dualism, then, is an application to human activity of Descartes' general distinction between mechanical and spiritual levels of the universe. In Descartes' psychology, the mind is a spiritual, immaterial entity, different from the body and easier to know than the body because of the first principle of self-reflection. The body is the physical entity that, in common with all animals, responds to the external world through the mechanics of physiology. The emotions are rooted in the body and represent movement or reflexes to the stimulation of sensory impulses by environmental sources. The relationship between the mind and the body is truly a psychophysical interaction. The human body with its mechanical operations is distinguished from other animals only because it is acted upon by the mind. The exact manner of this interaction is left unclear, although Descartes suggested that the site of the mind–body interaction may be the pineal gland of the midbrain because this organ is singular and situated between the hemispheres of the brain. Although Descartes' hypothesis about the pineal gland reflected the relatively primitive state of physiology, it is nevertheless important to recognize that Descartes consistently pointed to the role of the brain as the transitional agent between the spiritual energies of the mind and the physical forces of bodily mechanics. Descartes taught that the study of bodily processes was the province of physiology and the study of the mind belonged to psychology; thus the first modern philosopher firmly defined psychology's subject matter as the mind.

Descartes was dedicated to empirical observations, and his interest in laboratory study seemed to increase as he grew older. His dissections led him to speculate about a nervous system of hollow tubes through which flow animal spirits that account for voluntary movement. In analyzing vision, he studied the lens of the eye and also described the mechanisms that underlie ocular reflexes. In mathematics, he developed analytic geometry and made studies toward a primitive calculus. The consensus of these wide-ranging investigations confirmed his belief that mechanical laws govern the entire universe, except for God and the human soul. According to Descartes, if our knowledge were sufficient, we would be able to reduce all sciences—astronomy, chemistry, and physics—and all bodily operations—respiration, digestion, and sensation—to mechanical explanations. The single exception to this conclusion, directly from our own experience, is human reasoning.

Toward the end of his life, Descartes' system became widely known, gaining him both praise and condemnation. The Calvinist theologians were especially vehement against Descartes' support of free will as contrary to their rigid beliefs on predestination. However, the great and noble of Europe protected Descartes from harm by either Protestant or Catholic clerical authorities. Descartes agreed to the request of Queen Christina of Sweden to journey to Stockholm to tutor her in philosophy. Unfortunately, the chilly climate took its toll and he died a good Catholic on

February 11, 1650. Consigned to the same fate as Galileo's efforts, Descartes' works were placed on the Index of Forbidden Books by the Church in 1663, again ensuring their author at least notoriety, if not further success. Descartes' views on the supremacy of rationalism and on the derivation of all knowledge from the self won widespread acceptance throughout Europe, challenging the dominance of Scholasticism.

Three trends in psychology may be traced to Descartes. First, psychology as an introspective science investigating human consciousness finds support in the validity of the first principle of the mind espoused by Descartes. Second, psychology as a purely behavioral study is reinforced, although somewhat indirectly, by Cartesian dualism; that is, the interaction between mind and body indicates that overt, observable behavior is meaningful. Such activity reflects the mind, as the mind acts on the body, producing behavior. Finally, psychology as a physiological science is supported by Descartes' assertion that all human activities except thinking and feeling are related to bodily physiology and may be understood as truly psychophysiological. The successors to Descartes were able to find some support for differing orientations by selectively emphasizing certain aspects of Cartesian thought in their search for models of psychological inquiry. Thus, Descartes' importance lies in his directly stimulating the movement toward the founding of psychology.

In concluding this chapter, we can say that the sixteenth and seventeenth centuries witnessed several trends important to the eventual emergence of an empirical scientific psychology. The first concerned the products of scientific advancement, which clearly demonstrated the value of empirical inquiry. From their studies of the physical universe, scientists not only completed the Copernican revolution by providing empirical support for the theory of planetary motion, but also showed the impressive extent to which the physical world operates under specific lawful relationships. The orderliness of nature suggests that all of reality, including the operations of life, may conform to lawful relations revealed through scientific study.

The second trend, represented in the philosophical views of Spinoza and Descartes, offered an alternative to the Scholastic commitment to Christian morality. Both philosophers focused on the primacy of reason. Spinoza's philosophy was the more radical, in that it departed from belief in the guidance of a personal God and the advocacy of a monistic mind–body relationship. Spinoza's description of the unity of mind and body resulted in his emphasis on the uniquely human capacity of reasoning. Others followed Spinoza, taking his monistic concept, but they shifted the emphasis to either the materialism of the body or the spiritualism of the mind. Descartes was clearly a dualist in his interpretation of the mind–body interaction. However, his impressive discourses on the mechanics of bodily action opened the way for successive philosophers to reduce the mental aspect of experience to the physical mechanics of the body. Both Spinoza and Descartes provided a philosophical transition from Scholasticism and proposed a variety of assumptions about the nature of human experience, which later philosophers used as a basis for models of psychology.

CHAPTER SUMMARY

Two parallel trends prepared scholars for the investigation of the mind–body relationship so that a model of psychological inquiry could evolve. The first trend was methodological, characterized by the triumph of empiricism. Scientific innovations by Francis Bacon, Galileo, Kepler, and Newton were firmly based on careful observations and quantification of observables. Using inductive methods, moving from observed particulars to cautious generalization, empiricism stood in contrast to the deductive methods of the Scholastic philosophers. The second trend occurred in the attempt to develop conceptions on the nature of humanity and was more a philosophical enterprise. Spinoza taught that mind and body are manifestations of the same unity of the person. Human activity, although unique because of humanity's higher intellectual powers, is determined by the laws of nature. Descartes stated that the first principle of life is self-awareness of the idea, and all else that we know proceeds from self-reflection. His dualism of the interaction between mind and body distinguishes psychology from physiology. Descartes' views were developed in the French and British philosophical traditions; Spinoza influenced the German efforts to develop a model of psychology.

BIBLIOGRAPHY

Primary Sources

BACON, F. (1878). *Novum Organum*. In *The works of Francis Bacon* (Vol. 1). Cambridge, MA: Hurd & Houghton.

DESCARTES, R. (1955). *The philosophical works of Descartes* (E. Haldane & G. R. T. Moss, Trans.). New York: Dover.

NEWTON, I. (1953). *Newton's philosophy of nature* (H. S. Thayer, Ed.). New York: Hafner.

SPINOZA, B. (1955). *The chief works of Benedict de Spinoza* (R. H. M. Eleves, Trans.). New York: Dover.

Resources

BUTTERFIELD, H. (1959). *The origins of modern science: 1300–1800*. New York: Macmillan.

HALL, A. R. (1963). *From Galileo to Newton: 1630–1720*. London: Collins.

SARTON, G. (1957). *Six wings: Men of science in the Renaissance*. Bloomington: Indiana University Press.

Studies

BALZ, A. G. A. (1952). *Descartes and the modern mind*. New Haven: Yale University Press.

BERNARD, W. (1972). Spinoza's influence on the rise of scientific psychology. *Journal of the History of the Behavioral Sciences, 8*, 208–215.

ORNSTEIN, M. (1928). *The role of scientific societies in the 17th century*. Chicago: University of Chicago Press.

PIRENNE, M. H. (1950). Descartes and the body-mind problem in physiology. *British Journal of the Philosophy of Science, 1*, 43–59.

TIBBITTS, P. (1975). An historical note on Descartes' psychophysical dualism. *Journal of the History of the Behavioral Sciences, 9*, 162–165.

WATSON, R. I. (1971). A prescriptive analysis of Descartes' psychological views. *Journal of the History of Behavioral Sciences, 7*, 223–248.

❋ 6 ❋

Sensationalism and Positivism: The French Tradition

This chapter and the three that follow examine advances in science and philosophy during the seventeenth, eighteenth, and nineteenth centuries, advances that provide background for the emergence of modern psychology. The chapters are organized around respective developments in France, Britain, and Germany, because the structure of national movements in science and philosophy provided differing models for the building of psychology. Although these traditions overlapped, the teachings of Descartes and Spinoza were interpreted along characteristic national lines. In France, the mind–body dualism of Descartes was reduced to a materialism that focused on the mechanics of sensory processes to explain all psychological activity. The British tradition, while retaining the notion of the mind, stressed environmental input to explain the contents of the mind. The German tradition, following Spinoza more than Descartes, emphasized the self-initiating, dynamic qualities of mental activity that transcend both environmental stimuli and the mechanics of sensory physiology.

After the death of Descartes in 1650, France entered a golden era of political and cultural ascendancy under the long rule of two absolute monarchs, Louis XIV (reigned 1643–1715) and Louis XV (reigned 1715–1774). Although this was not a period of complete political tranquility, an intellectual enlightenment prevailed, and led to a flowering of literary, scientific, and philosophical achievements. France became the leading nation of continental Europe during this time (Map 7). Especially significant were the works of the dramatists Jean Baptiste Poquelin, better known as Molière (1622–1673), and Jean Baptiste Racine (1639–1699), as well as the writings of Pierre Corneille (1606–1684) and Jean de La Fontaine (1621–1695). Collectively, they made French the language of literary society.

The view of education as the responsibility of society became increasingly prevalent. The rational ideal of the age, reflecting belief in the limitless horizons of human knowledge, resulted in educational opportunities no longer being limited to the nobility. The acquisition of knowledge through education was viewed as the key to success and class mobility. The Church continued to provide early educational opportunities for most people, and the great philosophers of France, such as Denis Diderot (1713–1784) and Jean Jacques Rousseau (1712–1778), established the intellectual basis for universal education. It was François Marie Arouet, better known as Voltaire (1694–1778), who personified this Age of Reason. Voltaire's prolific writings on every aspect of life provided the foundation for transforming the conception of Western government from aristocratic obligation to modern social responsibility.

ADVANCES IN SCIENCE

Scientific inquiry developed rapidly during the seventeenth and eighteenth centuries, aided by advances in both mathematical and empirical disciplines. These developments are important for the history of psychology because they contributed to the supremacy of nineteenth-century science, upon which psychology was modeled. In France, as well as in England and Germany, mathematics and the physical sciences began to assume modern forms.

The French mathematician Joseph Louis Lagrange (1736–1813) was born of French parents living in Turin, Italy. After his preliminary education in Italy, he went to Berlin to study calculus under the mathematician Euler. During his 20 years in Berlin, Lagrange formulated his work *Mécanique Analytique (Mechanical Analysis)*, which gave physics a series of formulations for mechanical relationships based on algebraic proof and calculus. After the death of his benefactor Frederick the Great in 1786, he accepted an invitation to join the Académie des Sciences. His prestige allowed him to escape the excesses of the French Revolution. He was instrumental in rebuilding French educational institutions and played a leading role in devising and introducing the metric system. Through his long life of research and teaching, Lagrange produced a group of distinguished students who contributed to nineteenth-century mathematics, physics, and engineering. A contemporary of Lagrange, Jean Le Rond d'Alembert (1717–1783), published classic works on such problems of applied mathematics as the refraction of light and fluid mechanics.

The discovery of oxygen is an example of the growth of a true international scientific community. A Swedish scientist, Karl Wilhelm Schule (1742–1786), is credited with the pioneering research that anticipated the discovery of oxygen, and the English investigator Joseph Priestley (1733–1804) effectively claimed the discovery in a publication of

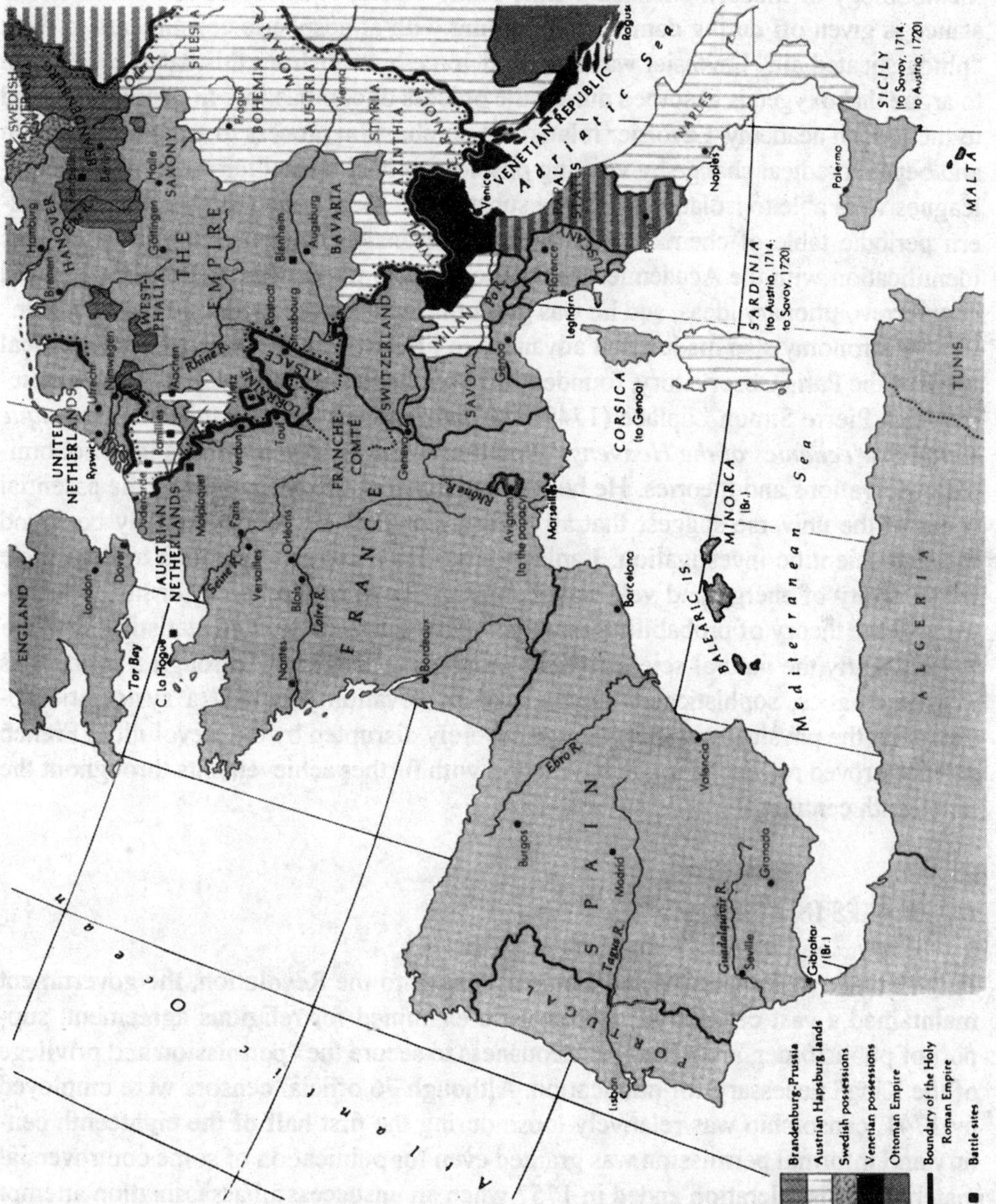

MAP 7. FRANCE AND NEIGHBORING STATES AT THE END OF THE REIGN OF LOUIS XIV, 1715.

1775. However, it was a group of French researchers, headed by Antoine Lavoisier (1743–1794), who actually named the element and went on to establish the scientific methodology of modern chemistry. Like many others, Priestley believed that a substance is given off during combustion, joining with atmospheric constituents to form "phlogisticated air." Lavoisier was able to divorce himself from this dated conception to argue that oxygen is absorbed during the process of combustion. In papers delivered to the French academy, Lavoisier related the combustion process to animal respiration and began a radical change by viewing physiology in chemical terms. He and his colleagues were able to isolate 32 "simple substances" that formed the basis of the modern periodic table of chemical elements. Unfortunately, Lavoisier's politics and his identification with the Académie des Sciences led to his condemnation for supposed counterrevolutionary ideas, and he was guillotined during the French Reign of Terror.

Astronomy also made great advances in France, centered around the empirical work of the Paris Observatory, founded in 1671. Between 1799 and 1825, the mathematician Pierre Simon Laplace (1749–1827) published his multivolume *Mécanique Céleste (Mechanics of the Heavens),* which summarized the advances in astronomical observations and theories. He believed that scientific discoveries of the essential order of the universe suggest that all explanations of life may eventually be found through scientific investigation. Laplace formulated classic equations to determine the intensity of energy and velocity of motion. He is remembered for his contributions to the theory of probability, establishing the basis of modern statistics.

Clearly the natural sciences were prospering in France. Through a solid basis in mathematics, sophisticated methods of observation provided a systematic approach to the physical world. Although severely disrupted by the Revolution, French science proved resilient enough to respond with further achievements throughout the nineteenth century.

ADVANCES IN PHILOSOPHY

In the France of Louis XIV and Louis XV, prior to the Revolution, the government maintained a vast censorship. Books were examined for religious agreement, support of public order, and moral righteousness to secure the "permission and privilege of the King" necessary for publication. Although 76 official censors were employed by 1741, censorship was relatively loose during the first half of the eighteenth century and informal permission was granted even for publication of some controversial materials. This toleration ended in 1757 when an unsuccessful assassination attempt on the life of Louis XV prompted severe restrictions of literature criticizing state or church. This repression succeeded in uniting all writers holding anti-Church or anti-state views. They were collectively called the *philosophes,* and they cooperated in clandestine publication within France and collaborated to have works printed outside of France to smuggle into the country. The net result was a widespread distribution system of censored works, which, aided by the learned academies, guaranteed the dissemination and success of the writings of the philosophes.

As individuals, the philosophes held wide-ranging views, agreeing only in their opposition to government repression. Through their antigovernment writings, the philosophes influenced some of the leaders of the American Revolution and contributed to the growing concerns that led to the French Revolution. The anti-Church writings of the philosophes had an important impact on psychology, most notably by developing a view that tried to purge the Christian (that is, the Scholastic) notion of soul from consideration in psychology.

Sensationalism

After Descartes, French thought on psychological issues concentrated on the sensory aspects of human experience. Specifically, a coherent theme of French thought involved the study of human nature based on sensations and perceptions. Accordingly, a group of philosophes examined the mechanisms of sensation, and in so doing gradually reduced mental activity to sensory mechanisms. Thus, the mind–body distinction, so carefully made by Descartes, became obscured in French thought.

Étienne Bonnot de Condillac. The first major figure in this trend of sensationalism, Étienne Bonnot de Condillac (1715–1780), was born in Grenoble and educated at a Jesuit seminary in Paris. Soon after his ordination to the Roman Catholic priesthood, he found his way into the literary and philosophical salons of Paris and steadily lost interest in his religious career. Condillac's early publications illustrated his appreciation of antecedent philosophical views on psychology, especially those of Descartes and the English philosopher John Locke, whom we consider in the next chapter. In his most famous work, *Traité des sensations* (*Treatise on Sensation;* 1754), Condillac radically departed from his predecessors and offered an interpretation of psychological activity based on sensory experience alone.

Condillac began by denying Descartes' notion that the mind is born with certain innate ideas. Rather, he argued that the entire complexity of the mind can be derived from a single sense capacity. To illustrate his argument, Condillac proposed the analogy of a statue, endowed as human beings are with an internal organization and a mind, empty of any ideas. The statue has the single sense of smell, and it is capable of realizing the difference between pleasure and pain. Condillac then attempted to demonstrate how complete psychological activity may be derived from this relatively simple statue as it gradually develops additional sensory capacities. With the first sense alone, attention is acquired by the compelling stimulation of sensory input. When a second sense is developed, judgment is acquired because the statue can now compare the input of two sensory modalities. Memory is a past sensation that is retrieved because of the stimulus of a present sensation, and imagination is an enhanced memory or a new combination of past sensations. Approach and avoidance behaviors are the active recall of pleasant or unpleasant sensations, and the will is a desire based on an exaggerated approach tendency toward an attainable object. Condillac stated that the capacity for self-reflection occurs as the result of alternations between memories and the objects of the will. Aspects of personality, such as

ÉTIENNE BONNOT DE CONDILLAC (1715–1780). (Courtesy, New York Public Library.)

the concept of the self, develop only gradually with the accumulation of experiences through memories and desires. Thus, Condillac formulated psychological functions from a single sense by adding the four other senses. The mind, then, is reduced to the roles of a receptor for sensory experience and a receptacle for memories. Moreover, the mind is shorn of any initiating functions.

The simplicity of Condillac's view was appealing. He also caused quite a furor in French intellectual circles. He was criticized for his deductive approach and lack of any inductive support from empirical evidence. However, in contrast to philosophers such as Descartes who postulated an active mind that required some kind of spiritual, or at least nonphysical, entity, Condillac relied solely on the physiologically based senses. In addition, Condillac introduced the notion of materialism to modern psychological thought. If the contents of the mind are reduced to their sensory bases, one does not have to go much further to equate the mind and the senses. Thus, the concept of a "mind" itself becomes superfluous. Indeed, Condillac's materialistic psychology was adopted in the school system reforms instituted by the French Revolution and was discarded only with the advent of Napoleon and the accompanying reaction against materialism.

Charles Bonnet. Born in Geneva, the most prominent French-speaking city of Switzerland, Charles Bonnet (1720–1793) studied plants and insects intensively and presented several experiments to the Académie des Sciences beginning in the 1740s. He studied reproduction in tree lice and reported that the female can reproduce fertile offspring without the male of the species, and suggested that sex may be not only for reproduction, but also for enriching offspring with the diverse characteristics of two parents. He was one of the first scientists of the eighteenth century to use the term *evolution,* although he meant by it the chain of life from simple atoms to human beings. His investigations of plants led him to conclude that plants are endowed with sensation, discrimination, and even judgment, which Bonnet viewed as evidence of intelligence. Accordingly, Bonnet's interpretation of the living world focused on the unity of living beings based upon the mediation of mechanical agents.

Bonnet extended Condillac's views by examining the physiological mechanisms of sensory processes. To continue with Condillac's statue analogy, Bonnet gave the statue a nervous system to accomplish sensation. He argued that the tracing of nerve fibers would explain not only sensory processes but also the psychological functions of attention, memory, and recognition. In so doing, Bonnet was one of the first scholars to mention specific nerve energies, wherein a given function is accommodated by a certain system of neural fibers. He viewed higher mental processes in terms of the association of sensations or memories through the commonality of some dimension, such as time, place, or meaning. For example, sensory event *A* may be linked to sensory event *B* by their simultaneous occurrence. Thus, Bonnet added to Condillac's view by establishing a more reasonable basis for psychological materialism through the nervous system, and the necessity of a special mental agency was further diminished.

Julien Offroy de La Mettrie. Julien Offroy de La Mettrie's (1709–1751) most famous work, *L'Homme Machine* (*Man, a Machine;* 1748), shook intellectual Europe because of its simple and clear statement of materialism. La Mettrie was the son of a wealthy merchant, who gave his precocious son a superb education. After receiving his doctorate in medicine, La Mettrie studied anatomy in Leiden, the Netherlands, and published several works that emphasized the role of the brain in human pathology. He eventually became a surgeon in the French army but continued his studies and writings.

La Mettrie's materialism held that matter has an active element, which is motion. He based this conclusion on sensory feelings found in the lowest animals and plants. This observation led him to propose a type of evolutionary hierarchy in the motion of matter. Thus, in the higher animals, the motion of matter allows the heart to beat and the brain to think. La Mettrie argued that psychology is ultimately physiology, and the dualism of Descartes was completely forsaken for the animal machine.

La Mettrie's views created problems for him with his military superiors, and he had to flee to Leiden for safety, but eventually, in 1748, he received an invitation from Frederick the Great to join the Berlin Academy of Sciences with a stipend. There La Mettrie developed his psychology further by asserting a motivational principle for human activity. This principle was hedonistic in that the seeking of pleasure was the ultimate force that propels the individual. In three publications he opposed Christian teachings and argued the importance of sensual pleasure. He established an ethic that judged the actions of people as determined by their desire for sensual gratification. Although La Mettrie's views were received with considerable scorn, he placed French psychology under the direction of the mechanistic laws of physiology. In his short, frantic life, La Mettrie succeeded in arguing against the need for a separate discipline of psychology. Faith in materialistic science was pushing psychology out of consideration only 100 years after Descartes first defined psychology by distinguishing it from physiology.

Claude Adrien Helvétius. Retreating from the extreme materialistic position of the French tradition, Claude Adrien Helvétius (1715–1771) retained some use for the concept of mind. The son of the queen's physician, he was born in Paris and educated by the Jesuits. As a tax collector, he became wealthy, married a beautiful countess, and retired to the countryside to live a contented life as a gentleman philosopher. The charm of his estate drew many of Europe's finest thinkers, and in 1758 Helvétius published his memorable work, *De l'Esprit (On Intelligence)*.

In this work Helvétius added a critical, complementary dimension to the French sensationalistic tradition. He concentrated on the environmental determinants of the individual. Although agreeing with La Mettrie on the basis of desire in pleasure seeking, he related this motivational principle to environmental influences. According to Helvétius, all people are born with equal capacities, but the environment acts differently on individuals, strengthening attention and widening perception in some people but not in others. This difference in capacity to deal with the environment is what Helvétius defined as intelligence. Believing that the key to success in the environment is the opportunity for enriching experiences, Helvétius argued for better educational benefits and more open social structures. Thus, although he did not disagree with the French sensationalists, Helvétius' emphasis on the environment reserved a place for psychology: Physiology may explain the mechanisms of psychological functions, but the mechanisms are still dependent on environmental context.

Pierre Cabanis. A final figure in the French sensationalistic tradition is Pierre Cabanis (1757–1808). Like Helvétius, Cabanis modified the extreme views of Condillac, Bonnet, and La Mettrie. A distinguished physician, he met the great thinkers who gathered in the literary salons of Paris. Although accepting the materialism of mechanical sensations, Cabanis nevertheless argued against the complete reductionism of his predecessors. Their view equated mental operations with their sensory input, logically leading to the discarding of the mind as unnecessary. Cabanis drew back from this position and proposed a central ego of the brain that acts as the integrator and synthesizer of sensory input. Cabanis' view, then, preserves the need for the concept of mind, even if described in terms of the physical brain. Moreover, he recognized levels of consciousness, including unconscious and semiconscious processes. According to Cabanis, sensations do not exist as pure forms; rather, sensations are part of an entire system, mediated by the central ego, or self, and sensations are known only through the integration of the entire system.

Cabanis' additions to Condillac's psychology rescued the mind, but tied it firmly to brain physiology. Unlike the British thinkers (considered in the next chapter), Cabanis disagreed with the view of the mind as passive and reactive, filled up by the accumulation of experiences. In contrast to German philosophers, especially Kant (see Chapter 8), Cabanis did not consider the mind as an entity having an integrity and independent processes divorced from physiology. Without necessarily attempting a compromise, Cabanis articulated a view that retained the need for the mind, recognized by the British and German scholars. Loyal to the French tradition, however, he embedded mental processes in the materialism of the nervous system.

To summarize briefly, although the major figures of the French sensationalistic tradition held differing views, they limited the concept of psychological processes to the level of sensory input. Emphasizing the critical role of sensory experience, they de-emphasized the need for the initiating central construct of the mind. Thus, their selectivity within Cartesian psychology tended to be one-sided, neglecting Descartes' defined province of psychology—the mind.

The Psychology of Maine de Biran

The renowned American philosopher and psychologist of the late nineteenth century, William James, referred to Maine de Biran (1766–1824) as the greatest psychologist of the eighteenth century. Biran began his writing committed to the French sensationalist tradition, but steadily moved beyond such restrictions to advocate a more complete, dynamic psychology. Although his writings reflect the interest of the sensationalists, he cannot be categorized in that group, as he personified the full gamut of eighteenth-century psychological views.

Biran was a soldier of the Garde du Corps of Louis XVI and witnessed the women's march on Versailles in 1789. During the Revolution he wisely retired to his country estate, reemerging to oppose the rule of Napoleon. He ended his political career as treasurer of the Chamber of Deputies after the restoration of Louis XVIII. During this politically tense time he continued his writings, which went through four rather distinct phases of intellectual evolution.

During the first phase, 1790–1800, Biran belonged to a group called the Ideologists, which had been founded by Cabanis to promote the teachings of Condillac. At this stage of Biran's thinking, he agreed that human understanding comprised the sum of the associations of the brain, caused by the stimulation of nerve fibers from motion in the environment. Accordingly, Biran believed in a physiological psychology explained by sensory processes. He broke with the Ideologists in 1805 and published *Mémoire sur la décomposition de la pensée (Essay on the decomposition of thought).* In this work he argued against the "fiber" psychology of the Ideologists that relegated human activity to the mechanistic atomism of sensory elements. Biran wrote that thought is a whole entity composed of distinct processes, but that it is not simply an aggregate of those processes. He focused on the will as an intentional activity that defines the essential character of the self. Thus, the will makes the individual more than the passive receptacle of sensations; it defines a spiritual force that explains life itself.

By 1810 Biran had moved into a third phase, and his conception of psychology took final form in *Essai sur les fondements de la psychologie* (*Essay on the fundamentals of psychology;* 1812). He concluded that psychology is the science of the immediate data of consciousness. To Descartes' "I think, therefore I am," Biran responded, "I will, therefore I am." Psychology's province is to study the intentionality of the self represented in consciousness. In terms of methodology, Biran insisted on the objective observation of the self through individual experience. Thus, the active self or ego is the central fact of psychology, so that the individual is intelligent to the extent that he or she is free. In his fourth phase, beginning in 1820, Biran turned to religious experience and attempted to integrate religious aspirations in life to his total concept of psychology.

Biran has been criticized for his changing views of psychology, ranging from physiological to mystical interpretations. However, the range of his opinions is fascinating. Indeed, Biran seems to have expanded his conception as he became dissatisfied with the limitations of fundamental explanations based on sensory physiology. His emphasis on the uniqueness of the individual dictated his intellectual evolution. Biran was impressed not with the commonality of physiological makeup or even psychological processes. Rather, his interest steadily centered on those aspects of human nature that result in creative, unpredictable activities fully expressive of the individual person. This same trend toward expanding psychology into a more comprehensive discipline aimed at explaining individual diversity is common to several figures in the history of psychology. Although Biran died at the comparatively early age of 58, he was able to accommodate an entire evolution in his thinking. Others who lived longer, such as Wundt, whom we shall consider in a later chapter, did not succeed in completing the cycle, although they were well on their way to the same goal that Biran achieved. Nevertheless, we can well understand James' appreciation of Biran for the breadth of his vision of psychology as well as for his anticipation of the variety of models that may be applied to psychology.

The Advent of French Positivism: Auguste Comte

By considering Auguste Comte (1798–1857) at this point, we are jumping ahead somewhat and leaving the historical sequence of psychology in a strict sense. Indeed, Comte's place in history is clouded by ambiguity. He expressed the scientific spirit that psychology adopted as it emerged as a formal discipline. At the same time, Comte's application of his own views resulted in an attempted utopia that proved embarrassing to those who tried to take him seriously.

The controversial life of Auguste Comte began at Montpellier, where he received his early education under Catholic auspices. He then studied at the École Polytechnique in Paris under some of the leading scientists of France. Expelled because of his republican sympathies, Comte remained in Paris and continued to study with the Ideologists. He secured a position as secretary to the social philosopher Saint-Simon (1760–1825), who advocated a reorganization of society under the guidance of emerging social science. Comte incorporated many of Saint-Simon's ideas into his own views. After a bitter quarrel, he parted with Saint-Simon and supported himself mainly by tutoring and giving lectures through private subscriptions. The lectures formed the basis of his most famous work, *Cours de Philosophie Positive (Course on Positive Philosophy)*, published in six volumes between 1830 and 1842. This monumental and revolutionary work took on the ambitious task of completely reorganizing intellectual conceptions of knowledge and applying this theory to the eventual reformation of social structures.

AUGUSTE COMTE (1798–1857).
(Courtesy, New York Public Library.)

Although Comte never gained a professorship, he did gather loyal and devoted students, and his views spread widely. The British philosopher-feminist Harriet Martineau (1802–1876) translated the *Cours* into English in 1858, and Comte carried on an extensive correspondence with the foremost spokesman of British psychology, John Stuart Mill. His precarious livelihood and seemingly reckless ventures soured many of his earlier admirers, including Mill. By the late 1840s, Comte's application of his theory took the form of a religion of humanity. The structure of his proposed society was remarkably similar to the hierarchical organization of the Roman Catholic Church, with humanity substituted for God and Comte substituted for the pope. This fanciful utopia based on reformulated social relationships tainted Comte's entire systematic thought.

However, Comte's earlier writings contained in the *Cours* are important, both for their consistency with the model of sensationalism in French thought and for their attempt to instill an objective method of science for psychology. We have already touched on Comte's notion of historical progress in Chapter 2. Briefly, he argued that explanations of life shift in focus from a theological to a metaphysical basis as human intellectual progress continues. A final shift from a metaphysical to a positivist basis defined the maturity of science for Comte. Whereas the metaphysical stage seeks causal explanations in nonphysical abstractions or universals, the positivist stage seeks to coordinate observable facts and find descriptive laws of natural events. By emphasizing description, Comte did not preclude causal relations in positivism, but he did argue against the preoccupation with the search for causality, which concerned so many previous philosophers. According to Comte, such a preoccupation led to artificiality because philosophers were susceptible to a preconceived notion of universals at the expense of observables, the true level of scientific enterprise.

Various sciences progress at different rates through these stages of intellectual development. Accordingly, science for Comte is relative knowledge, for positivism permits only a limited and changing view of nature. Comte listed six basic sciences: mathematics, astronomy, physics, chemistry, physiology or biology, and social physics or sociology. Interestingly, he omitted psychology, and placed the study of the individual under physiology, thereby agreeing with the sensory–physiological view of psychology advocated by Condillac and La Mettrie. The individual behaving in a group context is the subject matter of sociology for Comte. Elaborating on this "social psychology," Comte later added the science of ethics, which he meant not as the study of morals but rather as the study of observable social behavior aimed at finding laws of prediction for social planning.

It may be argued that Comte, writing before the advent of psychology's formal emergence, could not foresee the later coherence of psychology as a discipline. However, it appears that he recognized the trend of French sensationalism and saw disparity rather than unity. Accordingly, he was consistent with the French trend and simply carried to its logical conclusion the reduction of psychology, defined as sensation, to physiology. Comte's conclusions about psychology did not directly help the push toward its recognition as a discipline. However, his positivism indirectly helped identify a methodological strategy that helped psychology emerge as a recognized, separate discipline within the sciences. The emphasis on objective observation was clearer among British writers, whom we shall consider next. Moreover, Comte's positivism was resurrected in an updated form during the early part of the twentieth century and succeeded in establishing behaviorism as a dominant model in contemporary psychology.

This survey of two centuries of French thought identifies several influences on psychology. First, the benefits of a natural science were articulated, and they created an ideal model for psychology to emulate. Second, Descartes' dualistic conception of mind–body interaction was seriously challenged. Emphasizing materialism at the expense of mentalism, the main theme of French thought opted to restrict mental operations to sensory mechanisms, leading to the questioning of psychology's place by both Biran and Comte.

CHAPTER SUMMARY

The seventeenth and eighteenth centuries marked the ascendancy of French political power, literary success, and scientific achievement. In the natural sciences, such investigators as Lagrange, Laplace, and Lavoisier gave mathematical and empirical support to the modern basis of chemistry, physics, and biology. In a parallel movement, philosophical discourses on psychology led to a reinterpretation of Descartes' formulation so as to focus on sensation. Condillac, Bonnet, and La Mettrie progressively argued for the equation of mental operations with sensory input and worked to articulate the physiological mechanisms of sensation. In so doing, they logically reduced psychology to sensation. Helvétius and Cabanis attempted to back off from such extremism by asserting the mediating role of a central ego, although both remained committed to sensory physiology. Biran and Comte recognized the consequences of reducing psychology to mere sensory physiology, but each worked out separate solutions. Biran rejected sensationalism as completely inadequate, suggesting a total view of individuality based on the immediate data of consciousness expressing the dynamics of the will. In contrast,

Comte ultimately accepted the conclusions of sensationalism and dismissed psychology. Human activity of the individual should properly be studied by physiology; the individual behaving in a group is the province of sociology. Comte, however, advocated a spirit of objective observation that was eventually useful to psychology. Thus, the successors to Descartes in France left psychology in a somewhat tenuous position, removed from recognition as a formal discipline.

BIBLIOGRAPHY

Primary Sources

COMTE, A. (1858). *Cours de philosophie positive (1830–1842) [The positive philosophy of Auguste Comte]* (H. Martineau, Trans.). New York: Calvin Blanchard.

LA METTRIE, J. O. DE. (1912). *L'homme machine [Man, a machine]* (M. W. Calkins, Trans.). New York: Open Court.

MILL, J. S. (1965). *Auguste Comte and positivism*. Ann Arbor: University of Michigan Press.

RAND, B. (1912). *The classical psychologists*. New York: Houghton Mifflin.

General References

COPLESTON, F. (1960). *A history of philosophy, Vol. 4, Modern philosophy—Descartes to Leibniz*. Garden City, NY: Image Books.

COPLESTON, F. (1964). *A history of philosophy, Vol. 6, Modern philosophy, Part I—The French enlightenment to Kant*. Garden City, NY: Image Books.

COPLESTON, F. (1977). *A history of philosophy, Vol. 9—Maine de Biran to Sartre*. Garden City, NY: Image Books.

DURANT, W., & DURANT, A. (1965). *The age of Voltaire*. New York: Simon & Schuster.

DURANT, W., & DURANT, A. (1965). *Rousseau and revolution*. New York: Simon & Schuster.

DURANT, W., & DURANT, A. (1975). *The age of Napoleon*. New York: Simon & Schuster.

Studies

CHARLTON, D. G. (1959). *Positivist thought in France during the second empire*. Oxford: Clarendon Press.

DIAMOND, S. (1969). Seventeenth century French "connectionism": La Forge, Dilly, and Regis. *Journal of the History of the Behavioral Sciences, 5*, 3–9.

LEWISOHN, D. (1972). Mill and Comte on the method of social sciences. *Journal of the History of Ideas, 33*, 315–324.

MCMAHON, C. E. (1975). Harvey on the soul: A unique episode in the history of psychophysiological thought. *Journal of the History of the Behavioral Sciences, 11*, 276–283.

MOORE, F. C. (1970). *The psychology of Maine de Biran*. London: Oxford University Press.

STAUM, M. S. (1974). Cabanis and the science of man. *Journal of the History of the Behavioral Sciences, 10*, 135–143.

WOLF, A. (1939). *A history of science, technology, and philosophy in the eighteenth century*. New York: Macmillan.

7

Mental Passivity: The British Tradition

The strong affinity between American and British intellectual thought has been forged through four centuries, beginning with the colonial period and continuing through decades of sharing a common language and cultural inheritance. The British influence, more than any other European movement, was a primary determinant in the development of psychology in the United States. This influence is readily apparent when we consider both the content and the methodology during the dynamic growth of twentieth-century American psychology. For this reason, the earliest expressions of modern psychological inquiry in Britain are of special significance.

Britain of the seventeenth and eighteenth centuries was an exciting place of political and economic progress as the nation moved steadily toward its position as the dominant power of the nineteenth century. The seventeenth century witnessed the consolidation of national interests in the British Isles (Map 8) under the monarchy and the established Church of England, both controlled by Parliament. The eighteenth century saw the extension of British influence throughout the

MAP 8. THE UNITED KINGDOM OF ENGLAND, WALES, IRELAND, AND SCOTLAND AT THE ACCESSION OF WILLIAM AND MARY, 1689. The shaded area shows lands controlled by the English rulers as early as the twelfth century. Also shown are the major cities and university towns (circles) and historical battle sites (squares).

world, with the American Revolution the only setback in colonial expansion. The writings of the epic poet John Milton (1608–1674) successfully challenged the British licensing regulations for censorship, which were abolished in 1694 by William III, and freedom of the press became a reality in Britain. English literature then entered a period of achievement through the writings of John Dryden (1631–1700), Daniel Defoe (1659–1731), and Jonathan Swift (1667–1745). The cause of science was also advanced by the intellectual freedom of Britain, and even the restored Stuart monarch Charles II (reigned 1660–1685) favored scientists with approval and support. As we have already seen in the brief biography of Isaac Newton, the policy of the British government and society rewarded scientific achievement and encouraged such efforts, seeing them as national assets.

ADVANCES IN SCIENCE

In mathematics, the spirit of Newton continued in England with the full development of calculus. The British also made great strides in the application of mathematics in physics. Joseph Black (1728–1799), working at the University of Glasgow, did pioneer experiments in oxidation and discovered the exchange of heat in substances changing from liquid to gas and gas to liquid. A later scientist, James Watt (1736–1819), applied this principle in his improvement of the steam engine.

Although the frictional properties of electricity had been known since ancient Greek times, a British scientist, Stephen Gray (1666?–1736), did the precise experimental work on the conduction of electricity. The American scientist and statesman Benjamin Franklin (1706–1790) described the identity between electric sparks and lightning in a letter he sent to the Royal Society in 1750. Franklin's famous experiment harnessing electric power using a kite during an electric storm earned him membership and an award from the Royal Society in 1754.

The astronomers of Britain contributed to the country's developing naval superiority. Edmund Halley (1656–1742) published his first paper on planetary orbits at the age of 20 and was instrumental in building the observatory at Greenwich, which in turn established the definitive methods of calculating longitudes to aid British shipping. Halley is perhaps best remembered for his successful prediction of the comet that bears his name. James Bradley (1693–1762) succeeded Halley as the royal astronomer at Greenwich and studied the annual parallax of the stars. Bradley also influenced the adoption of the Gregorian calendar in 1750, after 170 years of British resistance to the papal reform. British astronomy reached a peak with William Herschel (1738–1822), who not only discovered the planet Uranus but also developed models of the motion of the solar system through space.

Of interest in the field of biology during this period is Erasmus Darwin (1731–1802), grandfather of the advocate of nineteenth-century evolutionary theory, Charles Darwin. Educated as a physician at Cambridge and the University of Edinburgh, Erasmus settled into the practice of medicine and joined the Lunar Society in Birmingham. This group of scientists, which included Priestley, provided a forum for Erasmus Darwin's thoughts on biology. He proposed a theory of plant and animal evolution based on the needs of the organism. His grandson took this concept and introduced the principle of natural selection, greatly changing the shape of the scholarly examination of species diversification.

British medicine made slow advances. The anatomical teachings of William Hunter (1718–1783) and the animal experimentation of his brother John Hunter (1728–1793) improved the quality of medical education. Epidemics of infectious diseases periodically resulted in cries to clean up the filth of the cities, and pioneering work in immunization helped control outbreaks. Smallpox inoculations were tried in England by Charles Martland in 1718 and in Boston by Zabdiel Boylston in 1721. However, quackery was still prevalent and bloodletting was the standard cure for varieties of illnesses. Nowhere were superstition and intolerance

THE ROYAL OBSERVATORY AT GREENWICH, OUTSIDE LONDON.
(Courtesy, New York Public Library.)

greater than in the treatment of the mentally ill. For a small fee, visitors could enter Bethlehem (Bedlam) Hospital for Lunatics in London to stare at the antics of the inmates chained with ankle and collar restraints. Patients were "treated" with bloodletting, enemas, or mustard plasters on the head. The first attempts to treat the mentally ill humanely and to recognize their maladies as a disease were accomplished by the Quakers of Pennsylvania, who founded asylums to care for such individuals.

ADVANCES IN PHILOSOPHY

The major theme of psychology pursued by British philosophers centered on a faith in empiricism. Empiricism has been generally defined as the view that experience is the only source of knowledge. Accordingly, the theme that prevailed throughout the British tradition emphasized the development of the individual psychological framework through the accumulation of experiences. As a major implication of this stand, British psychological inquiry studied the relationship between the sensory input of experience and the operations of the mind.

The Early Empiricists

The initial formulation of British psychological opinion derived from issues originally proposed by Descartes. As Cartesian dualism stimulated French thought and evolved into sensationalism, so too we may find in Descartes' writings the basic position that promoted the empirical basis of British thought.

Thomas Hobbes. Recognized as the most brilliant philosopher of his age, Thomas Hobbes (1588–1679) published on wide-ranging subjects that collectively advocated the submission of society and the Church to the orderly rule attained only under absolute monarchy. His views on psychology were likewise radical and started the British empiricist tradition.

A rich uncle provided Hobbes with an Oxford education, and he secured employment with an aristocratic family, allowing him some protection for his anti-parliamentary and anti-Church views as well as offering him some financial security. He became acquainted with the great scholars of his time, including Galileo and Descartes, and served briefly as secretary to Francis Bacon. During Cromwell's protectorate he lived in exile in France, tutoring the children of the aristocracy, including the future Charles II, but soon alienated the devout Anglicans among his fellow expatriates by insisting that the Church should be subordinate to the monarch. Upon the restoration of Charles II in 1660, Hobbes received an annual pension. He spent the remaining days of his long life defending his views.

His most famous work, *The Leviathan, or Matter, Form, and Power of a Commonwealth, Ecclesiastical and Civil* (1651), was primarily intended as a political treatise, but Hobbes also expounded his essential views on psychology. His first principle of psychology asserted that all knowledge is derived through sensations. Moreover, he went on to suggest that nothing exists, internal or external to us, except matter and motion, thus grounding his psychology firmly in materialism. Sensations, then, are reduced to motion in the form of change. For example, we know the sensory qualities of light and dark by their contrast; we could know neither alone or absolutely. Hobbes did not agree with Bacon's faith in induction, but rather believed that deduction from experience constitutes the only valid method of knowing.

By proposing that the motion of physical objects in the environment gives rise to sensations, Hobbes used the rules of mechanical association to derive ideas and memory. For Hobbes and his successors in the British tradition, the mind acquires knowledge through associations. Associations are organized into general principles that are usually mechanical in nature and that describe how the relationships between sensations are formed into ideas. For Hobbes, the contiguity in time or place of events provided the association of sensations to form the idea unit, which is then stored by the mind in memory. It is the association mechanism that determines the sequence of ideas, defined as thought. The motivational principle in Hobbes' psychology was desire, ultimately a physiological process governed by seeking pleasure and avoiding pain. Thought sequences, according to Hobbes, are directed by desire and based on external sensation. Hobbes argued that dreams are thought sequences unregulated by sensations. The determinants of associative mechanisms built into thought sequences precluded the notion of free will for Hobbes. Rather, he viewed the will as a convenient label for the alternating desire and aversion confronting the individual with respect to a given object in the environment.

Hobbes described the universe as an environmental machine of matter in motion. His psychology portrayed the individual as a machine operating in this mechanized world. Sensations arise from motion and result in ideas, following the laws of association. The nervous system accomplishes the transference of sensory motion to muscular motion, so that the mind is a physical process centered in the brain. The major inconsistency in Hobbes' view lies in consciousness. His sequence of thought implies an awareness of a cognitive content, but he was unclear on the manner of movement from physically based sensations to nonphysical thought. Despite this problem, Hobbes established the importance of associations in understanding the accumulation of experiences. His successors in the British tradition amplified the empiricist position.

John Locke. In addition to being the major leader of the British empiricist tradition, John Locke (1632–1704) was one of the most influential political philosophers of post-Renaissance Europe. Born near Bristol, England, Locke was educated in classics and medicine at Oxford. He remained at Oxford as a don, studying the writings of Descartes and assisting Robert Boyle in his laboratory experiments. In 1667 he became physician to the Earl of Shaftesbury and, through him, began close contact with the political turmoil of the 1680s. Because of his identification with Shaftesbury, he eventually had to flee to the Netherlands, where he remained until the revolution that deposed James II and brought William and Mary to the throne by the invitation of Parliament in 1688. Locke's political views asserted that individual abilities are determined not by heredity but by environment or experience and that the sole appropriate government is by authorization of the governed. These views justified the parliamentary invitation of new monarchs. Locke's political views also influenced some of the founding fathers of the American republic, such as Thomas Jefferson, John Adams, and James Madison.

JOHN LOCKE (1632–1704). (Courtesy, New York Public Library.)

Locke's psychological views were expressed in his *Essay Concerning Human Understanding* (1690). Extending Hobbes' first principle, Locke stated, "*Nihil est in intellectu nisi quod prius fuerit in sensu*—There is nothing in the mind that was not first in the senses." This principle was affirmed in Locke's description of the mind at birth in terms of a *tabula rasa*, or blank slate, upon which the accumulation of life's experiences are gradually impressed to constitute the entire contents of the mind. Locke rejected other sources of knowledge, innately endowed through God or otherwise built into our mental structure at birth. Rather, all knowledge, including our ideas of God or morality, is derived from experience. He distinguished between sensations, which are physical, and perceptions, which are the reflected products of sensations. The units of the mind, called ideas, are derived from sensations through self-reflection. Further, he affirmed that physical objects have inherent primary and perceived secondary qualities. The primary qualities are the properties of the objects as they exist—their volume, length, number, motion. The secondary qualities, however, are produced by us and attributed to the objects in the process of perceiving—sounds, colors, odors, tastes. This distinction led Locke into the dilemma of whether objects exist in themselves as substances. Locke concluded that there are two kinds of substances. Material substances exist in the physical world, but we know them only through their primary qualities. Mind substances exist as mental elements and are our perceptions of objects.

In contrast to the French sensationalists, who eliminated the need for the mind by equating it with sensations, Locke's empiricism has definite need for the concept of a mind. However, we can characterize this mind construct as predominantly passive; the denial of innate ideas coupled with the dependency on sensory ideas limits the mind's role to reacting to the environment. Locke, however, did reserve two important operations for the mind. The first was association. Although less associationistic than Hobbes, Locke believed that the mind links together sensations to form perceptions by the principles of logical position or chance. His notion of logical position was broader than Hobbes' contiguity principle; it meant that the contiguity, the contingent relationship, or the meaning of two or more events would result in the association of those events. Associations by chance are spontaneous linkages without an apparent logical position. They constitute what we today call superstitious reinforcement. The second type of mental operation was reflection. Through reflection the operations of the mind in themselves produce a new or compound idea based on the simple ideas derived from sensation. Locke's view is at odds with Hobbes' position because Locke believed that reflection may be viewed as an activity of the mind, only remotely related to the sensory level.

As we have already seen and will explore further below, Locke's views were highly influential. His psychology may be described as rational empiricism, as he succeeded in retaining the need for the mind construct while discarding the theological implications of the soul. Others such as Condillac, however, were able to take the basic teachings of Locke and, by dropping the reflective operations of the mind, use them to question the need for the mind construct. Nevertheless, Locke's environmental determinacy provided the staging for the remainder of the British empiricist movement.

George Berkeley. George Berkeley (1685–1753) was a fascinating character because his interest in Locke's notion of mental perception led him to deny reality. Berkeley was born in County Kilkenny, Ireland, and entered Trinity College, Dublin, at age 15. By age 29 he had completed three significant works, including *An Essay Towards a New Theory of Vision* (1709), which contained his important psychological views. Berkeley became an Anglican clergyman and in 1728 set out for the New World to establish a college for the spread of the Gospel among the "American savages." He reached Newport, Rhode Island, and spent 3 years among the leading New England intellectuals, including Jonathan Edwards. Whitehall, his home near Newport, is now a museum containing artifacts of Berkeley's stay in colonial America. However, the funds for his college never arrived from England, and he had to return to Britain. In 1734 he was appointed Bishop of Cloyne in Ireland.

According to Berkeley's view, if all knowledge is derived from the senses, reality exists only to the extent that the mind perceives it. Locke had tried to salvage reality by his notion of primary qualities possessed by objects themselves. However, Berkeley asserted that we have no way of proving the existence of primary qualities independent of the senses (that is, through secondary qualities), so he dismissed the notion and asserted that sensation and perception are the only reality about which we can be certain. Berkeley used the principles of associations to explain the accumulation of knowledge. Simple ideas of sensory origin are compounded or constructed to form complex ideas. This mechanical coupling adds nothing in the association process, so that complex ideas are directly reducible to simple elements. Berkeley's association principle, active during perceptual processes, allows us to acquire knowledge of the environment. Berkeley explained depth perception through associations. That is, two-dimensional perception is readily accommodated by the physiology of the retina. However, the third dimension of depth results from our experiences with objects of various distances and our movements toward or away from them. An association is formed between the ocular sensation and our experience, producing the perception of depth.

Berkeley's solution to the problem of reality was that God, not matter, is the source of our sensations, and God provides the necessary order to our sensations. Some critics saw Berkeley's position as absurd. The question of the existence of physical objects, independent of a perceiving mind, has sometimes been proposed in the form of the situation of a tree falling in a forest: Does the tree make any noise? Berkeley would state that the tree could not make any sound without a mind to hear it. Indeed, for Berkeley there would be neither falling tree nor forest without a mind. Nevertheless, Berkeley's views represent a progression from Descartes that fortified the empiricist position against dismissing the mind as the sensationalists had done. As Descartes asserted "*Cogito ergo sum*," Berkeley held that "*Esse est percipi*—To be is to be perceived." To paraphrase Boring's (1950, p. 184) summary of the progressive relationship among Descartes, Locke, and Berkeley: Berkeley saw the problem not as how the mind is related to matter (Descartes), nor how matter generates the mind (Locke), but how the mind generates matter.

David Hume. Although agreeing with Berkeley's conclusion that matter, independent of perception, cannot be demonstrated, David Hume (1711–1776) applied the same strategy to mind and denied its existence. He was born in Edinburgh of a moderately affluent family and reared in the Calvinistic tenets of Scottish Presbyterianism. He entered the University of Edinburgh at an early age but left after 3 years to devote himself completely to philosophy at the expense of his childhood religion. He took various positions as secretary and tutor to supplement his inherited income. His initial writings on psychology, politics, and religion gained him little attention, but gradually his attacks on established Christian beliefs earned him a controversial reputation as an atheistic political theorist. Finally, in 1752, he was elected librarian of the law faculty at Edinburgh. With access to a vast literary collection, he wrote his *History of England* (1754–1761) and secured praise for his admirable scholarship.

Hume's psychological works were contained in *A Treatise of Human Nature* (1739), later elaborated in *An Enquiry Concerning the Human Understanding* (1748). He accepted the basic empirical premise that all ideas are ultimately derived from sensation, and acknowledged the distinction between primary and secondary qualities proposed by Locke. However, he defined mind solely in terms of the sensations, perceptions, ideas, emotions, or desires of a person at any given point. In so doing, like Berkeley, he denied matter because we know our mental world only. Moreover, by confining "mind" to only ongoing sensory and perceptual processes, any additional spiritual characteristics of the mind are unnecessary. Accordingly, "mind" for Hume was the transitory collection of impressions. The mental operations of reflection suggested by Locke were dismissed. Associations are recognized as compelling links of sensations, links that are formed by the contiguity and similarity of events. Hume's skeptical account adopted a very passive view of association processes, far removed from Locke's notion of reflection. Even such basic relationships as cause and effect were illusory for Hume. As an example, Hume cited the perception of a flame followed by the perception of heat. Although we may attribute the heat to the causal agency of the flame, Hume insisted that all we have observed is a succession of events, and we have simply imposed the cause–effect relationship as derived from custom. Thus, Hume extended Berkeley's skepticism of matter to a denial of the traditional Cartesian notion of the mind. In its place he promoted the role of ideas to account for mental activity.

Personal freedom was also an illusion for Hume. Because we are determined by the momentary influx of sensory events, any subjective freedom is simply some idealistic concept again taught to us by custom or religion. The primary motivational construct for Hume was based on emotion or passion governed by the seeking of pleasure and avoidance of pain. Indeed for Hume it is the antagonism or tension between emotions that results in their control, or ethical constraint, not the pretense that reason, as a higher mental process, should control emotions. Hume believed that reason is slave to the emotions. The motivational states derived from emotional interplay are integrated and mediated by physiological mechanisms.

Thus Hume succumbed to reductionism. Following Berkeley's inadvertent conclusion about matter, Hume taught a most passive view of empirical psychology.

He viewed human activity as reactive and having little initiative or control of the environmental events impinging on the organism. By identifying mind solely with its functions and nothing else, Hume questioned the need for a mind construct.

David Hartley. David Hartley (1705–1757) was originally trained as a member of the clergy, but found biology more to his liking and became a physician. After spending considerable time collecting data, he published *Observations on Man* (1749), containing his views on psychology. Essentially, Hartley established a physiological basis for Hume's brand of empiricist psychology. Extending Hobbes' and Locke's principles of associations as responsible for the formation of ideas and storage in memory, Hartley advocated the explanation of all human activity, including emotion and reason, through the mechanism of association. For Hartley, associations were formed by the contiguity of events and strengthened by repetition. Further, he stated that fiber connections of the brain comprise the correlates of all mental operations. He believed that the vibrations of brain fibers form the basis of ideas. Hartley viewed nerves as solid tubes that are set in motion by external stimuli, causing vibrations that transmit the stimulation to various parts of the body. The neural vibrations in turn stimulate smaller vibrations in the brain, which Hartley argued were the physiological basis of ideas, thus proposing a physical mechanism that underlies a so-called mental operation.

Hartley's importance in the British empiricist movement was in his role as a synthesizer. He defined his psychology in the empirical mold suggested by Hobbes and fully elaborated by Locke. Accepting the material skepticism of Berkeley and the mental skepticism of Hume, Hartley took the latter's reliance on the association of ideas and built for it a physiological basis. According to Hartley, every mental activity has a concomitant physiological activity; the association of ideas is the mental aspect of the sensory association of events occurring together in time and place. Hartley's physiological psychology brought together trends that resembled the psychology of Condillac and his followers in France. However, Hartley made a significant distinction in retaining the need for some notion of mental activity.

To summarize briefly at this point, the early British empiricists presented a psychology resting firmly on experience. Sensory input constituted the first state of the mind. The critical mechanism relating the sensory level to higher mental processes was associations. Thus, what we might call learning occupied a critical position in early British psychology. The tendency to reduce such mental operations to simpler ideas or sensations was readily seen by Hume and Hartley. Such reductionism, as in French thought, is a problem for psychology because the logical implications of reduction eliminate the very need for psychology. The successors to the early tradition attempted to remedy the situation by qualifying radical empiricism.

Scottish Common Sense

The eighteenth century in Scotland was a period of intellectual activity centered around the universities in Edinburgh and Glasgow. We have already seen in Hume a major figure in the development of empiricism. However, Hume was rather atypical of the Scottish Enlightenment because he fits more appropriately into the British tradition. Most of the philosophers and literary contributors to the Scottish Enlightenment were more independent of British thought, perhaps as a reaction to British political domination or as a reflection of the traditional link between Scotland and France. At any rate, for psychology, the Scottish writers succeeded in shaking the foundations of British empiricism by highlighting the absurdity of skeptically denying the existence of matter and mind.

Thomas Reid. While teaching at Glasgow, Thomas Reid (1710–1796) wrote his *Inquiry into the Human Mind on the Principles of Common Sense* (1764), which became the cornerstone of his successors in Scotland. Reid took issue with the skepticism that had led Berkeley and Hume to extreme doubt and reductionism. Rather, acknowledging Locke's distinction between the primary and secondary qualities of physical objects, Reid argued that the primary qualities justify belief in the reality of physical objects. That is, he believed that we perceive objects directly; we do not perceive sensations arising from objects. He viewed secondary qualities not as projections of the mind but as mental judgments stimulated by the objects. Thus secondary qualities make sensations the product of a true interaction between physical objects and mental operations.

Reid proposed that these principles of common sense are instinctive parts of a person's constitution, taken for granted in daily life and their value continually confirmed. In contrast, he viewed the metaphysical discourses of Berkeley and Hume as intellectual games. Not only are objects present in reality, but ideas need a mind contained in the self. Thus Reid used his common sense to save empiricism from the sterile path that Hume had followed.

Thomas Brown. Another figure of importance from the Scottish Enlightenment was Thomas Brown (1778–1820), a student of Reid. Essentially, Brown emphasized the role of associations in mental operations and restored the importance of associative processes in empiricism. However, his views on associationistic processes were less mechanical than those of Hartley and Hume. Arguing that associations may be suggestions, he used associations to propose an explanation for mental consciousness. He introduced the notion of mental chemistry as a contrast to the reductionistic notion of mental compounding suggested by the early empiricists. Brown described two kinds of suggestion: simple and relative. Simple suggestion produces complete ideas; for example, the title of a musical work can evoke an entire thought sequence of melodies. Relative suggestion involves nonsensory input, resulting in exclusively mental operations. For example, multidimensional mathematics is studied in topology and is not represented through sensory experience. Thus Brown attempted to broaden the basis of associations by utilizing suggestion to explain the complexity of mental operations.

Scottish common sense was like a breath of fresh air for the empirical movement. By absorbing both the spirit and the content of the Scottish writers, the later British empiricists were able to broaden the scope of their consideration of the mind and to lay the foundation of modern psychology. Without the contributions of common sense psychology, empiricism might have stagnated and withered in the sterility of skepticism.

The Later Empiricists

The major focus of the later empiricists concerned principles of association. Recognizing the environmental determinacy of the early empiricists and bolstered by the common sense of Reid and Brown, they viewed the contents of the mind in terms of the acquisition of experiences by the individual. Association was the mechanism of acquisition, and an emphasis on learning and memory evolved in British psychology.

James Mill. A throwback to the earlier empiricists, James Mill (1773–1836) was educated at the University of Edinburgh and became a journalist in London. He began writing his *History of British India* in 1806 and completed it in 1818, providing an indictment against British colonial management. In 1808 Mill met Jeremy Bentham (1748–1832), the spokesman for the movement called utilitarianism in British political philosophy. Bentham's utilitarian views had a major impact on Mill's psychology. Briefly, Bentham dismissed the theological and metaphysical assumptions behind social institutions, such as divine law, natural law, and the "rights" of people. Rather, he held that the usefulness of an act for the individual determines its morality and lawfulness. Thus, the ultimate test for any action or law is whether it adds to the benefit and happiness of people. Bentham defined happiness in terms of the individual's seeking pleasure and avoiding pain. Although his work had its greatest impact on British legal and social institutions and led to many reforms, James Mill was sufficiently captivated to become the champion of Bentham's views in psychology.

Mill's major contributions to psychology were contained in his *Analysis of the Phenomena of the Human Mind* (1829). He held the extreme associationist position that ideas are the residual of sensations when the physical stimulating object is removed in the environment. His view of association postulated complete mental passivity; he saw contiguity between events as giving rise to associations. Mill argued that thought sequences are trains of successive or synchronous ideas, and they mimic the order of sensations. Moreover, complex ideas are simply aggregates of simpler ideas and reducible to them. Thus, Mill gets into the absurdity of reducing complex psychological constructs such as the self to constituent, additive components. Accordingly, his system leaves little room for any dynamic synthesis; rather, it sees the mind as only reacting to sensation.

Mill's background was humanistic, and his lack of appreciation of the physiological basis of sensory processes probably hindered his conceptualization of the possibilities of sensory mechanisms that could have admitted some flexibility into his psychology. Thus his views on the additivity of mental processes led him to reduce psychology to absurdity. Nevertheless, Mill saw the utility of associations as a means of explaining environmental determinacy. His son, John Stuart Mill, succeeded in moderating the extreme view of associations as mental combinations.

John Stuart Mill. Subjected to a severe regimen of education as a child, John Stuart Mill (1806–1873) lived timidly under his father's eye for 30 years, and then burst forth with independent opinions after the latter's death. We have already

JOHN STUART MILL (1806–1873).
(Courtesy, New York Public Library.)

noted that John Stuart Mill's empiricism found common points of agreement with Comte's positivism, and indeed much of that commonality stemmed from the influence of Bentham. Mill's major writings in psychology were contained in his *System of Logic* (1843), which was immediately popular and went through eight editions before Mill died. This work served as the standard scientific reference for many years.

Mill's empirical psychology was firmly based on induction. He argued that human thought, feelings, and actions are the province of psychology. The goal of psychology is to try to find underlying causality in human cognitive and emotional activity. Rather than viewing associations as mental combinations—as his father did—John Stuart Mill saw associations as governed by three principles:

1. Every experience has a corresponding idea.
2. Contiguity and similarity produce associations.
3. The intensity of an association is determined by the frequency of its presentation.

Moreover, in his views on habit formation, Mill recognized the subjective perception of relationships between events and agreed with the notion of mental suggestion offered by Brown. Thus Mill acknowledged the mind as generating the complex out of the simple.

Mill also noted contemporary advances in the neurophysiology of the brain, but was not prepared to settle into the materialistic basis of thought proposed by Hartley and, to some extent, his father. He argued that psychology, by virtue of the changing social context of humanity and concomitant individual differences, would not evolve laws to predict human activity. Rather, he was content to advocate the search for "empirical laws," which were expressions of systematic variation. We shall see in Chapter 10 that others, such as Galton, pursued empirical laws to develop the statistical techniques of systematic covariation, or correlation.

Alexander Bain. Educated at the University of Aberdeen, Alexander Bain (1818–1903) was impressed with the essential compatibility between philosophy and the natural sciences. Although his views on psychology, initially formed by 1855, were empiricist in approach and inductive in method, he later modified his system to conform with the evolutionary theory of Darwin. We shall consider the tremendous impact of Darwin at a later point; however, natural selection confirmed Bain's stress on the importance of the physiological correlates of psychological events. He argued for the concept of psychophysical parallelism, which holds that any given event has both a psychological side and a physical side. Bain believed that the body responds to the physical limitations of cause-and-effect relationships and conforms to quantitative laws of movement, or reflexology. The mind is not quantifiable, according to Bain, but possesses innate abilities or aptitudes.

Bain's major works in psychology were *The Senses and the Intellect* (1855) and *The Emotions and the Will* (1859), and he founded the philosophical journal *Mind*, which dealt almost exclusively with psychological issues. Bain's empiricist views of the mind relied on the association principles derived from contiguity, similarity, and agreement among environmental events. This last point was derived from the recognition that present experience is based on past events. Bain was quite aware of nineteenth-century advances in neurophysiology, and incorporated such findings into his work by asserting the possibility of spontaneous action of the nervous system. Thus, because of the biological makeup of the individual, Bain allowed for psychological activity independent of experience. Accordingly, he moved away from the sterile materialism of Hartley and James Mill, allowing British empiricism to finish the nineteenth century in a flexible position.

The entire course of British empiricism encompassed a variety of interpretations and emphases in describing the mind. However, all of the empiricists accepted the view that the mind is determined by individual experience. Further, they agreed that the predominant activity of the mind is associating sensations and ideas. Psychology as a form of scientific inquiry was seen as a legitimate and acceptable intellectual endeavor of British philosophy.

CHAPTER SUMMARY

The relative freedom and political stability of seventeenth- and eighteenth-century Britain produced an intellectual milieu amenable to advances in the natural sciences and philosophy. The major theme of British psychological thought was empiricistic, emphasizing knowledge acquired through sensation. The mechanism of this acquisition process was association. Founded by Hobbes but fully articulated by Locke, British empiricism retained the necessity of the mind construct while underlining the importance of sensations. Berkeley, Hume, and Hartley evolved skeptical positions concerning the reality of matter and mind that could have left the British movement in the same sterile position as French sensationalism. In addition, James Mill, although he was somewhat salvaged by the utilitarian influence, reduced associations to mental compounding. However, the Scottish common sense writers succeeded in restoring empiricism to a more flexible and open-ended position that recognized complex and integrative psychological phenomena. Thus, the later empiricism of John Stuart Mill, while adhering to scientific inductive methods, adopted a broadly based model of psychology that viewed mental operations and physiological processes as complementary and necessary dimensions of psychological inquiry. By the nineteenth century British philosophy was providing strong support for the study of psychology.

BIBLIOGRAPHY

Primary Sources

BERKELEY, G. (1963). An essay towards a new theory of vision. In C. M. Turbayne (Ed.), *Works on vision*. Indianapolis: Bobbs-Merrill.

HUME, D. (1957). *An enquiry concerning the human understanding* (L. A. Selby-Bigge, Ed.). Oxford: Clarendon Press.

LOCKE, J. (1956). *An essay concerning human understanding*. Chicago: Henry Regnery.

MILL, J. S. (1909). *Autobiography*. New York: P. F. Collier.

MILL, J. S. (1973). *Collected works*. Toronto: University of Toronto Press.

RAND, B. (1912). *The classical psychologists*. New York: Houghton Mifflin.

General References

BORING, E. G. (1950). *A history of experimental psychology*, 2nd ed. Englewood Cliffs, NJ: Prentice Hall.

COPLESTON, F. (1964). *A history of philosophy, Vol. 5, Modern philosophy: The British philosophers, Part I—Hobbes to Paley*. Garden City, NY: Image Books.

COPLESTON, F. (1964). *A history of philosophy, Vol. 5, Modern philosophy: The British philosophers, Part II—Berkeley to Hume*. Garden City, NY: Image Books.

DURANT, W., & DURANT, A. (1965). *The age of Voltaire*. New York: Simon & Schuster.

DURANT, W., & DURANT, A. (1967). *Rousseau and revolution*. New York: Simon & Schuster.

DURANT, W., & DURANT, A. (1975). *The age of Napoleon*. New York: Simon & Schuster.

MAZLISH, B. (1975). *James and John Stuart Mill: Father and son in the nineteenth century*. New York: Basic Books.

Studies

ALBRECHT, F. M. (1970). A reappraisal of faculty psychology. *Journal of the History of the Behavioral Sciences, 6*, 36–40.

ARMSTRONG, R. L. (1969). Cambridge Platonists and Locke on innate ideas. *Journal of the History of Ideas, 30*, 187–202.

BALL, T. (1982). Platonism and penology: James Mill's attempted synthesis. *Journal of the History of the Behavioral Sciences, 18*, 222–230.

Bricke, J. (1974). Hume's associationistic psychology. *Journal of the History of the Behavioral Sciences, 10,* 397–409.

Brooks, G. P. (1976). The faculty psychology of Thomas Reid. *Journal of the History of the Behavioral Sciences, 12,* 65–77.

Dreuer, J. (1965). The historical background for national trends in psychology: On the nonexistence of British empiricism. *Journal of the History of the Behavioral Sciences, 1,* 126–127.

Greenway, A. P. (1973). The incorporation of action into associationism: The psychology of Alexander Bain. *Journal of the History of the Behavioral Sciences, 9,* 42–52.

Heyd, T. (1989). Mill and Comte on psychology. *Journal of the History of the Behavioral Sciences, 25,* 125–138.

James, R. A. (1970). Comte and Spencer: A priority dispute in social science. *Journal of the History of the Behavioral Sciences, 6,* 241–254.

Miller, E. F. (1971). Hume's contribution to behavioral science. *Journal of the History of the Behavioral Sciences, 7,* 154–168.

Moore-Russell, M. E. (1978). The philosopher and society: John Locke and the English Revolution. *Journal of the History of the Behavioral Sciences, 14,* 65–73.

Mueller, I. W. (1956). *John Stuart Mill and French thought.* Freeport, NY: Books for Libraries Press.

Petryszak, N. G. (1981). Tabula rasa—Its origins and implications. *Journal of the History of the Behavioral Sciences, 17,* 15–27.

Robinson, D. N. (1989). Thomas Reid and the Aberdeen years: Common sense at the wise club. *Journal of the History of the Behavioral Sciences, 25,* 154–162.

Robson, J. M. (1971). "Joint authorship" again: The evidence in the third edition of Mill's Logic. *Mill's News Letter, 6,* 15–20.

Shearer, N. A. (1974). Alexander Bain and the classification of knowledge. *Journal of the History of the Behavioral Sciences, 10,* 56–73.

Smith, C. U. (1987). David Hartley's Newtonian neuropsychology. *Journal of the History of the Behavioral Sciences, 23,* 123–136.

Webb, M. E. (1988). A new history of Hartley's Observations on Man. *Journal of the History of the Behavioral Sciences, 24,* 202–211.

8

Mental Activity: The German Tradition

Advances in Science
Advances in Philosophy
  The Founders
    Gottfried Wilhelm von Leibniz
    Christian von Wolff
    Immanuel Kant
  The Psychology of the Self-Conscious
    Johann Friedrich Herbart
    Friedrich Eduard Beneke
    Rudolf Hermann Lotze
    Arthur Schopenhauer
    Eduard von Hartmann
Chapter Summary

The German philosophical basis of psychology took greater inspiration from Spinoza than from Descartes. The mind–body dualism of the latter gave rise to the distinction between physiological and psychological levels of study, which provided a conceptual framework contrasting the two realms of investigation. The French sensationalists blurred the distinction through reductionism; the British retained the distinction but allowed some mentalistic functions, such as associations, with a physiological basis. Spinoza conceived of physiological and psychological processes as descriptions of the same entity, which resulted in an emphasis on continuity in the activity of human functioning. Thus, rather than viewing physiology and psychology as contrasting areas of investigation, he viewed them as integrative aspects of human activity. The German model of psychology was not confounded by the contrast between sensations and ideas, as both were seen as aspects of the same active process. Before examining the specifics of the German model, we briefly consider the diverse intellectual climate of Germany.

German history has been characterized by political fragmentation. Surviving the Middle Ages and the Renaissance as a loose confederation of small kingdoms, principalities, and bishoprics, Germany entered the modern era under the nominal leadership of the Holy Roman Emperor, one of the last vestiges of the feudal political structure. Moreover, Germany was sharply divided by the Reformation and by the attempt of the Roman Catholic Church to regain ground lost during the Counter Reformation. The disastrous Thirty Years' War (1618–1648) was fought over the religious allegiance of the Protestant North and Catholic South of Germany.

Amid this political and religious confusion, the German state of Prussia developed in the northeastern portion of Germany. Modern Prussia evolved through the combination of the estates of the Teutonic Knights and the Brandenburg lands. In 1411, Frederick of Hohenzollern became ruler of Brandenburg, with headquarters in Berlin. His successors continued a steady policy of small acquisitions, so that by 1619 the Hohenzollerns ruled over Brandenburg as well as East Prussia. In the nineteenth century, the family presided over the unification of all German lands under their chancellor Otto von Bismarck (1815–1898) and the German Empire (Map 9) lasted until the last Hohenzollern emperor abdicated in 1918.

In the seventeenth and eighteenth centuries Prussia took the lead in German cultural activities, which reached a zenith under the versatile Frederick the Great (reigned 1740–1786). Under his leadership Prussia grew in wealth and power, and the population prospered as education spread and religious tolerance prevailed. Frederick ran an efficient government that was merciless in stamping out bureaucratic corruption. He fostered scientific societies, invited scholars from throughout Europe to Berlin, and even pursued a learned correspondence with Voltaire. University professors were appointed and paid by the government, and German replaced Latin as the language of instruction. German literature flowered and reached its fullest expression in the writings of Johann Wolfgang von Goethe (1749–1832). German music enjoyed a period of creativity unrivaled in history with the contributions of the family of Johann Sebastian Bach (1685–1750) and culminating in the genius of Austrian Wolfgang Amadeus Mozart (1756–1791) and Ludwig van Beethoven (1770–1827).

ADVANCES IN SCIENCE

As in France and Britain, seventeenth-century advances in science saw the triumph of mathematics and physics in Germany. Otto von Guericke (1602–1686) developed the barometer and invented an air pump that examined the physics of the vacuum. Gabriel Fahrenheit (1686–1736) proposed a system of temperature measurement with a column of mercury, leaving his name attached to the resulting scale. Ehrenfried von Tschirnhaus (1651–1708), who explored the basis of radiation from the sun, studied heat absorption.

Perhaps the greatest mathematician of the eighteenth century was Leonhard Euler (1707–1783), who was born in Basel and at the age of 26 became director of mathematics for the St. Petersburg Academy of Sciences. Later, he took a similar post in Berlin, but eventually returned to Russia. He applied calculus to light vibrations and determined the systematic relationship between density and elasticity. Moreover, he contributed much to establishing the modern forms of geometry,

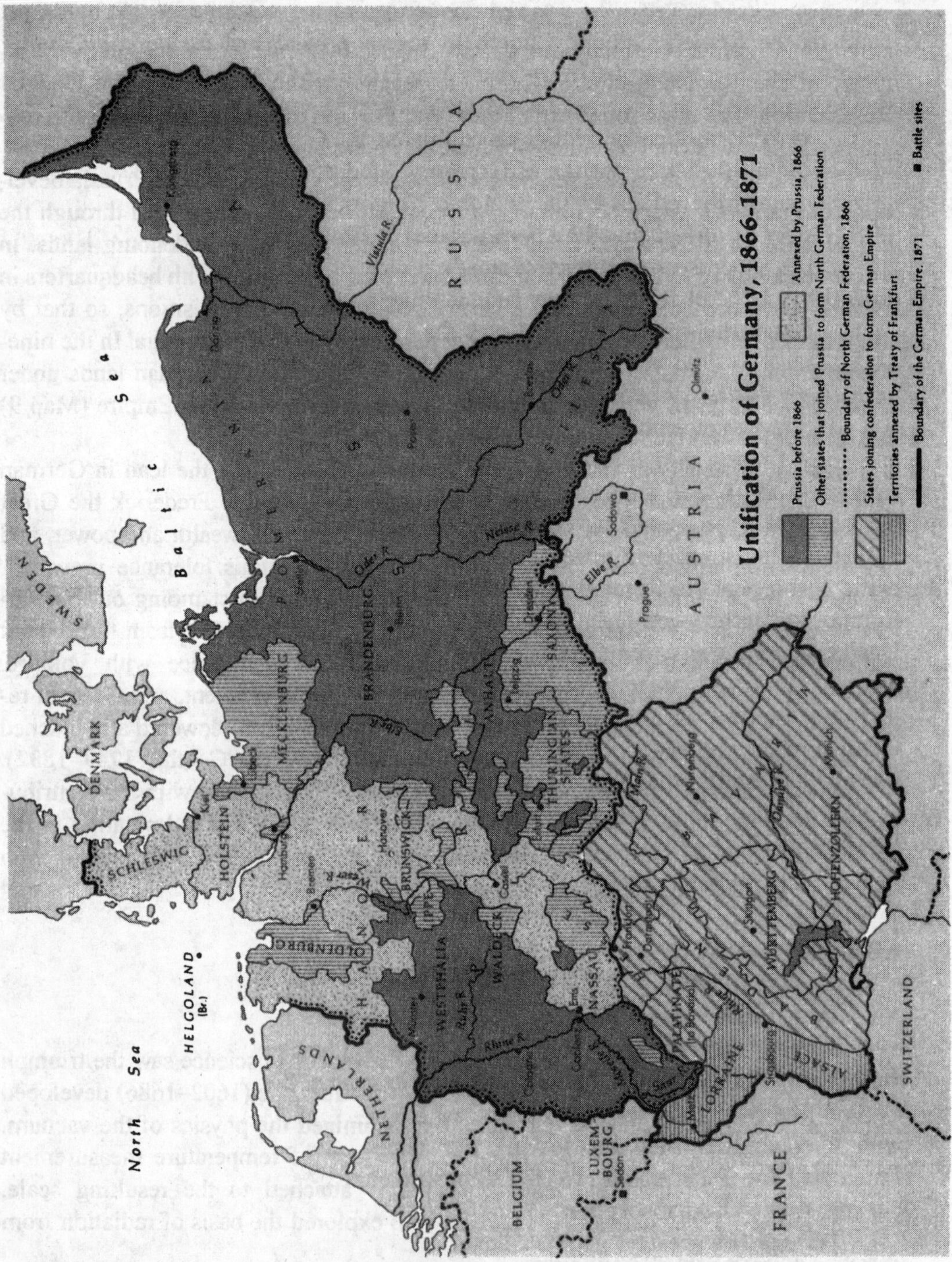

MAP 9. THE GERMAN STATES AT THE TIME OF UNIFICATION UNDER BISMARCK. At the end of the eighteenth century, Poland was occupied by Prussia, Russia, and Austria, and ceased to exist as an independent state.

trigonometry, and algebra. His work charting the planets and lunar positions gave a basis to determining longitude. Euler's genius, and his extensive travels, produced many students who spread his teachings throughout Europe.

Advances in electrical theories also occured. George Bose presented a paper to the Berlin Academy in 1742 arguing that the aurora borealis is of electrical origin, and then went on to show how electricity can be used in explosives. In 1745, E. G. von Kleist developed a battery capable of sustaining an electric charge for several hours, leading Daniel Gralath of Leiden to develop stores of potent electrical charges in 1746 by using jars placed in series.

Following in the tradition of the spokesman of modern botany, Swedish scientist Carolus Linnaeus (1707–1778), Philip Miller in 1721 wrote of plant fertilization by bees. In 1760, Josef Krölreuter reported extensive experiments in the physiochemistry of pollination. In 1793, Konrad Sprengel investigated cross-fertilization and postulated the basis of plant anatomy, and in 1791 Josef Gärtner finished an encyclopedic study of the fruit and seeds of plants, which became a classic work of nineteenth-century botany.

Turning to medicine, it is perhaps unfair to include the most famous quack of the eighteenth century, Franz Anton Mesmer (1734–1815), in a chapter on German culture. He was German only by virtue of his birth and education in German speaking Vienna, and spent his lucrative years catering to the idle rich in Paris before being forced to flee the Revolution to exile in Switzerland. However, we shall momentarily consider an active model of mental processing, and Mesmer's teachings are in clearer agreement with the dynamics of mental activity than are the passive views discussed earlier.

Mesmer's thesis for a doctorate at the University of Vienna revived speculation in astrological influences on personality, which he conceived as apparent in the form of magnetic waves. After opening an office for faith healing, he was exposed as a charlatan by the police, who gave him 2 days to leave Vienna. Arriving in Paris, he published *Mémoire sur la découverte du magnetisme animal* (*A statement on the discovery of animal magnetism*; 1779). Affluent patients soon arrived to be "mesmerized," a process wherein he touched them with a wand or stared in their eyes to the point of making them immobile and susceptible to suggestion. He even provided magnetic tubes filled with a hydrogen sulfide solution to effect his cures. After Mesmer's demise, others used similar curative techniques throughout the Continent and in Britain. Essentially, these quacks struck a responsive chord among a population serviced by primitive medicine and barbaric methods of treating behavioral abnormalities. It was not until late in the nineteenth century that more sophisticated scholars in France gave hypnotism an aura of respectability, prompting the young Sigmund Freud to study under them.

With the exception of Mesmer's work, science in Germany enjoyed success similar to that in France and Britain. Moreover, the Prussian government's efficiency and support of scientific enterprises provided a climate for the ascendancy of the German university system in the nineteenth century, and within that system psychology formally emerged.

ADVANCES IN PHILOSOPHY

The psychology that emerged within German philosophy differed from the sensationalism of France and the empiricism of Britain. Common to this German school was the essential activity of the mind. Whereas the other schools of thought looked to the environmental input of the mind, these German figures initially looked to the preexisting dynamics of the mind to order the environment.

The Founders

Descartes' writings were known and were influential in German philosophical circles. Unlike the French and British scholars, German philosophers emphasized Descartes' views on the activity of the mind, especially innate ideas. However, more than Descartes, it was Spinoza who served as the intellectual forebear of German philosophy. Spinoza's attempt to realign philosophical inquiry away from theological determinacy while retaining the dynamic activity of the mind found loyal followers among German thinkers.

Gottfried Wilhelm von Leibniz. As a statesman, mathematician, and philosopher, Gottfried Wilhelm von Leibniz (1646–1716) lived a full life that tended to minimize discord and exemplify optimism, to the point of his simultaneous profession of Catholicism and Protestantism. His father was a professor of moral philosophy at the University of Leipzig, which Gottfried entered at age 15. Later denied a doctoral degree because he was only 20 years old, he went to Nuremberg, and his thesis so impressed the faculty that he was offered a professorship. Rejecting the offer for more exciting prospects, he secured a diplomatic post as counselor to the Archbishop of Mainz, a position that allowed him to travel throughout France and Germany and brought him into contact with contemporary intellectual leaders. Working independently, he published his works on differential calculus (1684) and integral calculus (1686) before Newton, although Newton had completed his formulations by 1666.

Leibniz's views on psychology were initially undertaken as part of a commentary on the publication of Locke's *Essay Concerning Human Understanding*. He expanded the commentary to a discourse in dialogue form, *Nouveaux Essais sur l'Entendement Humain* (*New Essays on Human Understanding*), which was completed by the year of Locke's death in 1704, but not published until 1765. Leibniz viewed the mind not as a passive receptor of experiences, but rather as a complex entity that transforms the input of sensations by both its structure and its functions. Leibniz edited the empirical dictum to say: *Nihil est in intellectu quod non fuerit in sensu, nisi ipse intellectus*—Nothing is in the intellect that has not been in the senses, except the intellect itself. Acknowledging Locke's mental operation of reflection, Leibniz argued that the ultimate dependency of reflection on sense data is unsatisfactory. He stated that the mind itself possesses certain principles or categories, such as unity, substance, being, cause, identity, reason, and perception. These categories are keys to understanding and are innate to the mind; they are not in the senses nor in physical objects. Without these categories, we would be aware only of a succession of motions or sensations, so that for Leibniz all ideas were innate. Leibniz also added the notion of continuity in describing the activity of the mind. Thus, *thinking* was viewed as an incessant activity, and the thinking process allowed for both conscious and unconscious dimensions.

To this point, Leibniz elaborated on Spinoza's views in response to Locke's passivity of the mind. Leibniz's original contribution to psychology was in his agent of activity, the monad. Perhaps borrowing the term from its prevalent use to describe small seeds created by God from which all matter and life grow, Leibniz employed "monadology" to describe the essential activity of the mind. Considering the diversification of life expressed in plants and animals, the problem of defining life itself leads to absurd division. A field of grain is composed of living entities—individual plants that in turn have growing stalks with living seeds. The seeds themselves may be individually divided into the living structures of embryo, endosperm, and seed coat. With the aid of a microscope, we can conceivably divide these seed structures into components, and so the endless divisibility continues. As Democritus taught, life may be found in the smallest of atoms, but if we confine our view of life to the extension of matter, we are left with the puzzle of the continuing divisibility of the components of life. Leibniz dismissed the definition of life as infinite divisibility in the search for basic atoms. Instead he offered the concept of monads, which he defined as unextended units of force or energy. Each monad is a separate, independent force asserting its uniqueness against all other centers of force. All living beings are composed of monads that define individuality and reflect the universe. The monad of an individual human being is mind, to the extent that it has sensitivity and responsivity. The monad grows and develops throughout life; change occurs because of internal, individual striving. The elements of life in the individual are the result of a collection of various monads, each with a specific purpose and direction and with varying degrees of consciousness. This aggregate becomes the living harmonious organism of the person under the organizational direction of the dominating monad of the soul. Whereas Descartes advocated psychic and physical interaction in the person, and Spinoza denied this interaction because the physical and the psychic are two aspects of the same entity, Leibniz denied the interaction and yet asserted independent physical and mental processes. In place of the interaction, the harmony of personality is achieved by the purpose and direction of individual monads orchestrated by the organization of God.

A number of important themes may be extrapolated from Leibniz's psychological views. First, the individual is not at the mercy of environmental determinants. Rather, a person's mind is structured to act on the environment. Second, the concept of monadology, although perhaps vague and abstruse, does offer an explanation for the dynamics of mental activity. Processes such as attention, selective memory, and the unconscious are easily accommodated in ways not permitted in empiricist or sensationalist frameworks. With Leibniz, German psychology was committed to the mind construct and capable of fully exploring the implications of mental energy.

Christian von Wolff. The son of a tanner who rose to be a professor at the University of Halle, Christian von Wolff (1679–1754) published a total of 67 books in an attempt to scrutinize all knowledge under the guiding principle of reason. Consistent with Leibniz and the major theme of German thought, Wolff rejected Locke's assertion of knowledge as dependent upon sensory input, but drew back from some of the difficulties of Leibniz's alternative, monadology. Wolff served as a transition figure between Leibniz and Kant by emphasizing mental activity and bodily activity as two separate, noninteracting processes. Wolff was one of the most praised scholars of his time and was decorated by both French and Prussian academies. Exiled from Prussia out of fear that his writings might encourage rebellion, he was invited to return as chancellor of the University of Halle upon the accession of Frederick the Great.

His major works on psychology were *Psychologia Empirica* (1732) and *Psychologia Rationalis* (1734). As suggested by the separate titles, he elaborated two approaches to psychology. The first was the more limited and dealt with the sensory process, not unlike what we have seen in the British tradition. However, in his *Rational Psychology*, he argued for a full elaboration of mental activity within the framework of Leibniz; that is, he asserted the active role of the mind in the formation of ideas. Like Leibniz, Wolff taught that body and mind are known by action and idea, respectively. Action and idea are parallel and independent processes. The body and sensory level operate mechanically under purposive design. The mind is governed by the determinacy of cause and effect, and it controls the environment by its categories. In so arguing, Wolff's rational psychology may also be described as "faculty psychology" in which the capacities, or faculties, of mental activity form the proper area for study of human understanding. Psychology, then, is defined as the study of mental faculties, and the uniqueness of the human mind transcends all forms of life.

Immanuel Kant. German psychology received its permanent imprint in rationalism from the writings of Immanuel Kant (1724–1804), one of the most influential philosophers of post-Renaissance Europe. He never journeyed from his birthplace in the capital of East Prussia, Königsberg (since 1945 known as the Soviet, now Russian, city of Kalingrad). In 1740, he began his studies at the University of Königsberg, where he was introduced to the writings of Wolff, although he concentrated on the natural sciences. From 1755, when he received his doctorate, until 1770, when he finally received a professorship in logic and metaphysics after two earlier rejections, Kant supported himself through meager fees as a tutor or as a *privatdozent*, a lowly private teacher whose pay was determined by the students.

As was the tradition at this time, all new professors gave an inaugural address in Latin before the university community. Kant chose to describe the sensible world and intelligible world. The sensible world for Kant meant sense information or the world of appearances, whereas the intelligible world was conceived by the intellect or reason. To this distinction Kant added the basic position that the

IMMANUEL KANT (1724–1804).
(Courtesy, New York Public Library.)

dimensions of time and space are not properties of the objective environment, but rather perceptual forms innate in the mind. Thus the mind is not the passive agent produced by sensations, as the empiricists suggested. The mind is an active entity governed by innate laws and structures, and it translates sensations into ideas. Kant's position implied a psychology of mental operations that is not solely dependent on sensory experience.

After 12 years of contemplation, Kant formalized his psychological views with his monumental *Kritik der Reinen Vernunft* (*Critique of Pure Reason*; 1781). By pure reason, Kant meant knowledge requiring no experiential proof, which he called a priori knowledge. Kant admitted that he was prompted to this undertaking after reading Hume, who had written that all reasoning is based on the notion of cause and effect, which in turn is actually an observation of sequences but has no reality; such relationships are intellectual artifacts. Kant wanted to rescue causation by showing it to be independent of experience and a priori knowledge, and inherent to the structure of the mind. He began by dividing all knowledge into empirical knowledge, which depends on sense experience, and transcendental knowledge, independent of experience. Kant accepted that all knowledge begins with sensations insofar as they provide stimulation to activate the operations of the mind. However, once that stimulation has occurred, the experience is then molded by the mind's inherent forms of perception and conception. The perceptual forms then transform the experience as the external sense of space and the internal sense of time. Reminiscent of

Aristotle's teachings on mental categories, the forms of conceptualization for Kant are independent of experience and mold an experience through mental categories, summarized as follows:

> *Categories of quality*: limitation, negation, and reality
>
> *Categories of quantity*: plurality, totality, and unity
>
> *Categories of relation*: substance and quality, cause and effect, activity and passivity
>
> *Categories of modality*: possibility and impossibility, existence and nonexistence, necessity and contingency

Each perception falls into at least one of these categories, so that perceptions are sensations interpreted by the inherent forms of time and space. Knowledge, then, is perception molded into an idea of judgment. The subjective experience of the individual is not the passive processing of sense impressions but the product of the mind operating on sensation.

In 1788, Kant finished another work of importance to German psychology, *Kritik per Praktischen Vernunft* (*Critique of Practical Reason*). Kant wanted to extend the earlier work to a consideration of morality to show that values are not a posteriori social traditions, but a priori conditions of the mind. To do this, he had to examine the will. Kant asserted that every person has a moral consciousness that is determined not by experience but by the structure of the mind. This consciousness is absolute and basically follows the Golden Rule. According to Kant, in our subjective world of perceptions and ideas—the only world we know—we are free to make judgments that conform to our moral consciousness. It was Kant's intention to give society social responsibility, relying on something more than human reason without resorting to theological arguments. By linking a priori moral consciousness with the will, he elevated the notion of volition to a level of great psychological importance.

Kant's system held that the objective world is unknowable and that sense data are ordered by the mind. Thus, all knowledge exists in the form of ideas. The materialism that dominated French thought and influenced British empiricism was impossible for Kant. At the same time, in contrast to Hume, Kant did not dismiss the objective world, because its existence is confirmed by the stimulating and initiating functions of sense data in the formation of ideas. Accordingly, Kant included both empiricism and rationalism, although his major impact on psychology rests in the latter. Finally, Kant's emphasis on the primacy of the will, along with his rationalism, provided a dominant theme for the future of German psychology and added a critical dimension to the definition of mental activity.

To summarize briefly at this point, the founders of the German psychological tradition presented a new perspective relative to the French and British views. They opted for a model of the mind that was clearly active and dynamic. Mental activity was not a new hypothesis. However, the German movement, crowned by Kant, was developed in full light of and in response to other, alternative models. Thus this particular outline of mental activity formed a powerful argument for preconceived notions of the nature of people. This movement determined the immediate course of psychology in Germany. Moreover, the German model established a standard of mental operations with which all psychological models from that point had to deal.

The Psychology of the Self-Conscious

After Kant, the German tradition of mental activity elaborated and modified details of Kantian psychology, but retained the essential activity of the system. Thus, by the nineteenth century, psychological discussion within German philosophy was confined to the assumption of mental activity, just as psychological issues within British philosophy assumed mental passivity.

Johann Friedrich Herbart. The title of Johann Friedrich Herbart's (1776–1841) major work in psychology may hold the record for the most encompassing scope: *Psychology as a Science Newly Founded upon Experience, Metaphysics and Mathematics* (1824–1825). Born in Oldenburg, Herbart received his doctorate at the University of Göttingen and taught there in philosophy and education. Although in his doctoral thesis he was at odds with Kant in certain details, he nevertheless falls into the dynamic tradition of mental activity started by Leibniz. In 1809, Herbart was appointed to the chair of philosophy previously occupied by Kant at Königsberg. He remained there until 1833, when he returned to Göttingen to take a similar position.

For Herbart, psychology was a science based on observation. In contrast to Kant's, Herbart's psychology was empirically based in experience. However, psychology was not an experimental science, like physics, and the central province of psychology, the mind, was not subject to analysis. Reminiscent of Pythagoras,

JOHANN FRIEDRICH HERBART (1776–1841). (Courtesy, New York Public Library.)

Herbart asserted that psychology should use mathematics to move beyond simple description and expound the relations of mental operations. The basic units of the mind are ideas, which have characteristics of time, intensity, and quality. Ideas are active in terms of a tendency toward self-preservation against opposing ideas. Accordingly, the dynamics of self-preservation and opposition explain the flow between the conscious and the unconscious presence of ideas. Herbart viewed those dynamics as a type of mental mechanics, analogous to physical mechanics.

Herbart dismissed from psychology physiological considerations and the use of the experimental method. Moreover, his metaphysics of mental operations leading to a system of mental mechanics appears inconsistent with his objection to analysis. Nevertheless, Herbart successfully moved German thought away from the pure rationalism of Kant toward a better appreciation of empiricism. In addition, he must be credited with attempting to establish a psychology that was independent of philosophy and physiology.

Friedrich Eduard Beneke. Friedrich Eduard Beneke (1798–1854) was a contemporary opponent of Herbart. His major work, *Psychological Sketches* (1825–1827), was condensed and subsequently published as *Psychology as a Natural Science* (1833). Beneke's choice of the latter title was startling by German philosophical traditions, and would be considered somewhat misleading by a contemporary interpretation of natural science methodology. In contrast to Herbart, Beneke envisioned a psychology that included physiological data. Moreover, he held that psychology is not derived from philosophy, but rather is the basis of philosophy and all other disciplines. For Beneke, the mind was essentially active, and the psychological processes of knowing, feeling, and willing were mediated by both acquired and native mental dispositions.

Beneke was influenced by the association hypotheses of the British empiricists, and in opposition to Herbart's mathematical approach, Beneke favored the introspective methods of the British philosophers. Although he took exception to Kant's faculties of the mind, he did assert that mental dispositions exist and accomplish approximately the same functions. However, his importance lies in his recognition of physiological components of the experiential input to mental operations.

Rudolf Hermann Lotze. The son of an army physician, Rudolf Hermann Lotze (1817–1881) received his university education at Leipzig. He was a medical student and received scientific training from Weber and Fechner of the psychophysics movement (we shall consider this movement in Chapter 10), but he was drawn toward philosophy as a student. After briefly trying to practice medicine, he decided on an academic career and returned to Leipzig in a teaching post. In 1844, he succeeded Herbart at Göttingen, and remained there for 37 years. Lotze was not the founder of a novel and influential movement in psychology but rather, by virtue of his teaching and writing, succeeded in influencing a generation of German scholars active in the founding of the new discipline.

Lotze's contribution in psychology was titled, *Medical Psychology*, or *Physiology of the Soul* (1852). In it he attempted to blend the mechanical and the ideal through a synthesis of science and metaphysics, although he seems to have ended up firmly emphasizing the latter. He provided a multitude of data from physiology to determine empirically how the physical becomes psychic. He argued that objective, environmental events stimulate inner senses that are conducted by nerve fibers to the central agent. The soul, a term he retained, is affected unconsciously; a conscious reaction can occur, but the degree of the reaction is dependent on attention factors. To Lotze, then, the nervous system was simply a mechanical conductor of motion. Sensations themselves were experiences mediated by the central agency of the soul. In describing mental operations, Lotze rejected Herbart's mathematical speculation. Rather, he posited that the elements of experience are qualitative and require a qualitative, not a quantitative, methodology. Lotze, for example, viewed space perception as a process initiated by raw data entering the person via neural conduction and having intensity and qualitative dimensions only. Perceived space is inferred from conscious data through past experiences by means of a mental capacity. The entire process was labeled "the empiricistic intuition of space."

Lotze opposed materialism and entirely mechanical explanations. His inclusion of physiological data was limited to a portion of the total process of mental activity, and he was not proposing any reduction of mental processes to the initial physiological stage. For Lotze, the central agency of the soul provided mental processes and activity with an essential unity that preserved the integrity of the self in psychology.

The Kantian views on the will and the unconscious were elaborated by two additional figures of the German tradition, who in turn provide a direct link between Kant and Freud.

Arthur Schopenhauer. Arthur Schopenhauer (1788–1860), known for his philosophical position of decided pessimism, pursued the concept of the will, which he described as the functionally autonomous will-in-itself. Reacting against the idealism surrounding Kant's description, Schopenhauer noted that many forms of activity are not intellectual but still achieve rational results. Below the animal level, expressions of activity are clearly not intellectual. Accordingly, Schopenhauer depicted the will as an irrational striving to live, with its own force removed from intellectual understanding or even awareness. The will, then, is a fundamental impulse. Consequently, psychology must extend its subject matter beyond the purely rational level to encompass the full underlying motivation of human activity in the will.

Eduard von Hartmann. Eduard von Hartmann (1842–1906) postulated the unconscious as the fundamental universal principle, creatively synthesizing intellect and will. The unconscious is defined as instinct in action with purpose, although without knowledge of the result. In this sense, von Hartmann viewed the unconscious as teleological, or as the determining motivational principle of the self. Von Hartmann suggested three unconscious levels. The first is the physiological level, exemplified by such actions as reflexes. The second level is psychic and includes mental events not within the awareness of the individual. The third level he described as absolute; it represents the underlying principal force for all life. Thus von Hartmann was able to express one side of a paradox that suggests that the individual does not act through conscious reason but rather constructs reasons to explain her or his acts. The implications of this view were fully developed by Freud through his dynamic theory of unconscious determinacy in personality.

As might be expected, the assertion of mental activity, as opposed to passivity, opened the way for a variety of interpretations. Specifically, these interpretations of human activity produced models of psychology that focused on the uniqueness of human life, as exemplified by the issue of personal freedom, levels of consciousness, or moral attitudes. Moreover, these interpretations of human activity rejected the mechanical and reducible aspects of mental passivity, and adherents of such models had to search for methodological approaches outside the physical sciences. It is the wealth of this German tradition that provided much of the antecedent views of twentieth-century psychology.

CHAPTER SUMMARY

German science and culture of the seventeenth and eighteenth centuries benefited from the enlightened patronage of the Prussian King Frederick the Great. Moreover, the universities of Germany prospered and became centers of excellence in the West, especially in science. Advances in psychology by German philosophers focused primarily on mental activity. Discarding the environmental determinacy of British empiricism, Leibniz defended the active agency of the mind in molding sensory data to provide experience. The active principle of his monadology lent itself to a dynamic view of harmony between independent physical and psychic processes. The rationalism of Wolff was fully elaborated by Kant, who described pure reason as the formation of perceptions innately through time and space, and asserted an elaborate structure of the mind in terms of categories that order the environment. From these formulations, German psychology received a variety of models suggested by Herbart, Beneke, and Lotze. Further, the Kantian notions of the strivings of the will and the unconscious were explored more fully by Schopenhauer and von Hartmann. Collectively, the German tradition is diverse but united by the belief in the activity of the mind and its control of environmental influences.

BIBLIOGRAPHY

Primary Sources

KANT, I. (1965). *Critique of pure reason* (N. K. Smith, Trans.). New York: St. Martin's.

RAND, B. (1912). *The classical psychologists.* New York: Houghton Mifflin.

General References

COPLESTON, F. (1964). *A history of philosophy, Vol. 6, Modern philosophy, Part II—Kant.* Garden City, NY: Image Books.

COPLESTON, F. (1965). *A history of philosophy, Vol. 7, Modern philosophy, Part II—Schopenhauer to Nietzsche.* Garden City, NY: Image Books.

DURANT, W., & DURANT, A. (1965). *The age of Voltaire*. New York: Simon & Schuster.

DURANT, W., & DURANT, A. (1967). *Rousseau and revolution*. New York: Simon & Schuster.

Studies

BUCHNER, E. F. (1897). A study of Kant's psychology. *Psychological Review, 1* (monograph suppl. 4).

DOBSON, V., & BRUCE, D. (1972). The German university and the development of experimental psychology. *Journal of the History of the Behavioral Sciences, 8,* 204–207.

GOUAUX, C. (1972). Kant's view on the nature of empirical psychology. *Journal of the History of the Behavioral Sciences, 8,* 237–242.

LEARY, D. E. (1978). The philosophical development of the conception of psychology in Germany, 1780–1858. *Journal of the History of the Behavioral Sciences, 14,* 113–121.

9

Competing Models of Psychology

An Integration
Chapter Summary

To appreciate the emergence of psychology as an independent scientific discipline, it is critical to understand its philosophical precursors. Western European philosophers inherited from the Greeks various approaches to psychological inquiry. These were refined, passed on to Christian thinkers, and later combined or otherwise modified throughout the Renaissance. Finally, the native intellectual movements in France, Britain, and Germany combined philosophical advances with an essential faith in science to set the stage for the nineteenth century.

The philosophical developments that immediately preceded the nineteenth century have an importance that exceeds mere historical interest. They directly affect all students of psychology, because before undertaking any psychological study, one must assume a basic belief in the specific nature of life, which is a philosophical exercise. For example, if one believes that people are governed entirely by the mechanics of neural action and the "mind" is a superfluous pseudoconstruct, then the major data of psychological significance are necessarily confined to observations of overt organismic behavior. In contrast, if one holds that dynamic mental processes determine psychological activity, then observable behavior may have meaning both of itself and for its symbolic value. Each model dictates a different approach to psychological inquiry. Obviously, we are aided in our trust in a given model by our understanding of current psychological research that may either support or refute the assumptions behind a model. However, given psychology's contemporary diversity, there is no easy way to avoid making some choice from among competing models.

AN INTEGRATION

The philosophical positions surveyed in Chapters 6 through 8 take on essential significance for the foundations of modern psychology. Further organization and articulation of those diverse views may be informative as we now consider nineteenth- and twentieth-century psychology. The task of examining the relationships among the various philosophers may be facilitated by isolating certain themes prevalent in their views. Such an approach can generate dimensions along which philosophers may be evaluated relative to each other. This technique has been proposed by Watson (1967), who listed 18 dimensions, described in terms of contrasting labels, and used them as prescriptions of psychology's evolution as a discipline.

Prescriptive Dimensions of Psychology

1. *Conscious mentalism/Unconscious mentalism*: awareness of mental operations or activity versus unawareness of them
2. *Contentual objectivism/Contentual subjectivism*: psychological activity viewed as observable (e.g., behavior) versus nonobservable mental activity
3. *Determinism/Indeterminism*: psychological activity explained by antecedent events versus not explained by such events
4. *Empiricism/Rationalism*: emphasis on experience as the source of knowledge versus reason as the source of knowledge
5. *Functionalism/Structuralism*: psychological events described as activities versus such events described as contents
6. *Inductive/Deductive*: method of approach from particular to general versus general to particular
7. *Mechanism/Vitalism*: psychological events explained as physiologically based versus other than physiologically based
8. *Methodological objectivism/Methodological subjectivism*: methods repeated and verified by another scientist versus not able to be replicated by another
9. *Molecularism/Molarism*: psychological data described in terms of relatively small units versus relatively large units
10. *Monism/Dualism*: basic principle of life of one kind (e.g., materialism) versus two kinds (e.g., matter and mind)
11. *Naturalism/Supernaturalism*: psychological events explained solely in terms of the resources of the organism versus the need for some other active power (e.g., God)
12. *Nomotheticism/Idiographicism*: emphasis on finding general principles versus individual events
13. *Peripheralism/Centralism*: emphasis on psychological events occurring away from the presumed center of the organism (e.g., sensory processes) versus at the center (e.g., thinking)
14. *Purism/Utilitarianism*: emphasis on knowledge for its own sake versus emphasis on the usefulness of knowledge
15. *Quantitativism/Qualitativism*: data of psychology in measurable form versus differing in kind or type
16. *Rationalism/Irrationalism*: emphasis on intellectual or common sense determinants versus emotional or nonintellectual dominance
17. *Staticism/Developmentalism*: cross-sectional view versus changes with time
18. *Staticism/Dynamism*: nonchanging versus changing factors emphasized

(Based on Watson, R. I. [1967]. Psychology: A prescriptive science. *American Psychologist, 22*, 436–437.

These dimensions are somewhat redundant and their use may be artificial as identifying a single dimension for evaluation distorts the totality of any given model. Nevertheless, they can be of assistance in ordering the diversity of models in psychology. For example, Marx and Cronan-Hillix (1987) asked subjects to rate twentieth-century systems of psychology (associationism, structuralism, functionalism, behaviorism, Gestalt theory, and psychoanalysis) along Watson's 18 dimensions. They reported fairly reasonable discriminations among the systems, and the variability of ratings was quite low, justifying the value of this technique.

In another approach, Coan (1968) asked 232 psychologists to rate 54 figures influencing the development of psychology on 34 characteristics. Six factors, or dimensions, emerged that accounted for most of the variability in the responses:

1. Subjectivism versus Objectivism
2. Holistic versus Elementaristic
3. Transpersonal versus Personal
4. Quantitative versus Qualitative
5. Dynamic versus Static
6. Synthetic versus Analytic

The use of dimensions such as those of Watson and Coan seems to offer an effective way to discriminate some of the major figures and movements in psychology. Such an evaluative approach along several dimensions may be applied to the philosophers of the national movements surveyed in Chapters 6 through 8. In so doing, schematic arrangements of philosophical positions may be generated. Such arrangements are not quantitatively meaningful, but serve a useful purpose in organizing the philosophical positions in qualitative relationships. For example, the most pronounced differences in the national intellectual traditions relate to the concept of the mind. At one extreme, the mind is viewed as essentially active; at the opposite pole, the mind is a superfluous concept. In between is the position that the mind concept is needed, but its role is confined to a passive receptor of ideas and memories.

Figure 9–1 illustrates how the various philosophers might be arranged along this dimension. At the extreme active position on this dimension are the views of Leibniz, Wolff, Schopenhauer, and von Hartmann, a position reflecting their commitment to inner, determining mental activity. Kant and Biran are placed slightly away from the extreme because of their recognition of the stimulating role of sensation. At the opposite pole, La Mettrie's complete materialism places his views at the extreme, and the somewhat moderating positions of Condillac, Bonnet, Helvétius, and Cabanis reflect their allowance for minimal mental activity. Comte's positivism results in his position in the materialist cluster. Hobbes' initial statement of empiricism places him in the center of the dimension, along with James Mill, because of his additive association view. Locke's notion of self-reflection moves him slightly toward the active pole; Berkeley's dependency on creative associations places him even closer to the mental activity pole.

We find that views on the mind coming from separate models converge. The more dynamic empiricism of the Scottish philosophers Reid and Brown, along with

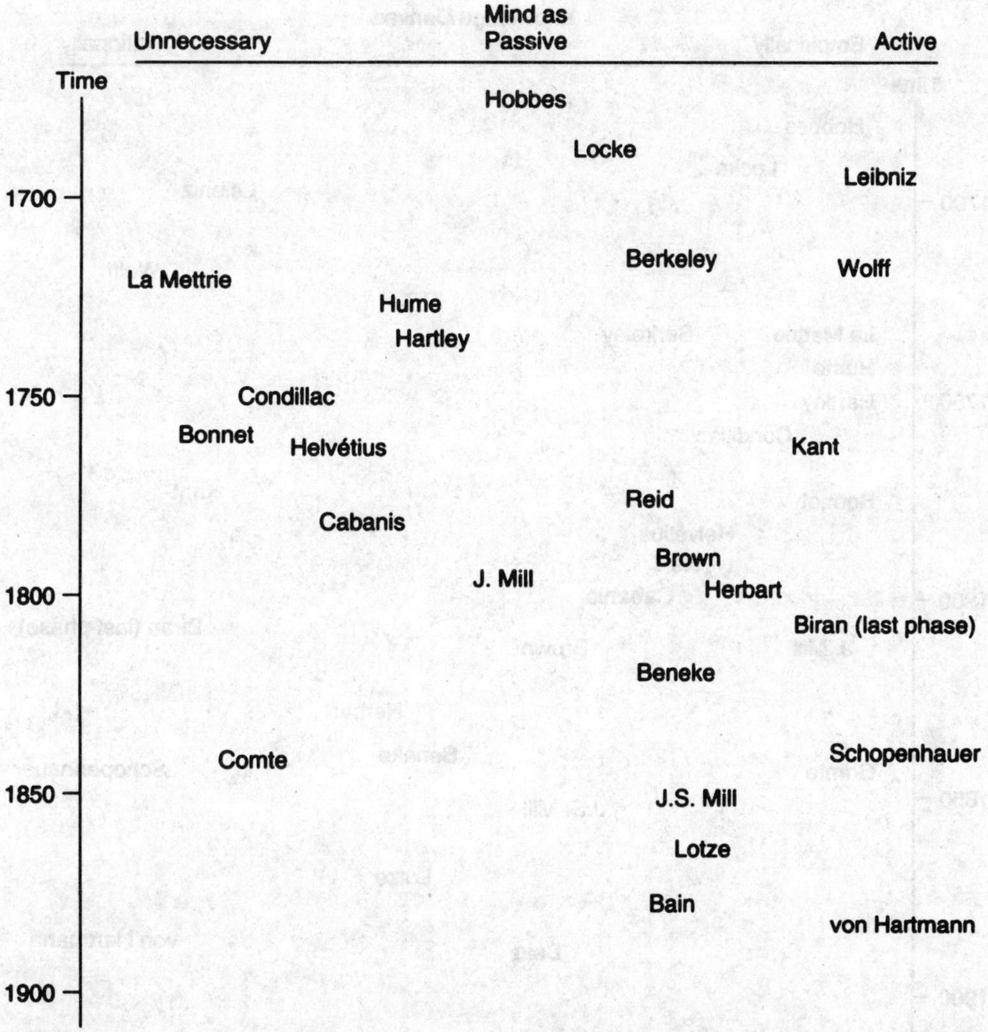

FIGURE 9–1 Psychological views of seventeenth-, eighteenth-, and nineteenth-century French, British, and German philosophers along the dimension of mental activity.

that of John Stuart Mill and Bain, is basically consistent with the views on mental activity tempered with a sensory basis offered by Herbart, Beneke, and Lotze. Thus, along this dimension of mental activity, the organizational scheme shows relatively clear distinctions among earlier expressions of the native intellectual movements. However, by the beginning of the nineteenth century, elements of the British and the German schools found common areas of agreement on mental activity.

Similarly, a dimension contrasting sources of knowledge as empirically or rationally based is illustrated in Figure 9–2. At the extreme pole, reflecting complete reliance on rational knowledge, are the Germans—Wolff, Schopenhauer, and von Hartmann. Biran joins them because of his emphasis on the will during his last phase.

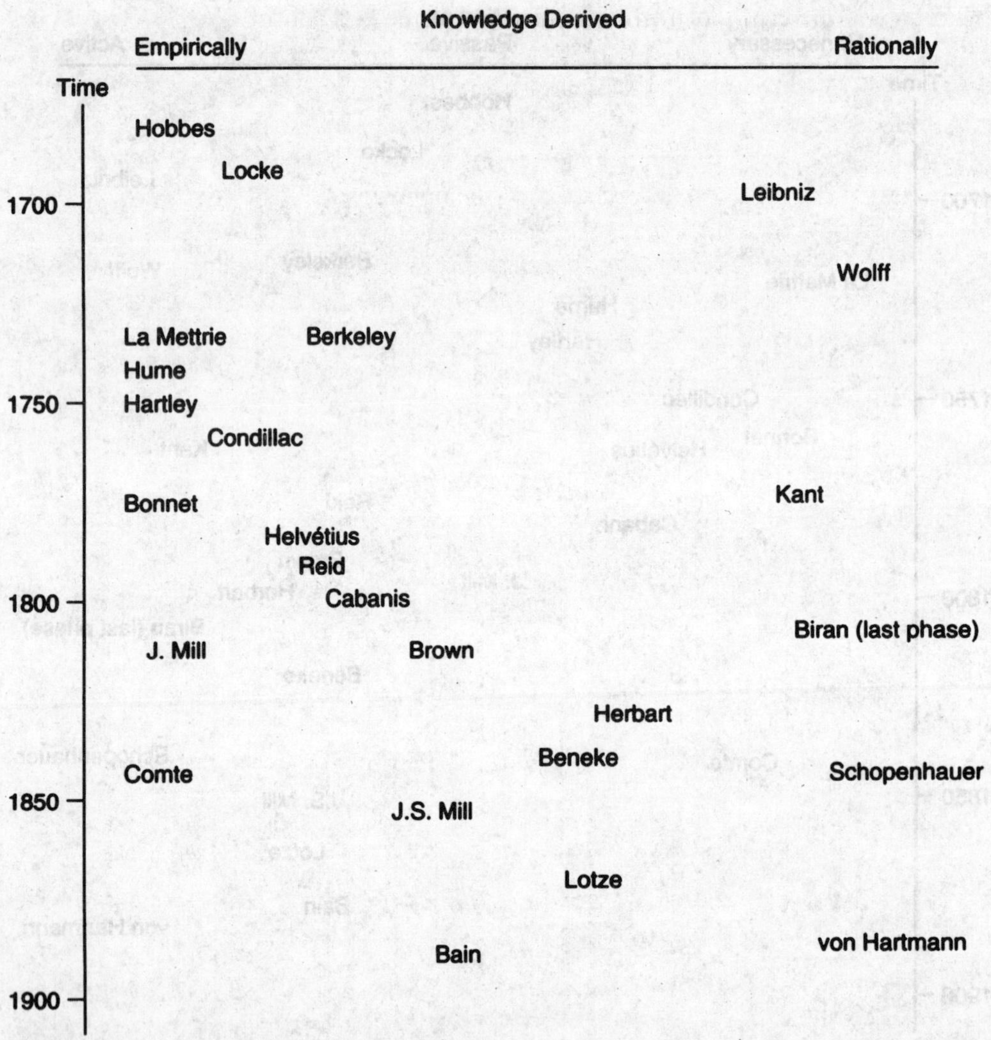

FIGURE 9–2 The views of French, British, and German philosophers placed on a dimension showing sources of human knowledge.

Leibniz and Kant are slightly removed from this extreme because of their acceptance of limited sensory input. At the other extreme, the empiricists relying exclusively on sensory knowledge include both British and French views. Locke's admittance of self-reflection pushes his views away from the extreme, as do the views of Berkeley, who, although mostly empirical, posited more generative activity in the mind. Condillac is also placed away from the extreme, but only because he did not carry his views to the conclusion reached by Bonnet and La Mettrie. Both the "common sense" of Reid and the moderating views of Helvétius and Cabanis recognized some nonsensory, rationalistic functions of the mind. Brown's notion of suggestion in his views on associations moves him still further away from extreme empiricism. Again, there is a nineteenth-century merger between British and German traditions. Thus the organization of this figure shows the early contrast between the isolated position of German rationalism and the united positions of British empiricism and French sensationalism. Following the Scottish influence on empiricism, however, the British view of the mental chemistry of associations approaches the modified rationalist tradition of those German figures who evolved to a position that recognized the significance of sensory physiology.

Another recurring theme concerned the opposing views of monism, which asserted a single materialistic foundation of psychology, and dualism, which retained the mind–body distinction. As arranged in Figure 9–3, both the British empiricistic and German rationalistic traditions retained the need for a mind construct, although for different purposes. In contrast, extreme French sensationalism reduced psychology to materialism. The conclusions of the additive associationistic interpretation in

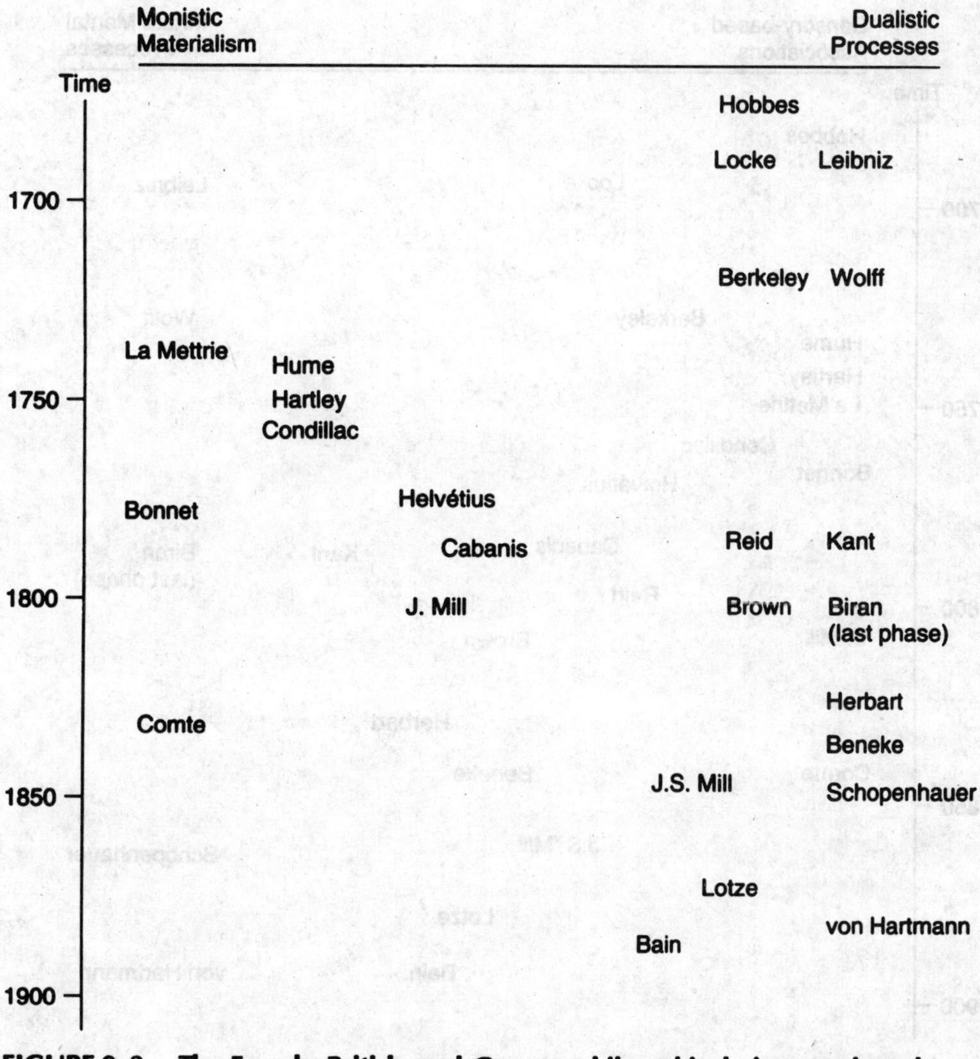

FIGURE 9–3 The French, British, and German philosophical views evaluated on a dimension of belief in monism versus dualism.

empiricism led to a denial of the need for a mind construct, firmly expressed by Hume and Hartley and, to a lesser extent, by James Mill. The moderation of sensationalism offered by Helvétius and Cabanis allowed some functions not adequately explained by sensory physiology only.

The method of acquiring knowledge may be placed on a dimension contrasting sensory associations with dynamic mental activity shown in Figure 9–4. The monadology of Leibniz, the extreme rationalism of Wolff, and the emphasis on the will of Biran, Schopenhauer, and von Hartmann all fit at the extreme mentalistic pole of this dimension. Again, Kant is removed from that extreme position because of his acceptance of the initiating role of sensory stimulation for the operations of the mind. At the other extreme pole of sensory-based associations, French materialism

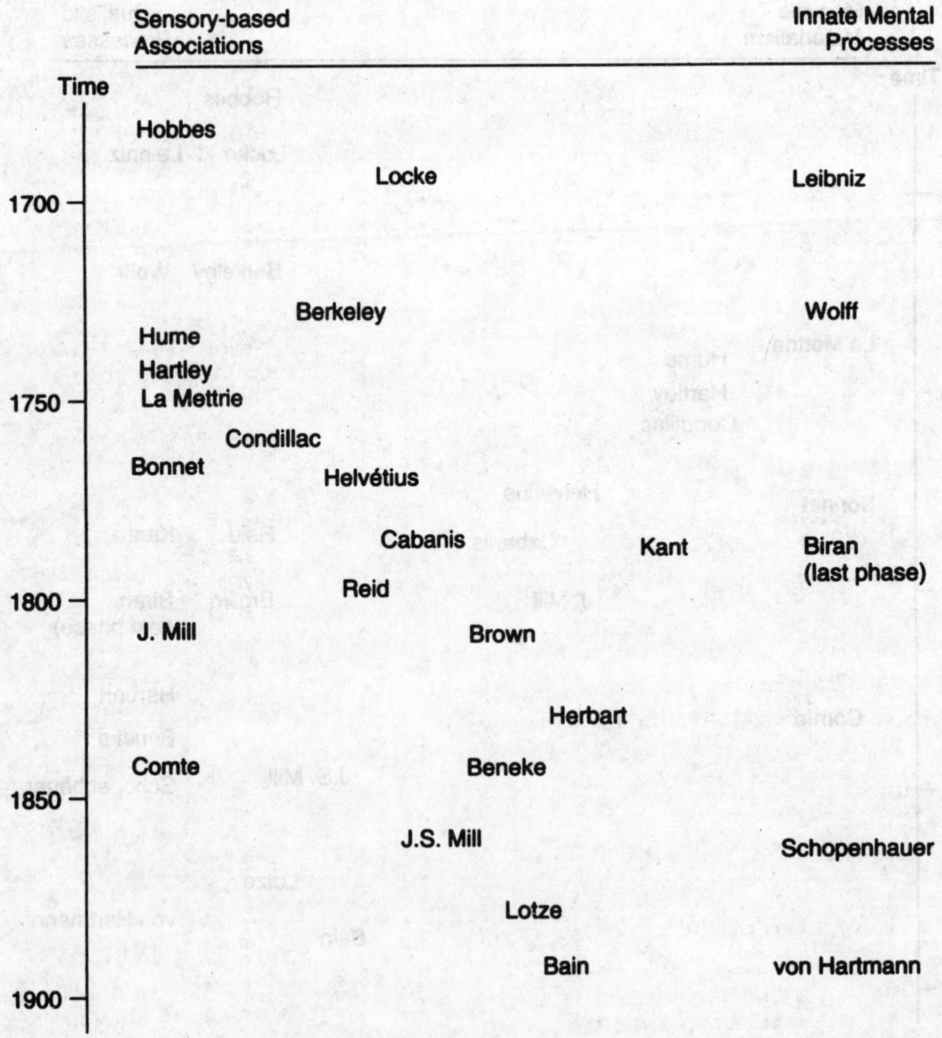

FIGURE 9–4 The views of French, British, and German philosophers on a dimension describing the formation of ideas.

and British empiricism merge, because both relied on the additive properties of associations in the formation of ideas. Locke's position admitted nonsensory associations in his notion of self-reflection, whereas Berkeley's antimaterialistic view of the mind removes him from the extreme pole. The mental chemistry of Reid, John Stuart Mill, and Bain, and Brown's notion of suggestion, allow their views to approach the recognition of sensory elements proposed by Herbart and Beneke.

These schematic organizational arrangements, and others that may easily be constructed, have value only to the extent of the qualitative, relative positioning of the major figures of the philosophical movements. As quantitative statements, they need further validation. Nevertheless, within such limitations, this technique provides insight into the flow of ideas that preceded the emergence of psychology. All four representations affirm the static position of French materialism; that is, with the moderating exceptions of Helvétius and Cabanis, the reduction of psychological processes to sensory elements left the French tradition with little room for a psychology separate from physiology. However, the clear differences between the German and British traditions allowed some eventual accommodation based on their mutual acceptance of dualism. It is important to qualify this statement, however, because the apparent agreement on single-dimensional illustrations of later German and British thought may mask very real differences. The German and British figures of the nineteenth century approached the common ground of agreement from very different positions. Mental chemistry and sensory-based mentalistic activity may serve similar functions, but they are modifications of different perspectives on psychology. Accordingly, although we may acknowledge the fluid character of British and German positions relative to the French, we must cautiously avoid interpreting areas of agreement as points of equality.

CHAPTER SUMMARY

The 25 figures selected to describe the place of psychology within the British, French, and German philosophical movements reflect important subtleties. Watson's prescriptive dimensions, Coan's factors, and the schemes diagrammed in Figures 9–1 through 9–4 are all useful in the organization of these diverse views. As psychology emerged in the nineteenth century, various forms of the new discipline reflected the underlying philosophical models developed over three centuries of post-Renaissance thought. To a great extent, the assumptions of those models are still relevant to the contemporary study of psychology.

BIBLIOGRAPHY

COAN, R. W. (1968). Dimensions of psychological theory. *American Psychologist, 23,* 715–722.

MARX, M. H., & CRONAN-HILLIX, W. A. (1987). *Systems and theories in psychology* (4th ed.). New York: McGraw-Hill.

WATSON, R. I. (1967). Psychology: A prescriptive science. *American Psychologist, 22,* 435–443.

10

Nineteenth-Century Bases of Psychology

This chapter considers three nineteenth-century scientific movements that had a direct impact on both the founding of psychology and its subsequent expression in the twentieth century. First, in physiology, research on nervous activity provided an empirical basis for many human functions that were previously considered functions of the mind. Second, a German development called psychophysics attempted to find the quantitative basis for the mind–body relationship, but it went beyond Herbart's psychological writings by employing an empirical approach. Finally, the writings of Charles Darwin in Britain affirmed a theory of evolution based on the empirical evidence of natural selection. All three movements directly helped to establish the formal study of psychology.

ADVANCES IN PHYSIOLOGY

Empirical research in physiology made important strides in the nineteenth century. The study of nervous activity, sensations, and brain physiology confirmed the benefits of careful systematic empirical strategies. For psychology, these benefits pointed to the possibility of elucidating the physiological basis of mental operations.

General Physiology of the Nervous System

The distinction between sensory and motor nerves was independently demonstrated by the experimental work of Charles Bell (1774–1842) and François Magendie (1783–1855). Bell was born in Edinburgh and achieved fame as an anatomist in London; Magendie was a widely esteemed professor and member of the French Academy. Their collective work, encapsulated in the Bell–Magendie Law, was based on their discovery that the posterior roots of the spinal cord contain sensory fibers only, whereas the anterior roots contain motor fibers. Thus, neural fibers were no longer considered either "hollow tubes" transmitting "spirits" of activity or all-encompassing fibers communicating both sensory and motor functions through "vibrations" arising from sensory stimulation. Rather, neural fibers were specific in function, and neural conduction seemed to occur predominantly in one direction.

The work of Bell and Magendie received systematic elaboration in the writings of Johannes Müller (1801–1858), who set the tone for nineteenth-century physiology. His exhaustive *Handbuch der Physiologie des Menschen* (*Handbook of Human Physiology*; 1833–1840) became the classic compilation of contemporary physiology. After receiving his doctorate from the University of Bonn in 1822, Müller served there as a professor until 1833, when he was called to be professor of physiology at Berlin. Many of the foremost European physiologists of the nineteenth century were his pupils, and still more were certainly influenced by the *Handbuch*. Based on the work of Bell and Magendie, Müller fully articulated the so-called doctrine of specific nerve energies. He described the specific qualities of neural transmission, formulating them under ten laws. The major implication of Müller's doctrine is the explicit statement that our awareness is not of objects, but rather of our nerves themselves. Accordingly, the nervous system serves as the intermediary between sensed objects and the mind. Müller asserted that five kinds of nerves each impose their own quality on the mind. As a physiological parallel of Kant's philosophical notion of categories of the mind, Müller's work stimulated the study of localization of functions in the brain, which is considered in the following sections.

The understanding of sensory physiology took a major step when neural conduction was discovered to be basically an electrical process, putting to final rest the traditional view of nerve fibers containing "animal spirits." We already noted in Chapter 8 that Gralath of Leiden was able to store electrical charges in a series of jars. An Italian physiologist, Luigi Galvani (1737–1798), used Leiden jars as an electrical source to do the classic experiment eliciting reflex action in the leg of a frog with a partially intact spinal cord. Galvani correctly concluded that nerves are capable of conducting electricity, although in attempting to fit his discovery into existing views, he thought he had isolated a unique substance—"animal electricity"—which is transported via a fluid from the nerve to the muscles. It was a student of Johannes Müller, Emil Du Bois–Reymond (1818–1896), who broke away from the traditional view of "animal spirits" and established the modern basis of neural transmission by describing the electrical properties of the neural impulse.

The speed of a nerve impulse was measured by another of Müller's students, Hermann von Helmholtz (1821–1894), whom we consider later in this chapter. In his

Handbuch, Müller had acknowledged, although somewhat skeptically, one of the major implications of the "animal spirits" view of neural transmission—namely, the speed of a nerve impulse is too rapid to be observed and studied empirically. However, Helmholtz devised a method of measuring temporal duration between the application of an electrical stimulus to a frog's nerve and the twitch of the muscle. In the frog he found reaction times of 0.0014 second and 0.0020 second for 60 millimeters and 50 millimeters of nerve fiber, respectively, yielding limits of 42.9 and 25.0 meters per second. Using the same method to measure reaction time in humans, Helmholtz stimulated a subject on the toe and thigh and calculated differences in reaction times. He found that the rate of transmission for sensory impulses is between 50 and 100 meters per second. Although others, such as Du Bois–Reymond, would later report more accurate calculations, Helmholtz had succeeded in empirically demonstrating neural transmission, increasing faith in the efficacy of empirical science. Moreover, because Helmholtz reliably measured the effects of stimulation through the overt behavioral responses (that is, the reaction time) of subjects, the reaction time experiment served as a prototype of empirical psychology.

Physiology of the Brain

Perhaps the most dramatic reflection of the major advances in brain physiology during the nineteenth century occurred in 1906 with the joint award of a Nobel Prize to an Italian neurologist, Camillo Golgi (1844–1926), and a Spanish anatomist, Santiago Ramón y Cajal (1852–1934). In 1873, Golgi published a paper reporting his use of silver nitrate to stain nerve cells, revealing under a microscope the structural details of nerves. Ramón y Cajal, a professor of neuroanatomy at the University of Madrid, later used this staining technique to make his discovery of the neuron, the basic unit of the nervous system. Their work, concluding a century that began with the prevailing view that the workings of the nervous system were analogous to those of the circulatory system, demonstrated the value of empirical strategies in the study of nervous activity.

At the beginning of the nineteenth century, the dominant interpretation of brain functions was contained in the doctrine of phrenology, expressed by Franz Joseph Gall (1758–1828) and his student J. G. Spurzheim (1776–1832). To a large extent, phrenology and similar movements in brain physiology were a logical consequence of the mentalistic model embodied in the "faculty" psychology championed by Wolff and Kant. Specifically, phrenology attempted to find a physiological localization of mental faculties. Gall began as a lecturer in Vienna, but in 1800 was pressured into leaving by the Austrian government and spent his remaining years in Paris. Gall and Spurzheim suggested that there are 37 mental powers corresponding to the same number of brain organs, and the development of these organs causes characteristic enlargements of the skull. Accordingly, they developed a pseudoscience that claimed extreme localization of brain functions. Phrenology held that the degree of a mental faculty or trait possessed by an individual is determined by the size of the brain area controlling that function and that this can be evaluated by measuring the overlying skull area.

Gall's phrenology forced the question of brain localization into the forefront of physiological investigation. One scientist whose work led him to reject phrenology and substitute better evidence of brain localization was Luigi Rolando (1773–1831). In 1809, Rolando published his research efforts in Italy, and in 1822 they were reviewed in French. Using pathological observations, Rolando argued that the cerebral hemispheres are the primary mediators of sleep, dementia, melancholia, and mania. Sensory functions are localized in the medulla oblongata. Although Rolando's experiments were primitive, he found that electrical stimulation elicited more violent muscle contractions as the stimulation point was moved to higher brain centers. Similarly, Pierre-Paul Broca (1824–1880), a French scientist, performed a postmortem examination of a man who had speech aphasia. Broca found damage in a specific area of the frontal cortex (now called Broca's area), which he interpreted in support of the thesis of localization of function by describing the area as the physiological basis of expressive language.

The study of brain physiology took on definitive form from both the precise methodology and the coherent interpretations offered by Pierre Flourens (1794–1867). After studying sensory physiology in Paris, Flourens secured a professorship in comparative anatomy and was elected to the French Academy for his clear and concise refutation of phrenology, summarized in *Examen de la Phrénologie* (*An Examination of Phrenology*; 1824).

Rather than relying on pathological clinical evidence observed during postmortem examinations, Flourens perfected the more controlled method of extirpation. Essentially, in this procedure an area of the brain of a living animal is isolated, then removed surgically or destroyed without damaging the remainder of the brain. After recuperation, the animal is observed for loss of function and recovery of function. Flourens assumed that six separate areas exist in the brain and, using his surgical skills, was able to identify the important functions of each area:

1. *Cerebral hemispheres*: willing, judging, memory, seeing, and hearing
2. *Cerebellum*: motor coordination
3. *Medulla oblongata*: mediation of sensory and motor functions
4. *Corpora quadrigemina* (containing inferior and superior colliculi): vision
5. *Spinal cord*: conduction
6. *Nerves*: excitation

Flourens noted the essential unity of the nervous system by stressing the common action of the various parts in addition to their specific functions. Although his anatomical approach reflected the localization stressed by the phrenologists, his emphasis on the common unity of the entire system represented a move away from the extremism of Gall. Moreover, his methodological innovations resulted in data that clearly anticipated the future of neurophysiological research.

The culmination of nineteenth-century advances in brain physiology, which formed the basis of modern neurophysiology and allied approaches in electrophysiology and histology, was reached by Charles S. Sherrington (1857–1952). His long career may be viewed in two parts. In the earlier phase, lasting until 1906, Sherrington carried to its conclusion the nineteenth-century work of such scientists as Müller, Bell, Magendie, and Flourens, which led to modern neurophysiology. This work established the neuroanatomical basis of reflexology—that is, the physiological causality underlying overt behavioral responses to environmental stimuli. Sherrington's research, summarized in his classic work *The Integrative Action of the Nervous System* (1906), paved the way for the behavioristic psychology of the twentieth century, initiated by the Russian physiologist Pavlov and the American psychologist J. B. Watson. During the second half of his career, crowned by his award of the Nobel Prize in 1932, Sherrington continued his prolific experiments and educated a future generation of neurophysiologists at Oxford University. Thus, he not only established the foundations of neurophysiology, but also continued to build on those foundations, which resulted in tremendous strides in the understanding of the physiological basis of psychological events.

Sherrington's early research into reflexes was dominated by his analysis of spinal-level activity and the reciprocal action of antagonistic muscles. To describe his findings, he developed a terminology that is now basic to the neurosciences. Such terms as *nociceptive, proprioceptive, fractionation, recruitment, occlusion, myotatic, neuron pool*, and *motoneuron* were coined by him to describe his observations. His neuroanatomical contributions, published in the 1890s, consisted of mapping motor pathways, identifying sensory nerves in muscles, and tracing the cutaneous distribution of the posterior spinal roots. These studies revealed the dynamics of nervous coordination, which he described as the "compounding" of reflexes constructed by the interacting of reflex arcs around common pathways. Sherrington concluded that underlying this reflex activity are the critical processes of inhibitory and excitatory actions at the regions between nerve cells; he labeled these junctions *synapses*.

Sherrington used the method of extirpation in his studies, and his 1906 work fully explored the potential for neurophysiology based on the integrative properties of the nervous system. In this work, complex reflexes were described in terms of the synaptic chain of converging pathways. It is difficult to overemphasize the impact of Sherrington's work and its significance for contemporary psychology. Sherrington's concept of excitatory and inhibitory processes forms a central place in our understanding of brain–behavior relationships and comprises the cornerstone of conditioning theory. His views have been greatly expanded, but essentially confirmed, during the intervening time, most notably by his brilliant students, especially John C. Eccles (1903–1997), who opened the possibility of entirely new interpretations in psychology.

Physiology of Sensations

A related nineteenth-century movement attempted to study sensations from the perspectives of physics and anatomy. The anatomical properties of the organ of reception (for example, the eye) were examined in terms of the physical properties of the stimulus (light), and the resulting psychological experience—sensation—was analyzed in terms of the combined physical and physiological processes.

This approach had been used by the English scientist Thomas Young (1773–1829), who is also known as one of the first translators of Egyptian hieroglyphics. Young attempted to extend Newton's work in optics, and successfully developed a theory of color vision. In papers published in 1801 and 1807, Young argued that there are three primary colors—red, yellow, and blue—that have characteristic wavelengths and differentially stimulate specific areas of the retina. This trichromatic theory was later bolstered with better evidence by the German psychophysicist Helmholtz (who shall be considered later) and is now known as the Young–Helmholtz theory of color vision. The physiologist Müller also contributed to sensory physiology by his description of direct subjective experience of neural action, not description of the environment, which we can know only indirectly. In addition, Müller attempted, less successfully, to develop a theory of hearing.

Perhaps the most interesting researcher of nineteenth-century sensory physiology was the Czech scientist Jan Purkinje (1787–1869). His varied investigations made him famous as a physiologist. He permitted subjective experience in his methodological approach relating the physical and physiological components to sensation. As a child, he was intended by his parents for the priesthood, but his own advanced study of contemporary philosophers led him to reject this direction. Instead, he supported himself by tutoring and eventually received a scientific education at Prague. From 1823 to 1850 he was a professor of physiology at the University of Breslau (now the Polish city of Wrocław), where he founded the first institute of physiology in any European university. In 1850, he returned to Prague, where he succeeded in having Czech accepted as a language of instruction along with German. During his last years he was active in the revival of Czech political life and the general uplifting of Slavic culture.

In his early research on sensory physiology Purkinje used himself as a subject because of lack of funds. In studying his visual reactions through meticulous self-observations, he was impressed that certain events, such as perceptual errors, discrepancies between stimulus intensity and perceptual strength, and uncaused sensory experiences, were not random. Rather, they were governed by the systematic relationship between the structure of the eye and the neural connection to the brain. In 1825, he published his observation, known as the Purkinje effect, that the relative luminosity of colors in faint light differs from that in full light. This difference between scotopic and photopic vision was later explained by the separate mediation of rods and cones of the retina. Purkinje also noted the inability to differentiate colors in the periphery of the retina.

Others, such as the celebrated German Romantic poet and dramatist Johann Wolfgang von Goethe, had made similar self-observations of perceptual illusions. As a scientist, Purkinje saw these phenomena in terms of their physiological value. He proposed a corresponding objective, physiological basis for all subjective sensory phenomena, and showed how these subjective phenomena may be used as an appropriate tool to explore the objective bases. Thus, Purkinje admitted a method of self-observation or self-description as a valid investigative approach. Moreover, he suggested several procedures for its utilization. Purkinje's substantial contributions, as well as his methodological approach, were recognized by later psychophysicists and incorporated into one of the first formal models of psychology.

Purkinje also worked extensively in neurophysiology, as reflected by his identification of certain cells of the cerebrum (Purkinje cells) and in the structure of the heart (Purkinje fibers). His recognition of the need for experimentation and self-observation in physiological research made a great impact on the methodological direction of psychology; he allowed for the study of subjective experience in addition to more objective physical and physiological components in the understanding of sensory processes. We now consider the movement called psychophysics, an immediate precursor of modern psychology, which owed a debt to Purkinje.

PSYCHOPHYSICS

The label *psychophysics* is given to a type of sensory physiology that emphasized subjective experience in the study of the relationship between physical stimuli and sensations. As a group, psychophysicists examined sensations from several perspectives. They considered sensations as a reflection of the mind–body problem, rather than as a situation for anatomical and physical study alone. At the same time, however, these psychophysicists were not psychologists, because they did not seek a new and comprehensive discipline. Rather, they remained inside the traditional disciplines of their training—physiology, physics, or natural philosophy. Indeed, only with the hindsight knowledge of the subsequent emergence of psychology does psychophysics take on coherence as a movement. Nevertheless, psychophysics served as a critical transition between the study of the physiological and physical components of sensation and the emergence of psychology itself. Thus, the scholars of the psychophysical movement were the immediate precursors of modern psychology.

Ernst Heinrich Weber

Ernst Heinrich Weber (1795–1878), the first person who may be categorized as a psychophysicist, was professor of anatomy and physiology at Leipzig from 1818 until his death. The University of Leipzig became the dominant institution for both the psychophysical movement and the emergence of a psychology modeled after the natural sciences. Weber's contributions included an exhaustive investigation of the sense of touch. He established a methodological orientation that seemed to demonstrate the possibility of quantifying mental or psychological operations.

His major work in psychology, *De Tactu: Annotationes Anatomicae et Physiologiae* (*On Touch: Anatomical and Physiological Notes*), was published in 1834 and contained extensive experimental work. He distinguished three manifestations of the sense of touch: temperature, pressure, and locality sensations. Temperature was dichotomized into positive and negative sensations of cold and warm, which Weber felt were analogous to the light and dark sensations of vision. In his investigations of pressure, Weber developed a methodological innovation known as the two-point threshold. Briefly, he used a compass with two points and attempted to measure cutaneous sensitivity by the smallest detectable distance between the two points that could be sensed by a subject. Weber found that this threshold of detectable difference between the two points varied with the places of stimulation, a variation he explained by postulating differential densities of nerve fibers underlying the skin's surface. This method led him to a study of weight discrimination and eventually to the formulation of "Weber's Law," named for him by his colleague Gustav Fechner, who is considered in the following discussion. Weber found that the smallest detectable difference between two weights can be expressed by the ratio of the difference between the weights relative to the absolute value of the weights, and that this ratio is independent of the absolute values of the weights. He extended his research to other senses and found general validity for the ratio of the smallest detectable difference between two stimuli. The last touch sensation, locality, was viewed by Weber to be more than a sensory dimension. Rather, he believed locality was more dependent on perception, which he interpreted as mental activity.

Weber succeeded in using a quantified approach to sensations, an approach that was adopted by his successors. However, in his interpretation of mental action on these sensations, he relied on the prevailing philosophical system of Germany—namely, Kant's views of the mind. In other words, Weber viewed perceptions as governed by mental categories of time and space, and did not speculate further.

Gustav Theodor Fechner

Gustav Theodor Fechner (1801–1887), the major proponent of psychophysics, attempted to explore more fully the relationships between sensations and perceptions. He labeled this movement through his *Elemente der Psychophysik* (*Elements of Psychophysics*; 1860), which was designed to be an exact science of the functional relations between the body and the mind. Moreover, Fechner's psychophysics was constructed as an attack on materialism. This goal is of interest because of the implied assumptions behind his psychophysics. Specifically, he did not believe that the notions of science and the mind are necessarily mutually exclusive; there is no compelling reason to reduce the mind to materialism (as in physiology) in order to study mental operations scientifically. Rather, in the tradition of German philosophy, he acknowledged the essential activity of the mind and proposed an empirical science of the mind that allows the relative increase of bodily, sensory stimulation to serve as the measure of the mental intensity of experiences.

Fechner was born in a small village in southeastern Germany, the son of the local church pastor. At age 16, he began to study medicine at the University of Leipzig and received his degree in 1822. Fechner's interest shifted to physics, and he remained in Leipzig to study, supporting himself by translating, tutoring, and giving occasional lectures. In 1831, he published a paper on the measurement of direct current, using the relationships published by Georg Ohm in 1826. Fechner was appointed professor of physics at Leipzig in 1834, and his future seemed secure. His interests began to move toward problems of sensations, and by 1840 he had published research on color vision and subjective afterimages. At about this time Fechner had what might be called today a nervous breakdown. He had overworked, exhausting himself, and he also damaged his eyes by gazing at the sun during his research on afterimages. Fechner's collapse seemed total, and he resigned his position at the university to live in seclusion for 3 years.

Fechner recovered, but his illness and confinement had a profound effect on him. He emerged from his crisis committed to the spiritual aspects of life, and renewed his religious convictions. He was convinced of the existence of both mind and matter, and believed that the materialism of science, exemplified by prevailing sensory physiology, is a distortion. For the rest of his life, he published on wide-ranging topics. In addition to psychophysics, he attempted to formulate an experimental esthetics, and even proposed a solution to the problem of determining the shape of angels.

His contributions to psychophysics are his most important works. After publishing two short papers on the subject, his *Elemente* appeared in 1860. This work was not widely recognized at first, but did attract the attention of two important leaders in German psychology—Helmholtz and Wundt. Any overview of Fechner's psychophysics must begin with the concept of *limen*, or threshold, which originated with Herbart and was developed by Weber. The notion of threshold is a quantitative expression that has two applications. The first usage refers to the minimal amount of physical energy needed in a stimulus for it to be detected by the observing subject, which was termed the *absolute threshold*. The second usage refers to the minimal amount of change in physical energy required for sensory detection.

Fechner began with the relationship expressed in Weber's Law:

$$\frac{\Delta R}{R} = K$$

Here, using the German symbols (R = *Reiz* = stimulus), Fechner expressed Weber's findings that the ratio of the change in stimulus value (ΔR) to the absolute value of the stimulus (R) is equal to a constant. This constant is a measure of the second usage of threshold, which Fechner called the *just noticeable difference* (jnd.) in stimulus intensity detected by the subject. Fechner then related the magnitude of an experienced sensation (S) to the magnitude of the stimulus using the jnd., or k, by the following relationship:

$$S = k \log R$$

Quadrant B of Figure 10–1 shows Fechner's empirically derived function for the relationship between the magnitude of the stimulus value (ordinate axis) and the strength of the sensation (abscissa). It is possible to extend Fechner's reasoning beyond his empirical demonstration, and quadrants A, C, and D of Figure 10–1 attempt to represent some of the hypothetical relationships within Fechner's approach. For example, the relationship between stimulus intensity and sensation that could occur in quadrant A would describe the nondetection of stimuli present, which relates to the subthreshold of attention. Quadrant C depicts possible sensory experiences in the absence of physical stimulation, and quadrant D describes nonsensory experiences of nonstimuli. The former (C) might define hallucinations; the latter (D) could be a definition of dreams. Although this interpretation may be pushing Fechner's conceptualization beyond his intention, it is fascinating that his views of the relationship between sensations and stimuli include a complete framework within the active model of the mind prevalent in German philosophy. Indeed, Fechner was very much a part of the intellectual climate of his contemporary Germany.

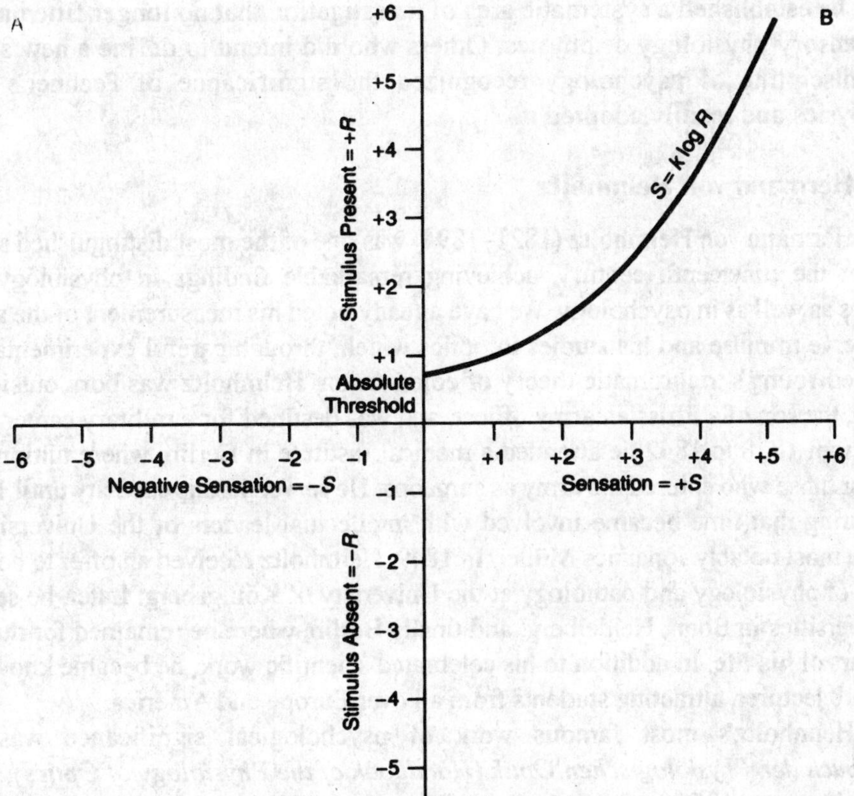

FIGURE 10–1 The possible relationships between stimulus intensity and the magnitude of sensation. The abscissa and ordinate axes are ordered in arbitrary units. Quadrant B shows the empirically derived relationship between the intensity of a stimulus and the magnitude of the sensation described by the function $S = k \log R$. In quadrant A, the nondetection (negative sensation) of a stimulus value would appear. Quadrant C describes a sensation in the absence of a stimulus. Quadrant D shows the nondetection of a nonstimulus.

Fechner proposed three fundamental methods to determine thresholds. The first was called the *method of just noticeable differences*, wherein the subject is asked to detect or respond to minimal change in stimulus values. The second was called the *method of right and wrong cases*, or the *method of constant stimuli*, in which the subject has to judge repeatedly which of two stimuli is the more intense. The third, the *method of average error*, requires subjects to adjust stimuli until they are equal. These techniques effectively estimate the major variables in psychophysical studies, and similar procedures are still employed in psychological investigations.

To judge by Fechner's stated goals of antimaterialism, he probably would not be very pleased with his own contributions to psychology, which have been largely methodological. However, without trying to begin a new disciplinary study, he established a systematic area of investigation that no longer fitted neatly into sensory physiology or physics. Others who did intend to define a new scientific discipline of psychology recognized the significance of Fechner's psychophysics and readily adopted it.

Hermann von Helmholtz

Hermann von Helmholtz (1821–1894) was one of the most distinguished scientists of the nineteenth century, achieving remarkable findings in physiology and physics as well as in psychology. We have already noted his measurement of the speed of a nerve impulse and his studies in optics which, through careful experimentation, affirmed Young's trichromatic theory of color vision. Helmholtz was born outside of Berlin, the son of a Prussian army officer, and was destined for a military career himself. From 1838 to 1842 he attended a medical institute in Berlin, where tuition was free for those who entered the army as surgeons. He served in this capacity until 1849, but during that time became involved with intellectual leaders of the University of Berlin, most notably Johannes Müller. In 1849, Helmholtz received an offer to be professor of physiology and pathology at the University of Königsberg. Later, he served at universities in Bonn, Heidelberg, and finally Berlin, where he remained for the last 23 years of his life. In addition to his celebrated scientific work, he became known as a superb lecturer, attracting students from all over Europe and America.

Helmholtz's most famous work of psychological significance was his *Handbuch der Physiologischen Optik* (*Handbook of the Physiology of Optics*), published between 1856 and 1866. In addition, he published his *Tonempfindungen* (*Tonal Sensations*; 1863), containing his resonance theory of hearing, which proposed that the transverse fibers of the basilar membrane act as a tonal analyzer and selectively respond to varying tonal frequencies. In contrast to Fechner, Helmholtz placed greater reliance on environmental or physical determinants of sensory activity. To some extent his approach to sensory physiology was closer to the British philosophical tradition than to the German—that is, Helmholtz agreed that experiences explain perceptions and not vice versa. Although not denying innate knowledge, and indeed recognizing the existence of instincts, he argued that the development of perceptions can be adequately explained from experiences.

Helmholtz postulated a perceptual doctrine of unconscious inference that may seem inconsistent with his empirical outlook. However, he proposed it as a perceptual response based on accumulated experiences. Helmholtz recognized that certain perceptual experiences are not readily accounted for by available elements in an actual stimulus presentation. For example, the age-old problem of depth perception cannot be completely explained by sensory stimulation alone. Helmholtz argued that we infer perceptual characteristics as a result of repeated experiences over time; the inferences are unconscious to the extent that we make them instantaneously, without conscious calculation or solution. He described unconscious inferences as "irresistible" because, once formed, they cannot be consciously modified. Moreover, he described this process as inductive, as the brain is unconsciously capable of generalizing an inference, once acquired, to other similar stimuli in the environment.

HERMANN VON HELMHOLTZ (1821–1894). (Courtesy, New York Public Library.)

Methodologically, Helmholtz stressed the importance of observing sensations as opposed to objects sensed—that is, the critical level of observation is the experiencing person, not characteristics of the stimulating object. Accordingly, he had high regard for the work of Purkinje, who obtained interesting data from his innovative observational techniques. Helmholtz's overview of perceptual processes clearly emphasized the dependency of sensory patterns on central functions such as unconscious inference and imagination. Thus, he advanced the study of psychophysics because he used an empiricist methodological approach to define perception as being more than sensory physiology.

This brief description of the major psychophysicists reveals quite different orientations. On the one hand, Fechner studied sensory and perceptual events from the perspective of the underlying mental activity characteristic of the German tradition. On the other, Helmholtz studied the same phenomena and developed interpretations consistent with an empirical orientation, related more to the British tradition. However, both scientists succeeded in pointing to an area of investigation not easily accommodated in physics, physiology, or natural philosophy alone, and that was the emerging subject matter of psychology.

EVOLUTION

The publication of Charles Darwin's *On the Origin of Species* (1859) represented for science a triumph of advances that began with Copernicus. The questioning of theologically based authority started with Copernicus and continued as various other thinkers chipped away at the province of theology throughout post-Renaissance development. The sciences of physics, physiology, and chemistry and early empiricist psychology provided reliable answers to the perplexing issues of life without reverting to explanations involving divine mediation. Darwin's theory of evolution, although not a complete innovation, provided convincing evidence that shocked theologians. First, if humans and apes derive from common ancestry, then the traditional privileged position of humanity, created in God's image, is unwarranted. Second, if all life evolved by the principle of natural selection, the role of God as the final cause of creation is unnecessary. For psychology, Darwin's theory of evolution represented the third movement of the nineteenth century (the other two being the sophistication of physiological research and the development of psychophysics) that not only allowed the formal study of psychology to emerge as a discipline but, indeed, made its establishment unavoidable and compelling.

Charles Darwin

We have already seen that Erasmus Darwin suggested an evolutionary concept in eighteenth-century England. His grandson Charles Darwin (1809–1882) was the fifth child of Robert Darwin, a successful physician in Shrewsbury in western England. In 1825, Charles was sent to the University of Edinburgh to study medicine, but the practicalities of clinical medicine were not to his liking. However, during his stay at Edinburgh, Darwin was exposed to the evolutionary teachings proposed by the French naturalist Jean Baptiste Pierre Lamarck (1744–1829). Briefly, Lamarck argued that changes in animals occur through the efforts of the species to adapt to its environment; that is, giraffes adapt to better food sources by developing longer necks. Thus, the acquired characteristics of environmental adaptation are passed from generation to generation. While at Edinburgh, Darwin was also introduced to the methodology of naturalistic biology and geology.

In an effort to find a new career for him, Darwin's father sent him to Cambridge University to prepare for the Anglican clergy. Darwin received his degree in 1831, and through his Cambridge contacts secured an invitation to join an expedition sponsored by the British Admiralty. He was to serve as an unpaid naturalist in a survey of the coasts of Patagonia, Tierra del Fuego, Chile, Peru, and some of the Pacific islands. After overriding objections from his family, Darwin set sail on December 27, 1831, aboard the H.M.S. *Beagle*, under the command of Robert Fitzroy.

The 5 years of the voyage of the *Beagle* had profound effects on Darwin, and he used the observations and evidence collected during those years as the basis for his subsequent writings. The places visited by the *Beagle*—from the Cape Verde Islands off the west coast of Africa to the isolated Galápagos Islands in the Pacific Ocean off South America—provided not only a living laboratory for Darwin but also a preserved history of tremendous significance. He was able to journey on land and observe primitive forms of human life and innumerable species of animal life. His work during these 5 years provided convincing botanical, geological, and anatomical evidence to support the theory of natural selection.

CHARLES DARWIN (1809–1882).
(Courtesy, New York Public Library.)

Darwin's theory of evolution differed from Lamarck's on several important points. Darwin proposed that, first, species variation over time results from chance and not by the adaptive effort of animals. Second, natural selection is an inherent struggle for species survival, a view Darwin found consistent with the economic views of the British philosopher Thomas Malthus (1766–1834). Darwin's argument for evolution by natural selection relied on several compatible lines of evidence. He postulated that the numbers of members of species are relatively constant, but also noted the overproduction of pollen, seeds, eggs, and larvae, leading him to conclude that there is a high mortality rate in nature. In addition, he gathered ample evidence to prove that all members of a given species are not identical but, rather, show variability along anatomical, behavioral, and physiological dimensions. He concluded that in the same species some members are better able to adapt than others, and they will tend to have more offspring, who will in turn reproduce. Finally, he pointed to the resemblance between parents and their offspring, concluding that subsequent generations will not only maintain but also improve their adaptability to environmental conditions. As environmental conditions vary, the criteria of natural selection differ, and over time, divergent generations gradually arise from common ancestry. For Darwin, the missing piece of his evolutionary theory was the exact nature of hereditary transmission, which he could not supply. A rather obscure Czech monk, Gregor Johann Mendel (1822–1884), completed Darwin's theory through his botanical experiments showing the inheritance of particular characteristics, and thus founded the study of genetics.

The implications of evolution by natural selection for psychology were addressed in two later works of Darwin, *The Descent of Man* (1871) and *The Expression of the Emotions in Man and Animals* (1872). Darwin argued that the essential difference between humans and the highest primates is one of gradation, not quality. He pointed to the full gamut of activities, ranging from self-preservation to cognition to emotions, shared by all animals, including humans. Moreover, Darwin included the evolution of moral attitudes in his framework, pointing out the survival value of moral development. An admirer of Darwin, George Romanes (1848–1894), pursued the comparative value of cross-species study in his work *Animal Intelligence* (1882). Romanes presented evidence to establish common dimensions of evolution between human and infrahuman activity and offered a primitive form of comparative psychology. The methodology of Romanes' work was somewhat loose and anecdotal in nature, leading to the criticism that his conclusions were anthropomorphic. Another early comparative psychologist, Lloyd Morgan (1852–1936), attempted to counteract the anthropomorphism of Romanes by urging a parsimonious convention, known as "Lloyd Morgan's Canon," in comparative studies: If a particular animal behavior could be explained by any one of several functions, the simpler, and presumably phylogenetically lower, explanation should be chosen. The validity of a comparative approach to psychological study was firmly established in Britain as a direct by-product of Darwin's teachings.

Herbert Spencer

A more comprehensive view of psychological studies with social implications derived from evolutionary theory was contained in the writings of Herbert Spencer (1820–1903). Spencer has been described as an "evolutionary associationist." His writings on the association of ideas, advocating associations as the principal mediating experiences, place him clearly in the British empirical tradition. Moreover, Spencer's writings come almost full circle to the Ionian physicists of ancient Greece, who were searching in nature for the basic substance of life that accounts for change. In a similar manner, Spencer used evolution as the basic principle and applied this interpretation of change in life to the individual in society.

Spencer stressed that the relations between feelings are based on the association principle of similarity. His evolutionary perspective led him to suggest that associations made repeatedly are passed along through heredity. Accordingly, Spencer's view of the inheritance of acquired associations led him to conclude that instincts become an inherent part of our ethnic and racial heritage. Indeed, it was Spencer, in applying evolution on the human social level, who coined the phrase "survival of the fittest," which not only distorts Darwin's theory but ultimately falls into redundancy. Spencer's evolutionary associationism, for all the empirical aura embodied in the concept of associationism from the British tradition, clearly supports a perspective more akin to the German philosophical position because of its proposal for inherited dispositions.

Francis Galton

A final figure in the nineteenth-century evolutionary context of Britain was Francis Galton (1822–1911). Both Darwin and Galton were grandsons of Erasmus Darwin, the former through Erasmus' first wife and the latter through his second wife. Galton's primary interest was human evolution and the inheritance of specific traits. Surely, his own family offered a ready example of the inheritance of intelligence. His two major works of psychological significance were *Hereditary Genius* (1869) and *Inquiries into Human Faculty and Its Development* (1883). Both works examined the inheritance of mental abilities with the goal of racial improvement. Indeed, the latter book took on an intense fervor as Galton argued the benefits of a belief in the attainment of human progress through evolutionary theory at the expense of religious approaches to human betterment.

Galton was a person of many talents, and his time spent in psychological inquiry was so significant that many psychologists point to him as the founder of experimental psychology in Britain. Perhaps because of the way in which nineteenth-century British philosophy accommodated psychological inquiry, the founding of modern psychology as a discipline separate from philosophy is looked upon as a German event. Another reason for attributing the separation of psychology from philosophy to the Germans is that psychology needed visible champions in the "hostile" climate of German philosophy. In this context, Galton's role as an early advocate of psychology tends to be overlooked. However, his methodological rigor and his emphasis on long-term adaptation in terms of species improvement had an impact on early American functional psychology. This is described in Chapter 12.

To assess human abilities, Galton developed a methodological strategy that rested on the statistical analyses of mental tests. Galton's tests were designed to measure individual achievement on mental exercises; they were brief to permit a wide sampling of subjects. Accordingly, Galton placed great emphasis on the measurement of individual differences, and attempted to study systematically varieties of mental activities ranging from motor behavior to mental imagery. Later, he devised a variety of apparatus to measure such characteristics as olfactory discrimination and space perception. He opened a laboratory where people could pay a small fee and receive a battery of tests, enabling him to sample over 9,000 people along many dimensions of intellectual and motor performance.

Galton started a movement in psychology that emphasized the value of testing and an associated statistical approach to defining population trends. This movement gained momentum during the beginning of the twentieth century, when psychologists, most notably in France and America, began to use mental testing on an increasingly wide scale. Moreover, Galton demonstrated the practicality or utility of translating Darwin's views on evolution from a biological abstraction into a mechanism for bettering society.

In summary, three nineteenth-century developments—neurophysiology, psychophysics, and evolutionary theory—formed an intellectual climate that demanded the founding of the new discipline of psychology. These movements were the immediate precursors of psychology and overlapped with the early development of modern psychology. We next turn to Germany, where the volatile intellectual climate of the nineteenth century produced the foundation of modern psychology, presented to the world in the context of two opposing models of psychological inquiry.

CHAPTER SUMMARY

Three movements in the nineteenth century formed the intellectual background from which psychology emerged as a discipline separate from the natural sciences and philosophy. In physiology, great advances were made in the understanding of the nervous system. The specific functions of nerve fibers were described by Bell and Magendie. Müller's systematic analysis of neural conduction led researchers such as Du Bois–Reymond and Helmholtz to describe the nature of the nerve impulse. As a reaction against Gall's phrenology, the localization of brain functions was studied through neuroanatomy and histology, reaching a culmination in the works of Flourens and Sherrington. The strides in physiological investigations were combined with advances in knowledge of physics to examine sensations, and Young, Helmholtz, and Müller all contributed theories of sensory processing. The methodological integrity of subjective sensory experience was justified by Purkinje.

The second nineteenth-century intellectual backdrop to modern psychology was psychophysics. This movement differed from sensory physiology by proposing that the integrity of sensory experience is not completely reducible to physics and physiology. Although Weber contributed both methodologically and substantively to psychophysics, its clearest expression is found in the work of Fechner. The quantitative analysis of sensory and perceptual experiences marked the need for a disciplinary approach not accommodated by the natural sciences. This view received strong support from the experiments of Helmholtz, especially in his doctrine of unconscious inference in perception—clearly a mental construct.

The final movement centered around Darwin's theory of evolution by natural selection. Darwin's writings completed the Copernican revolution in science and established the primacy of scientific empiricism in our quest for knowledge. Spencer applied Darwin's writings to evolutionary associationism, and Galton, also influenced by Darwin, made an intensive examination of individual differences through mental testing. All three movements aptly demonstrate the supreme position of empirical science in the nineteenth century. The scientific ideal represented the appropriate framework for the pursuit of psychological inquiry.

BIBLIOGRAPHY

Primary Sources

DARWIN, C. G. (1868, 1875). *The expression of the emotions in man and animals*. London: Murray.

DARWIN, C. G. (1871). *The descent of man, and selection in relation to sex*. New York: Appleton.

DARWIN, C. G. (1964). *On the origin of species by means of natural selection, or The preservation of favoured races in the struggle for life* (1859). Cambridge: Harvard University Press.

DENNIS, W. (Ed.). (1948). *Readings in the history of psychology*. New York: Appleton-Century-Crofts.

FECHNER, G. (1966). *Elements of psychophysics* (H. Adler, Trans.). New York: Holt, Rinehart & Winston.

GALTON, F. (1869). *Hereditary genius*. London: Macmillan.

RAND, B. (1912). *The classical psychologists*. New York: Houghton Mifflin.

ROMANES, G. J. (1883). *Animal intelligence*. London: Kegan Paul.

SHERRINGTON, C. S. (1906). *The integrative action of the nervous system*. New Haven: Yale University Press.

Studies

BOAKES, R. A. (1984). *From Darwin to behaviourism*. London: Cambridge University Press.

BUSS, A. R. (1976). Galton and the birth of differential psychology and eugenics: Social and political forces. *Journal of the History of the Behavioral Sciences, 12,* 47–58.

BUSS, A. R. (1976). Galton and sex differences: An historical note. *Journal of the History of the Behavioral Sciences, 12,* 283–285.

DE BEER, G. (1964). Mendel, Darwin and Fechner. *Notes and Records of the Royal Society, 19,* 192–226.

DENNY-BROWN, D. (1957). The Sherrington school of physiology. *Journal of Neurophysiology, 20,* 543–548.

DEWSBURY, D. A. (1979). Retrospective review: An introduction to the comparative psychology of C. Lloyd Morgan. *Contemporary Psychology, 24,* 677–680.

ERICKSON, R. P. (1984). On the neural bases of behavior. *American Scientists, 72,* 233–241.

FACTOR, R. A., & TURNER, S. P. (1982). Weber's influence in Weimar Germany. *Journal of the History of the Behavioral Sciences, 18,* 147–156.

FULTON, J. F. (1952). Sir Charles Scott Sherrington, O. M. *Journal of Neurophysiology, 15,* 167–190.

FROGGATI, P., & NEVIN, N. C. (1971). Galton's law of ancestral heredity: Its influence on the early development of human genetics. *History of Science, 10,* 1–27.

GILLISPIE, C. C. (1959). Lamarck and Darwin in the history of science. In B. Glaco, O. Temkin, and W. L. Straus (Eds.), *Forerunners of Darwin, 1745–1859*. Baltimore: Johns Hopkins University Press, 265–291.

GILMAN, S. L. (1979). Darwin sees the insane. *Journal of the History of the Behavioral Sciences, 15,* 253–262.

GREENBLATT, S. H. (1984). The multiple roles of Broca's discovery in the development of the modern neurosciences. *Brain and Cognition, 3,* 249–258.

GREENE, J. C. (1959). *The death of Adam. Evolution and its impact on Western thought*. Ames: University of Iowa Press.

GRUBER, H. E. (1983). History and creative work: From the most ordinary to the most exalted. *Journal of the History of the Behavioral Sciences, 19,* 4–14.

HURVICH, L. M., & JAMESON, D. (1979). Helmholtz's vision: Looking backward. *Contemporary Psychology, 24,* 901–904.

KRUTA, V. (1969). *J. E. Purkyne (1787–1869) Physiologist: A short account of his contributions to the progress of physiology with a bibliography of his works*. Prague: Czechoslovak Academy of Sciences.

MACKENZIE, B. (1976). Darwinism and positivism as methodological influences on the development of psychology. *Journal of the History of the Behavioral Sciences, 12,* 330–337.

MACLEOD, R. B. (1970). Newtonian and Darwinian conceptions of man, and some alternatives. *Journal of the History of the Behavioral Sciences, 6,* 207–218.

MOIRE, J. R. (1979). *The post-Darwinian controversies: A study of the Protestant struggle to come to terms with Darwin in Great Britain and America, 1870–1900.* Cambridge: Cambridge University Press.

PASTORE, N. (1973). Helmholtz's "popular lecture on vision." *Journal of the History of the Behavioral Sciences, 9,* 190–202.

RICHARDS, R. J. (1977). Lloyd Morgan's theory of instinct: From Darwinism to neo-Darwinism. *Journal of the History of the Behavioral Sciences, 13,* 12–32.

SOHN, D. (1976). Two concepts of adaptation: Darwin's and psychology's. *Journal of the History of the Behavioral Sciences, 12,* 367–375.

STROMBERG, W. J. (1989). Helmholtz and Zoellner: Nineteenth-century empiricism, spiritism, and the theory of space perception. *Journal of the History of the Behavioral Sciences, 25,* 371–383.

TURNER, R. S. (1977). Hermann von Helmholtz and the empiricist vision. *Journal of the History of the Behavioral Sciences, 13,* 48–58.

WARREN, R. M., & WARREN, R. P. (1968). *Helmholtz on perception: Its physiology and development.* New York: Wiley.

WASSERMAN, G. S. (1978). *Color vision: An historical introduction.* New York: Wiley.

WOODWARD, W. R. (1972). Fechner's panpsychism: A scientific solution to the mind-body problem. *Journal of the History of the Behavioral Sciences, 8,* 367–386.

ZUPAN, M. L. (1976). The conceptual development of quantification in experimental psychology. *Journal of the History of the Behavioral Sciences, 12,* 145–158.

11

The Founding of Modern Psychology

By the last quarter of the nineteenth century, European science had achieved widely recognized prestige as the optimal form of intellectual activity. The inductive method, successfully employed by Copernicus and subsequently nurtured through three centuries of philosophical endeavors, acquired a sense of reliability that evoked a faith in the scientific approach throughout the nineteenth century. Indeed, the dramatic advances in biology, chemistry, and physics, with their demonstrated implications for the betterment of society, provided ready justification for trust in scientific methods.

As psychology began to break away from the provinces of religion and speculative philosophy, the prevailing dominance of nineteenth-century scientific inquiry had profound importance. Specifically, if psychology was to embody a collection of knowledge derived from authoritative sources other than religious belief, science offered the most promising direction of pursuit. The findings of British natural philosophers and German psychophysicists had already demonstrated the viability of scientific methodology for certain psychological issues. Thus, by the end of the nineteenth century, the *Zeitgeist* ("spirit of the times") in the intellectual climate of Europe showed a readiness to accept the formalized study of psychology. The immediate problem was the specific scientific model for psychology to emulate. As had occurred since the time of early Greek philosophy, several models of scientific inquiry, derived from varying underlying assumptions about the nature of people and their concomitant psychological processes, competed to express the definition and form of modern psychology.

This chapter concentrates on the emergence of psychology in Germany. In a sense, it is curious that psychology first emerged as a formal discipline in Germany, as the intellectual climate of Britain was actually more amenable to its acceptance. As we have seen, the relatively homogeneous model of empiricism was widely accepted in Britain, and natural philosophers there investigated mental associations as the mediational agent of cognitive and emotional processes. Moreover, Darwin's evolutionary theory by natural selection had an impact that resulted in further acceptance of the possibility for psychology's studying the full range of animal activity. However, it was precisely this tolerant atmosphere that prevented the emergence of psychology from natural philosophy; there simply was no need for a separate discipline. In other words, the philosophical traditions of Britain were quite open to the study of psychological issues and ably accommodated new questions and methodological approaches. In contrast, the intellectual climate in Germany was varied; German philosophy reflected this diversity and was uncommitted to a single model of psychological inquiry. As seen earlier, the predominant underlying commonality of German philosophical views on psychology was the essential activity of the mind. Mental activity was expressed in the logical and metaphysical systems of rationalism proposed by Kant and Wolff, in the unconsciously motivated strivings inherent in the views of Schopenhauer and von Hartmann, and in the mechanical model of Herbart. Such diversity precluded the ready acceptance of scientific psychology within the province of German philosophy. Indeed, the preliminary attempt to study sensory and perceptual processes scientifically in psychophysics came about through the work of physiologists and physicists, not philosophers.

Accordingly, it was Germany that provided the setting for the emergence of psychology. The diverse intellectual climate of Germany was perhaps the most exciting in Europe. The nation had been recently united under the Hohenzollern dynasty, and the strong tradition of Prussian support for universities was extended throughout Germany. In all fields of science, philosophy, and literature, the German intelligentsia achieved international recognition. In psychology two separate models emerged. Although neither was successful in establishing the definitive framework of the new science, both gave psychology its push into science and served as anchor points for the discipline's progress.

PSYCHOLOGY AS A NATURAL SCIENCE

The label *natural science* as applied to psychology is meant to describe a framework of psychology that emulates the methodology and analytic goals common to biology, chemistry, and physics. It implies that psychology should be studied by defining psychological events in terms of variables and submitting such variables to the analytic scrutiny of the experimental method. Accordingly, this model of psychology was a radical departure from the German philosophical establishment of metaphysical psychology. This natural science conceptualization, in its attempt to set psychology off from the prevailing philosophical systems of Germany, limited both the scope and methodology of psychology to a level that confined its growth and led to its eventual rejection.

Structural, or Content Psychology

Psychology as defined in this system is the analytic study of the generalized adult human mind through the method of introspection. This approach originated with Wilhelm Wundt and was championed in the United States by his student Edward Bradford Titchener. Their conclusions about psychology were highly compatible, and our overview will consist of a synopsis of their collective writings. Because psychology was to study the contents of the mind, this system is sometimes called *content psychology*. In addition, Titchener, writing in 1898, emphasized mental structures and named this system *structural psychology*. The system, whatever the name given it, had as its goal the analysis of the human mind through the careful application of the experimental method of introspection carried out by trained scientists. By analogy, this system aimed to develop the "chemistry of consciousness."

In the following description of structural psychology, it is important to put into proper perspective the relative contributions of Wundt and Titchener. It should be emphasized that structural psychology was Wundt's invention. Titchener was only one of his many students, and probably a minor student at that, in comparison to some of the major figures in European and American psychology who trained with Wundt. But it was Titchener who strove to bring his own inflexible sense of Wundt's psychology to the American audience. Unlike other followers of Wundt in America, the British-born Titchener remained impervious to the influences that caused schools of psychological thought to develop uniquely American styles. Perhaps one reason for Titchener's unique place in the history of American psychology was that most of Wundt's other students in America were native Americans, or at least quite adapted to American life. Nevertheless, structural psychology for the most part reached American scholars through Titchener's teaching and writing, and this may have placed Titchener in an artificially preeminent position. We should give Wundt his rightful credit as the founder, but we must recognize Titchener's role as the prime spokesperson for structural psychology in the United States.

Wilhelm Wundt. Wilhelm Wundt (1832–1920) was born in the southwestern German province of Baden, the son of a Lutheran pastor. During his childhood and adolescence, he was allowed only a strict regimen of learning, with little time for

WILHELM WUNDT (1832–1920).
(Courtesy, The Granger Collection.)

play or idleness. This upbringing produced a rather dour person, totally committed to intellectual endeavors of a systematic type, and a prolific nature. Wundt eventually studied at the University of Heidelberg, intending to become a physiologist, but turned to medicine because of the practicalities of earning his own support. After 4 years of study, he recognized that he was definitely not interested in being a physician. In 1856, he went to Berlin to study in Johannes Müller's institute of physiology, where he also worked with Du Bois–Reymond. After this brief but stimulating experience, Wundt returned to Heidelberg, finished his doctorate in medicine, and took a minor teaching post in physiology. In 1858, Helmholtz came to Heidelberg, and for the next 13 years he and Wundt worked in the same physiological laboratory. When Helmholtz left Heidelberg for Berlin in 1871, Wundt was passed over as his successor. Wundt left Heidelberg in 1874 for a year as professor of inductive philosophy at Zurich, and then in 1875 accepted a professorship of philosophy at Leipzig, where he remained for the rest of his long career.

Wundt was attracted to psychological study after having acquired a firm basis in physiology. As a result, he transferred his appreciation for science—and especially for the experimental method—to his developing interests. In 1873 and 1874 he published in two parts his systematic call for a new discipline of psychology, *Grundzüge der Physiologischen Psychologie* (*Principles of Physiological Psychology*). This work, which went through six editions in Wundt's lifetime, attempted to establish the paradigm, or framework, of psychology as an experimental science of the mind, to be studied through its processes. In addition, Wundt envisioned a type of ethnic psychology in which the scientific study of human nature could reveal higher mental processes through an anthropological approach, relating to areas of child psychology and animal psychology.

Wundt started a laboratory at Leipzig in 1879, which may be viewed as the first laboratory dedicated exclusively to psychological research. In 1881, he founded *Philosophische Studien* (*Philosophical Studies*), a journal to report the experimental studies of his laboratory. The list of Wundt's students includes the names of many of the founders of psychological systems in Germany, throughout Europe, and in America. The vast majority of his students deviated from Wundt's conception of psychology to varying extents.

Edward Bradford Titchener. Edward Bradford Titchener (1867–1927) was one of Wundt's students who carried away from his studies a fixed view of Wundt's system, which he imported to America. Although Titchener studied with Wundt for only 2 years, that brief period made an indelible impression on him, and he adhered strictly to his interpretation of Wundt's system during his career at Cornell University in New York.

Titchener was born in southern England to a family of old lineage but little money. He entered Oxford University in 1885 on a scholarship to study philosophy, and became interested in Wundt's writings, translating the third edition of the *Principles of Physiological Psychology*. However, the new psychology of Wundt was not enthusiastically received at Oxford, so Titchener resolved to go to Leipzig and work directly under Wundt. There he took his doctorate after completing a dissertation in 1892 on binocular effects of monocular stimulation. After unsuccessfully searching for a position in England, Titchener accepted a professorship at Cornell, which had opened up when Frank Angell, another American student of Wundt, went to the newly founded Stanford University. For 35 years Titchener presided over psychology at Cornell, where he was a formidable institution unto himself, advocating a rigid version of structural psychology and tolerating no dissent.

Titchener's views are appropriately considered here with the founding of psychology in Germany because he never joined the mainstream of early American psychology, which is considered in the next chapter. Titchener's major works include *Outline of Psychology* (1896), *A Primer of Psychology* (1898), *Experimental Psychology* (1901–1905), *Psychology of Feeling and Attention* (1908), *Experimental Psychology of the Thought Processes* (1909), and *A Text-Book of Psychology* (1909–1910). These works are scholarly and systematic, almost encyclopedic in their scope. However, because Titchener would not admit applied aspects of psychology, he removed himself from the major theme of American psychology, which was the study of such topics as child psychology, abnormal psychology, and animal psychology. Titchener was solely concerned with the experimental analysis of the normal adult human mind, not with individual differences. In addition, he quarreled often with his American colleagues and founded his own organization to rival the fledgling American Psychological Association because of a dispute with members of the latter group. Although Titchener supervised a large number of students in early twentieth-century American psychology, his system (and therefore Wundt's) died with him in 1927.

Structuralism. The structural psychology of Wundt and Titchener had a threefold aim: to describe the components of consciousness in terms of basic elements, to describe the combinations of basic elements, and to explain the connections of the elements of consciousness to the nervous system. Consciousness was defined as immediate experience—that is, experience as it is being experienced. Mediate experience, in contrast, is flavored by contents already in the mind, such as previous associations and the emotional and motivational states of a person. Thus, immediate experience was presumed to be unprejudiced by mediated experience. Structural psychology attempted to defend the integrity of psychology by contrasting it with physics: Physics studies the physical, or material world, without reference to the person, through observational methods of carefully controlled inspection. Psychology studies the world, with reference to the experiencing person, through the observational method of controlled introspections of the contents of consciousness. The proper subject of structural psychology is the process of consciousness, free of associations. As such, Wundt and Titchener argued, psychology must be kept free of the forces of metaphysics, common sense, and utilitarian or applied interests, which would destroy its integrity.

The experimental method proposed to secure appropriate analysis of mental contents was introspection. This technique of self-report is the ageless approach to describing self-experience. We have seen earlier that Augustine employed it with remarkable clarity in his *Confessions*. An introspective approach in nineteenth-century German science was endorsed through the elegant work of Purkinje, described in the previous chapter. However, introspection as defined by Wundt and Titchener was far more rigorous and controlled. Moreover, the credibility of structural psychology rested on the proper use of introspection—that is, the emphasis on immediate experience (rather than mediate experience) as the subject matter of psychology dictated a reliance on the method of assessing such pure experiences. Accordingly, introspection was considered valid only if done by exceptionally well-trained scientists, not naive observers. Introspection depends on the nature of consciousness observed, the purpose of the experiment, and the instructions given by the experimenters. The most common error made by untrained introspectionists was labeled the "stimulus error"—describing the object observed rather than the conscious content. Stimulus errors, according to Titchener, result not in psychological data but in physical descriptions. Not surprisingly, making introspection the only acceptable method of psychological investigation was seriously questioned because no facts or principles could be derived from the introspective method. Unfortunately, there was no group agreement among trained introspectionists on the properties of sensory experiences.

Most of the major findings of this system were seriously challenged. In terms of higher mental processes, Titchener called "thought" a mental element that is probably an unanalyzed complex of kinesthetic sensations and images. Moreover, he perceived what we call the "will" as an element composed of a complex of images that form ideas in advance of action. As a result, thought and will are linked through mental images. According to this analysis, thought must be accompanied by images. This imperative gave rise to the "imageless thought controversy," in which others, most notably Külpe, Binet, and Woodworth (who are discussed later), argued the possibility of thought processes without discrete mental images. Such an interpretation was unacceptable to Titchener because it contradicted his analytic view of thought, which required description of elements consisting of images. Instead, it substituted a more holistic or phenomenal view of thought processes, unanalyzed into constituent elements.

In the 1890s Wundt developed a three-dimensional theory of feeling. Essentially, Wundt thought that feelings vary along three dimensions. Titchener agreed with only one:

Wundt	***Titchener***
Pleasant–unpleasant	Yes
Strain–relaxation	No
Excitement–calm	No

Titchener's acceptance of only the first dimension led him to relegate emotions to organic visceral reactions. Wundt's wider interpretation led him beyond the unconscious inference of Helmholtz to posit apperception as the creative process of the components making up total perception. That is, apperception is the focus of attention within the field of consciousness at a given point in time. As such, apperception was a cognitive activity that recognized the logical bond between mental contents; feelings were viewed as the product of apperception of sensory contents. Thus, Wundt's theory of feeling as a reflection of mental apperception approached a phenomenologic interpretation of higher mental processing. Titchener did not accept this more holistic direction in Wundt's thought but instead adhered to a more reductionistic view. Titchener did propose a theory of meaning suggesting that the context in which a sensation occurs in consciousness determines its meaning. Accordingly, a simple sensation has no meaning by itself, but acquires meaning by association with other sensations or images. Thus, Titchener described the mind in terms of formal elements with attributes of their own, connected and combined by the mechanism of associations. Wundt believed that associative combinations occur by fusion, as in tonal melodies; by assimilation, as in the integrating contrast and similarity of optical illusions; or by complication, defined as the bond formed by sensations of several modalities.

Structural psychology, in its attempt to adhere strictly to a natural science model, tended to overlook psychological processes and activities that did not easily fit into its methodological framework. In addition, the overreliance on the questionable, strict methodology of introspection led structural psychology into a dead end. In a sense, structuralism was caught between the empiricism of the British tradition and the nativism of the German tradition. In other words, Wundt and Titchener articulated a view of the mind as determined by the elements of sensation; at the same time they recognized mental activity and attempted to deal with activity through such constructs as apperception. Besides the inadequacies of introspection, structuralism failed to accommodate conflicting philosophical assumptions about the nature of the mind. Accordingly, the contributions of structural psychology are somewhat mixed. First and most important, this system gave psychology a push into science. Despite all its shortcomings, Wundt did proclaim a formal discipline of psychology based upon scientific formulations, and psychology was recognized as a science. Second, structural psychology put introspection as a method to the test, which it rather dismally failed. Finally, structuralism served as an anchor of opposition to subsequent developments within psychology. Boring (1950), who was a student of

Titchener, conceded that the Wundtian influence was most apparent as a negative force, motivating scientists to discredit both the substance and the methodology of this system. Perhaps this reaction to structural psychology in America was the product of a misunderstanding of Wundt's writings or of a reliance on Titchener's version of Wundt. Nevertheless, by 1930, just three years after Titchener's death, structural psychology had ceased to be a viable force in psychology.

Other Expressions of the Natural Science Model

Structural psychology holds a unique place in the German development of the natural science model for psychology. Specifically, the writings of Wundt and Titchener constitute a systematic attempt to start a coherent science encompassing all that they considered to be psychological. As such, structural psychology was a system of psychology. However, other scientists in Germany, contemporary with Wundt, responded to the same *Zeitgeist* and wrote on psychology. They wrote as individuals, though, not as system builders, and within the limits of a natural science approach to psychology, they all rejected the extremism of Wundt and Titchener, in terms of both the substance and the methodology of structuralism. These scientists were experimentalist in the sense that they were guided in their progress not by the framework of a preconceived system, as were Wundt and Titchener, but rather by the results and implications of their laboratory studies.

Ewald Hering. Ewald Hering (1834–1918), who succeeded Purkinje as professor of physiology at Prague in 1870, did extensive work in vision and touch. He proposed a three–substance, six-color theory of color vision, involving the dichotomous contrast of red–green, yellow–blue, and white–black receptors, which produce substances of differential retinal sensitivity. In addition, he opposed the empirical views of Helmholtz on visual space perception and cited evidence to support a more nativistic interpretation consistent with Kant's philosophy. Both at Prague and, after 1895, at Leipzig, Hering refrained from grand theorizing or systematic development of a complete psychology.

Georg Elias Müller. A similar approach may be seen in the long career of Georg Elias Müller (1850–1934). Born near Leipzig and educated in history, his lifelong interest, Müller's service in the Franco–Prussian War (1870–1871) convinced him of the narrow perspective of traditional historical study, and he returned to Leipzig to study the natural sciences. In 1872, he went to Göttingen to study under Lotze, whose antimechanical, metaphysical views of psychology exerted a consistent influence on Müller. Müller succeeded Lotze as professor of philosophy at Göttingen when Lotze went to Berlin in 1881. Müller's 40 years of research at Göttingen were marked by consistent experimentation, mostly in psychophysics, and many students from both Europe and America came to study in his laboratory. He became a leader in German psychology because of his commitment to data gathering at the expense of a determining overview, and he left not a system but rather an accumulation of experimental results. Such a true experimental attitude led to some unexpected directions; for example, Müller's student David

Katz published a paper in 1909 that attempted to describe color perception without the prevailing analytical sensory approach and anticipated several major tenets of Gestalt psychology.

Hermann Ebbinghaus. Another formidable figure of German psychology known for his individual experimentation rather than for system building was Hermann Ebbinghaus (1850–1909). Educated at the University of Bonn, Ebbinghaus wrote his doctoral dissertation on von Hartmann's views of the unconscious. After receiving his degree, Ebbinghaus spent 7 years in England and France, tutoring to support himself. While in Paris he came across a copy of Fechner's *Elemente der Psychophysik*, which became an intellectual stimulant for him. Essentially, he started to study memory in the same manner as Fechner had studied sensations. Having become acquainted with the classic associationistic philosophers of Britain, Ebbinghaus viewed the law of repetition as the key to the quantification of memory. He used the nonsense syllable as a means to measure the formation of associations. Basically, he presented the subject with a series of three-letter syllables, typically a vowel separating two consonants. He deliberately chose ones without meaning that might confound memorization (for example, MEV, LUS, PAQ). He used this method on himself to measure time to mastery and retention savings over time. His work *Ueber das Gedächtnis* (translated into English as *Memory*), published in 1885, described his methodology and findings, including the famous retention curve showing forgetting over time from initial acquisition. This work was widely acclaimed, not only because of its range of topics, completeness of data, and clarity of writing, but also because Ebbinghaus documented a full experimental attack on higher mental processes that was precluded by Wundt's system.

Ebbinghaus held professorships at Berlin, Wrocław (Breslau), and Halle, and attracted many students. He started a psychological journal with a national audience, *Zeitschrift für Psychologie und Physiologie der Sinnesorgane* (*New Writing for the Psychology and Physiology of Sense Organs*; 1890) that surpassed the provincial character of Wundt's journal. From the study of memory he moved on to study color vision, and also developed early versions of intelligence tests, anticipating by several years the work of the French psychologist Binet.

His reputation was enhanced with publication of his text of general psychology, *Grundzüge der Psychologie* (*Foundations of Psychology*; 1897–1902), which became the standard text in German universities, much as William James's *Principles of Psychology* (1890) was in American universities. Although noted for his work on memory, Ebbinghaus should be best remembered as an exponent of careful experimentation in psychology. Like the others whose work is described in this section, Ebbinghaus did not leave a "school" but rather contributed to the intellectual atmosphere that established psychology as a scientific enterprise.

Ernst Mach and Richard Avenarius. A final movement within the German natural science model for psychology contributed a philosophical justification for the scientific bases of psychology. The major figures were Ernst Mach (1838–1916) and Richard Avenarius (1843–1896). Their writings have been labeled radical empiricism or, more simply, Machian positivism. We have already seen positivism in Comte's attempt to explain intellectual progress through a reliance on observable events, undercutting metaphysical explanations that go beyond such direct observations, and we will again discuss positivism when behaviorism is presented. Behaviorism includes a modern expression of "logical positivism," a scientific attitude that defines scientific events in terms of the operations that produce them. The positivism of Mach and Avenarius was consistent with the skepticism of Hume, which held that causality is only the observed covariation of events and is valid only to that extent. Moreover, all events are reducible to the psychological and physical components of observations, defined as the process of sensation. Thus sensations and sensory data form the critical essence of science, so that the introspection of the scientist forms the basis of all methodology.

Mach, who was a professor of physics in Prague during his most productive years, published his most important work for psychology in 1886, *Analyse der Empfindungen* (*Analysis of Sensations*). In asserting that sensations are the data of all science, Mach pushed psychology further from metaphysics and established a standard of scientific criteria for psychology to emulate. According to Mach, the only certain reality is our own experience. His analysis of space and time viewed these processes as sensory and not the mental categories suggested by Kant. Avenarius, a professor of philosophy at Zurich, essentially agreed with Mach's conclusions, although his writing did not have the clarity and compelling justification of Mach's.

Both Mach and Avenarius presented complex views on science in general and psychology in particular. Their importance lies in their positive views on the sensory basis of science, which gave psychology a position alongside physics, adding greatly to its integrity as an independent discipline. Mach moved to Vienna in 1895 and influenced a new generation of philosophers of science. They in turn reformulated positivism at a time when behaviorism was being proposed as the revised definition of psychology, a view we consider later.

To summarize briefly, psychology emerged as a recognized discipline within a natural science model. The most coherent and systematic expression of psychology was by Wundt, although others contributed to the credibility of psychology as a natural science. Whereas this "founding" was somewhat of a false beginning, these German psychologists did succeed in establishing a scientific ideal for psychology. We now turn to a competing model of psychology, which, while not as well known as the natural science model, provided an alternative for a broad-based science of psychology.

PSYCHOLOGY AS A HUMAN SCIENCE

At the time of psychology's modern founding in the 1870s, science was largely equated with the study of physical events in the natural sciences of biology, chemistry, and physics. Structural psychology emerged from the joining of sensory physiology and the British empirical assumptions about the mind, methodologically combined by the psychophysicists and finally expressed systematically by Wundt. However, the strict definition of the scientific method in terms of experimental study confined the scope of psychology.

As it turned out, certain contemporaries of Wundt, including those who immediately and critically reacted to his views, did not agree with the restrictions that Wundt and Titchener imposed on psychology. This group of psychologists did not present a common alternative or a coherent rival school to structural psychology. Rather, they differed as individuals with both the scope and methodology of Wundt's formulations. However, they agreed that psychology should not be bound to a single method of science, and that science itself encompasses more than just the experimental method. Moreover, their conclusions on psychology fit the German philosophical assumption of mental activity better than Wundt's psychology, and subsequently more systematic expressions of psychology were derived from their teachings.

Act Psychology

Although varying interpretations of act psychology may be found in both historical and contemporary expressions of psychology, the defining character of this movement centers around the inseparable interaction of the individual and the environment. As such, psychological events are often defined as *phenomena*; that is, events that cannot be reduced to component elements without losing their identity. The form of act psychology we now examine was a contrast to the elementarism of structural psychology. However, contemporary forms of act psychology currently exist in contrast to the atomistic reductionism of stimulus-response behaviorism.

Franz Brentano. The person who came closest to Wundt, both in influence and in temporal contiguity in late nineteenth-century Germany, was Franz Brentano (1838–1917). Whereas Wundt's long career was characterized by systematic scholarship in a stable intellectual environment, Brentano's career was marked by controversy and upheaval. However, his relatively few works were significant and represented a viable alternative to Wundt's dominating conception of psychology. Indeed, Titchener, writing in 1925, named Brentano's act psychology as one of the major threats to psychology's integrity, inadvertently giving Brentano's views enhanced stature.

Brentano, the grandson of an Italian merchant who immigrated to Germany, was born in Marienberg near the Rhine. The family was known for literary achievements. Brentano's aunt and uncle were writers in the German romantic tradition, and his brother Lujo won the Nobel Prize in 1927 for his work in intellectual history. At age 17, Franz began studying for the Catholic priesthood in Germany.

He joined the same order of priests, the Dominicans, as did the great Scholastic Saint Thomas Aquinas, and his study of Scholasticism may have influenced his choice of a dissertation topic—*On the Manifold Meaning of Being According to Aristotle*—for which he was awarded a doctorate in philosophy from the University of Tübingen in 1862. During the next 2 years he finished his studies in

FRANZ BRENTANO (1838–1917). (Courtesy, The Granger Collection.)

theology and was ordained a priest at Würzburg. There he continued his work in philosophy, and in 1866 he became a docent at Würzburg in recognition of his study on the psychology of Aristotle, a work acclaimed as the most scholarly presented to the philosophy faculty since the turn of the nineteenth century. While at Würzburg, Brentano became known as an effective teacher because of the clarity of his presentations in philosophy and mathematics as well as because of his love of research. However, Brentano's revision of scholastic logic, his appreciation of British empiricism, and his favorable study of Comte published in 1869 earned him the sharp criticism of the Catholic establishment. Nevertheless, on the strong recommendation of Lotze and over the objections of the Austrian emperor and the archbishop of Vienna, Brentano was appointed professor of philosophy at Vienna in 1874. There he remained as a popular but controversial teacher until 1894, and enjoyed his most productive period. His many students included Karl Stumpf; Edmund Husserl, the founder of modern phenomenology; and Sigmund Freud, who took his only nonmedical courses from Brentano at the University of Vienna between 1874 and 1876.

The controversy that surrounded Brentano from 1870 onward revolved around his criticism of the Church. During this period the Church felt threatened by the forces of intellectual liberalism and political nationalism in Italy. Viewing these developments from Würzburg and Vienna, Brentano became increasingly disturbed by the anti-intellectualism of the Church. Finally, in April 1873, he left the priesthood and openly attacked the reactionary attitudes of the Church hierarchy and the doctrine of papal infallibility, which had been declared in 1870 by the First Vatican Council. In 1880, wishing to marry and being prohibited from doing so by Austrian law, he resigned his professorship and his Austrian citizenship so that he could marry legally in Savoy. He returned to Vienna in the lesser post of docent, which meant he could not direct doctoral students. However, the more conservative forces of the theology faculty continued to exert pressure against him and he severed all relations with the university in 1894, eventually finding a home in Florence. As a pacifist, he protested Italy's entrance into World War I by moving to Zurich, where he died in 1917.

Brentano's most important work in psychology, *Psychologie vom empirischen Standpunkt* (*Psychology from an Empirical Standpoint*), appeared in 1874 and was intended as the first of a multivolume explication of psychology's scope and methodologies. He never finished the later volumes, so we have only an outline of his views on psychology. Nevertheless, Brentano's proposals for psychology's progress stand in sharp contrast to Wundt's. Brentano defined psychology as the science of psychic phenomena expressed as acts and processes. This definition contrasts with psychology viewed in terms of physical reductionism, consciousness, or associationism. Brentano viewed consciousness in terms of a unity expressed by acts. Thus, structuralism's inherent goal of finding the elements of consciousness was meaningless for Brentano because such study destroys the essential unity of consciousness, and such elements, if they exist, do not have psychological meaning. Rather, according to Brentano, only the product of consciousness—the acts and processes—are truly psychological. Brentano subscribed to a physiological or biological substrate of psychological acts, which supplies information about, but is not identical with, psychological acts. Moreover, Brentano recognized two levels of psychological study: pure and applied. Pure psychology studies physiological considerations, individual differences, personality, and social levels. Applied psychology consists of the value of psychology for other sciences. Thus, to Brentano, psychology is the pinnacle of science and is differentiated from other sciences by its study of intentionality, or the ability of people to reach for some object–goal beyond themselves. The psychological act is directed; it is intentional, and this characteristic is unique and purposive.

Brentano argued for hierarchical levels of classes of psychic phenomena. At the representational level is mere awareness; it corresponds to the nonmediated experiences that Wundt saw as psychology's entire subject matter. However, Brentano taught that beyond the representational level is a cognitive class, which he described as a level of judgment. Finally, there is a level of personalization of psychic phenomena, a type of assimilation that individualizes experience, which he labeled the class of interest. As an empirical science, psychology is studied through observation, but it is not reduced to elementary components. Rather, Brentano allowed for various empirical methods that are adaptable to the subject matter of psychology. Perhaps most important was the method of inner perception of ongoing acts. This method was not introspection in the Wundtian sense, but rather the naive reporting of evident psychic phenomena. Other methods described by Brentano were the objective observation of past psychic acts in memory, the observation of the overt behavior of people, and the observation of antecedent and physiological processes concomitant to psychological acts. Accordingly, Brentano's empiricism was open-ended but firmly based on observation.

Brentano's later views did not benefit from the intellectual stimulation and interaction of a stable academic environment. Nevertheless, he moved toward the development of a phenomenological method for psychology. Specifically, he argued that phenomenology is a descriptive method that is explanatory and leads to understanding. This method is based partly on the a priori science that examines the ways we come to knowing and partly on empiricism. By using a personalistic orientation with the self as the point of reference, Brentano hoped for a psychological method that allows psychological acts to be described in terms of the subjective, experiencing person. Accordingly, objects in the environment may be described as part of the process of perceiving. For example, the physical light stimulus, the visual sensory mechanisms, and perceptual levels are interrelated in a psychological mode, best termed "seeing." It was left to Brentano's students, most notably Husserl, to develop this methodology further.

Brentano's psychology did not have the recognized impact of structural psychology. Indeed, the relatively scant outline of his views leaves the reader somewhat confused, because his works contain a framework that, although interesting, is rather vague. Various parts of Brentano's teachings were influential in psychology's subsequent developments. The Gestalt movement, the third force movement of phenomenological psychology, and even the eclectic orientation of American functionalism all owe a debt to Brentano.

Karl Stumpf. Karl Stumpf (1848–1936) was a major figure in German psychology and led the way to the acceptance of psychology by the European academic establishment. Moreover, he engaged Wundt in a dispute over the introspection of music that drew sharp distinctions between their views. However, Stumpf was not an originator of psychological views. Rather, his importance lies in his expression of the tremendous personal influence of Brentano, as well as in the achievements of his many students.

Stumpf was born in Bavaria, in southwestern Germany, the son of a court physician. Through his grandfather, he received a superb early education in the classics and the natural sciences. He also showed an early talent for music, and began composing at age 10. By the time he was an adult, he had mastered five musical instruments. In 1865, Stumpf entered the university at Würzburg, where he met Brentano and was captivated by the vitality of his teaching and his love of scholarship. Brentano sent him to Lotze at Göttingen to complete his degree, and there he studied physiology, physics, and mathematics. Stumpf's first psychological work, published while he was a docent at Göttingen, was a nativistic theory of space perception. This work won him a professorship at Würzburg. For the next 20 years, he moved among various universities in Germany and Prague, until he was appointed to the prestigious professorship at Berlin in 1894. During the time preceding this appointment, Stumpf published his *Tonpsychologie* (*Psychology of Tones*; 1883, 1890), in which he was able to blend his love of music and science. Also at this time, Wundt and Stumpf conducted a public argument over the proper description of melodies: through introspection or through the trained ear of the musician. Stumpf obviously favored the latter, but the importance of his argument rested on his emphasis on the essential unity of musical experience. In other words, whereas the introspectionist claimed a melody is reducible to its constituent sensory elements—that is, the individual notes—Stumpf held that the melody is a unity itself, noting that a key transformation, which would actually change the individual notes, would not change the perception of the melody. This interpretation is consistent with a phenomenological view, both reflecting Brentano's influence and anticipating Husserl's development after he finished his degree with Stumpf.

In bringing phenomenology into psychology, Stumpf followed the classification of levels of experience articulated by Brentano. The first level concerns the phenomena of sensory and imaginal data of experience. The second classification involves the psychic functions of perceiving, desiring, and willing, equivalent to Brentano's acts. Finally, there is the level of relations, a cognitive classification somewhat akin to Brentano's interests. Stumpf passed along his version of act psychology and his phenomenology to a generation of students. Köhler and Koffka, two of the three founders of the Gestalt movement in psychology, received their degrees under Stumpf at Berlin. Accordingly, Stumpf accomplished what Brentano was unable to do. He offered an alternative to Wundt's structural psychology, and as psychology progressed in Germany, it was Stumpf's students who came to dominate.

Christian von Ehrenfels. A student of Brentano, Christian von Ehrenfels (1859–1932) held views that actually represented a bridge between the natural and the human science models. He took Mach's notion of form in space and time and suggested that form is more than the sum of the parts. In a paper published in 1890, Ehrenfels introduced the concept of form quality, *Gestaltqualität*, as a new identity that appears when elements are brought together. Further, Ehrenfels distinguished between temporal and nontemporal form qualities. The former include sensations that are time related, such as a musical melody. Nontemporal form qualities are usually spatial and include the perception of movement. Following Brentano's lead, Ehrenfels pursued an empirical (but not necessarily an experimental) demonstration of form qualities. For example, as evidence of the presence of form qualities, he cited the reports of subjects indicating the persistence of form despite changes in the elements of the stimuli evoking the sensation.

Although dissatisfied with Wundt's system, Ehrenfels nevertheless retained an emphasis on the elements of perception. It was left to the Würzburg school to extend the progression from Mach to Ehrenfels and to pave the way for Gestalt psychology, which successfully challenged Wundt's system in Europe.

Alternative Scientific Approaches

Before continuing with the next expression of psychology as a human science, it is appropriate to mention briefly some issues for psychology in the philosophy of science. The human science model for psychology basically questions the equating of the methods of the natural sciences with the notion of science itself. Whereas the developing phenomenology of Brentano and Stumpf proposed methodological alternatives within psychology, other writers were also questioning the natural science model from a more general perspective.

Wilhelm Dilthey. A German philosopher, Wilhelm Dilthey (1833–1911) objected to the dominance of the natural science approach and proposed a view that emphasizes the person perceived in terms of historical contingency and change. In seeking understanding of the human situation, Dilthey argued that understanding is a matter of finding meaning—a mental operation just as perception and reasoning are mental operations. He actually used the term "human science" to propose an appropriate criterion for evaluating human understanding that is not distorted by artificially trying to conform to natural science criteria. Thus he viewed historical evaluation as a humanistic enterprise based on the meaning of a person's place in time. Natural science techniques, whether in experiments or through introspection, are far too narrow to assess the meaning of humanity adequately (see also Chapter 16).

Henri Bergson. Henri Bergson's (1859–1941) work is somewhat similar to Dilthey's. Bergson, a major figure in French philosophical thought, wrote an exhaustive treatment of the metaphysical problems of knowledge and time. He argued that the methodology of the natural sciences distorts time, motion, and change by interpreting them as static concepts. According to Bergson, the progress of life should be evaluated by appropriate criteria, and such criteria are not represented in natural science methodology. He defined "true empiricism" as finding the dynamics of becoming through participating in it. By using a method of intuition, metaphysics can provide the appropriate perspective to secure the meaning of life. Bergson concluded that the key to understanding life is found by viewing life as a process of creative evolution through the subjective consciousness arising in each individual.

This brief outline of the views of Dilthey and Bergson does not begin to touch on their depth and complexity. However, both philosophers questioned the prevailing scientific methodologies. As we shall see when we come to twentieth-century psychology, shades of these varying interpretations of science are found in the works of others. At this point it is important to note that the dominance of natural scientific methodology was beginning to erode, and in many respects, Brentano and Stumpf were furthering that process.

The Würzburg School

A final expression of a human science model for psychology came from the Würzburg school, associated with Oswald Külpe (1862–1915). Essentially, the Würzburg school investigated two areas, with dramatic results. Its first major finding was that thoughts do not necessarily have accompanying images, giving rise to a confrontation with the basic tenets of structural psychology. Second, it contended that thinking cannot be fully accounted for by associationism. The Würzburg school's short-lived productivity seriously challenged structural psychology on its own grounds. The Würzburg psychologists were not nearly as radical as Brentano, and accepted many of the proposals of Wundt's structural psychology. However, using the same framework as Wundt, they succeeded in doing far greater damage to belief in the validity of structural psychology.

Külpe was born in Latvia of German ancestry and received his early education there before journeying to Leipzig to study history. His contact with Wundt left him undecided about continuing to study history or changing to psychology. After studying both disciplines at several universities, he returned to Wundt and took his doctoral degree in 1887. He largely adhered to Wundtian psychology until he was appointed professor at Würzburg and his interests led him to investigate thought processes. In 1901, two of Külpe's students published a paper on associations, using empirical methods that went beyond introspection, accepting self-reports on thought processes. For the next 10 years Külpe and his associates produced data questioning the interpretation of thought processes inherent in structural psychology. They did not resolve the problem of imageless thought, but the very existence of the problem was enough to suggest that there are contents other than sensory elements in consciousness. Moreover, the Würzburg workers published data on thought processes, suggesting that such activities as judgment and willing are not the orderly and logical sequences that association theory proposed. Rather, spontaneous and extraneous patterns are present in thought processing, seriously undermining assumptions about the structures of the mind.

When Külpe left Würzburg for a new position at Bonn in 1909, the Würzburg school ceased. At Bonn, Külpe turned to the relationship between psychology and medicine. The burst of activity at Würzburg was an incomplete movement. Although the school's experiments seriously challenged the legitimacy of structural psychology on its own terms, the Würzburg school did not break out to provide an alternative system for German psychology. Rather, the break with the past was left to another movement—Gestalt psychology, to be considered later.

In summary, neither dominant figure of each model of psychology, Wundt or Brentano, succeeded in establishing contemporary psychology in a definitive way. With the hindsight of history, we may conclude that Brentano was the more successful, despite his lower profile, because his views were transmitted intact and not completely repudiated. However, in a very real sense, psychology was forced to go through sequences of being reestablished in the twentieth century.

CHAPTER SUMMARY

Psychology emerged in Germany during the 1870s as a recognized scientific discipline. The recurrent theme in German philosophy of the essential activity of the mind provided the exciting intellectual setting that made a compelling case for psychology's founding, and also gave rise to competing models of the proposed substance and methodology of psychology. One model, an outgrowth of studies of sensory physiology and psychophysics, has been labeled structural, or content, psychology, and Wundt and Titchener were its major spokesmen. Under this natural science approach, psychology was defined as the experimental study of the data of immediate experience through the method of trained introspection. The goal of psychology was to reduce the contents of consciousness to constituent elements of sensory origin. Both its restricted subject matter and ambiguous methodology led to structural psychology's being seriously challenged as the definitive framework for the new science. Nevertheless, structural psychology secured recognition of psychology as a new science, and others, such as Müller, Hering, and Ebbinghaus, attempted to modify structural psychology to accommodate more sophisticated psychological issues. Moreover, philosophers such as Mach and Avenarius bolstered the justification for the natural science approach to psychology.

An alternative, described as a human science model, proposed more open methodologies empirically based on observation but not necessarily experimental. Within this context, Brentano's act psychology defined its subject matter as processes of psychological events inseparable from the environment and consciousness. This phenomenological view offered a greater scope and several accepted methodologies for psychology. The works of Stumpf and Külpe, supported by the philosophical critiques of the natural science methods proposed by Dilthey and Bergson, fall into the human science model. However, these men's individual views did not offer a coherent or systematic theory able to compete successfully with structural psychology. Nevertheless, subsequent developments in psychology established viable alternatives. In many respects, the "founding" of modern psychology was a false beginning. Neither dominant model, as expressed by Wundt and Brentano, was successful in establishing a lasting framework for psychology. It was left to the immediate successors of these German psychologists to rethink the formulation of psychology's scope and method.

BIBLIOGRAPHY

Primary Sources

BERGSON, H. L. (1910). *Time and free will: An essay on the immediate data of consciousness* (F. L. Podgsen, Trans.). New York: Macmillan.

BRENTANO, F. (1973). *Psychology from an empirical standpoint (1874)* (O. Krauss, A. C. Rancurello, D. B. Terrell, & L. L. McAlister, Trans.). Atlantic Highlands, NJ: Humanities Press.

EBBINGHAUS, H. (1948). Memory. In W. Dennis (Ed.), *Readings in the history of psychology*. New York: Appleton-Century-Crofts, 304–313.

TITCHENER, E. B. (1898). A psychological laboratory. *Mind, 7*, 311–331.

TITCHENER, E. B. (1898). Postulates of a structural psychology. *Philosophical Review, 7*, 449–465.

TITCHENER, E. B. (1899). Structural and functional psychology. *Philosophical Review, 8*, 290–299.

TITCHENER, E. B. (1910). *A textbook of psychology*. New York: Macmillan.

TITCHENER, E. B. (1925). Experimental psychology: A retrospect. *American Journal of Psychology, 36*, 313–323.

WUNDT, W. (1907). *Principles of physiological psychology (I)* (E. B. Titchener, Trans.). New York: Macmillan.

WUNDT, W. (1912). *An introduction to psychology*. London: George Allen.

WUNDT, W. (1916). *Elements of folk psychology*. London: Allen & Unwin.

WUNDT, W. (1969). *Outlines of psychology*. Leipzig: Englemann, 1897; reprinted: St. Clair Shores, MI: Scholarly Press.

WUNDT, W. (1973). *The language of gestures*. The Hague: Mouton.

Studies

ANDERSON, R. J. (1975). The untranslated content of Wundt's *Grundzüge der Physiologischen Psychologie*. *Journal of the History of the Behavioral Sciences, 10,* 381–386.

BLUMENTHAL, A. L. (1975). A reappraisal of Wilhelm Wundt. *American Psychologist, 30,* 1081–1088.

BLUMENTHAL, A. (1979). Retrospective review: Wilhelm Wundt—the founding father we never knew. *Contemporary Psychology, 24,* 547–550.

BORING, E. G. (1927). Edward Bradford Titchener. *American Journal of Psychology, 38,* 489–506.

BORING, E. G. (1950). *A history of experimental psychology,* 2nd ed. Englewood Cliffs, NJ: Prentice Hall.

BRINGMANN, W. G., BALANCE, W. D. G., & EVANS, R. B. (1975). Wilhelm Wundt 1832–1920: A brief biographical sketch. *Journal of the History of the Behavioral Sciences, 11,* 287–297.

BROZEK, J. (1970). Wayward history: F. C. Donders (1818–1889) and the timing of mental operations. *Psychological Reports, 26,* 563–569.

COPLESTON, F. (1974). *A history of philosophy, Vol. 9, Maine de Biran to Sartre, Part I—The revolution to Henri Bergson*. Garden City, NY: Image Books.

DANZIGER, K. (1979). The positivist repudiation of Wundt. *Journal of the History of the Behavioral Sciences, 15,* 205–230.

EVANS, R. B. (1972). E. B. Titchener and his lost system. *Journal of the History of the Behavioral Sciences, 8,* 168–180.

EVANS, R. B. (1975). The origins of Titchener's doctrine of meaning. *Journal of the History of the Behavioral Sciences, 11,* 334–341.

FANCHER, R. E. (1977). Brentano's psychology from an empirical standpoint and Freud's early metapsychology. *Journal of the History of the Behavioral Sciences, 13,* 207–227.

HENLE, M. (1971). Did Titchener commit the stimulus error? The problem of meaning in structural psychology. *Journal of the History of the Behavioral Sciences, 7,* 279–282.

HENLE, M. (1974). E. B. Titchener and the case of the missing element. *Journal of the History of the Behavioral Sciences, 10,* 227–237.

HINDELAND, M. J. (1971). Edward Bradford Titchener: A pioneer in perception. *Journal of the History of the Behavioral Sciences, 7,* 23–28.

LEAHEY, T. H. (1979). Something old, something new: Attention in Wundt and modern cognitive psychology. *Journal of the History of the Behavioral Sciences, 15,* 242–252.

LEARY, D. E. (1979). Wundt and after: Psychology's shifting relations with the natural sciences, social sciences, and philosophy. *Journal of the History of the Behavioral Sciences, 15,* 231–241.

LINDENFELD, D. (1978). Oswald Külpe and the Würzburg school. *Journal of the History of the Behavioral Sciences, 14,* 132–141.

PILLSBURY, W. B. (1928). The psychology of Edward Bradford Titchener. *Philosophical Review, 37,* 104–131.

POSTMAN, L. (1968). Hermann Ebbinghaus. *American Psychologist, 23,* 149–157.

RANCURELLO, A. C. (1968). *A study of Franz Brentano.* New York: Academic Press.

ROSS, B. (1979). Psychology's centennial year. *Journal of the History of the Behavioral Sciences, 15,* 203–204.

SABAT, S. R. (1979). Wundt's physiological psychology in retrospect. *American Psychologist, 34,* 635–638.

SHAKOW, D. (1930). Hermann Ebbinghaus. *American Journal of Psychology, 43,* 505–518.

STAGNER, R. (1979). Wundt and applied psychology. *American Psychologist, 34,* 638–639.

SULLIVAN, J. J. (1968). Franz Brentano and the problems of intentionality. In B. Wolman (Ed.), *Historical roots of contemporary psychology.* New York: Harper & Row, 248–274.

TINKER, M. A. (1932). Wundt's doctorate students and their theses, 1875–1920. *American Journal of Psychology, 44,* 630–637.

WOODWORTH, R. S. (1906). Imageless thought. *The Journal of Philosophy, Psychology and Scientific Methods, 3,* 701–708.

❋ 12 ❋

American Functionalism

When the new German psychology of Wundt was introduced to the United States, it immediately took on a particular American character. With the exception of Titchener, who remained a strict adherent to Wundt's formulations, the American psychologists who had been trained in Germany imposed a functional interpretation on structural psychology when they returned to America. Briefly, functionalism was

an orientation in psychology that emphasized mental processes rather than mental content and that valued the usefulness of psychology. Ironically, it was Titchener who in 1898 coined the term *functional psychology*, to distinguish such views from his own "true" structural psychology.

Functional psychology was not a formal system of psychology in the way represented by structural psychology or later systems of Gestalt psychology, behaviorism, or psychoanalysis. Functional psychology did not provide a comprehensive view of psychological activity with underlying philosophical assumptions and prescribed research strategies and goals. Rather, it differed from structural psychology in a spirit or an attitude that emphasized the applications and usefulness of psychology. As Boring (1950) suggested, it was not so much that the functional psychologists did different experiments than the structural psychologists. Rather, it was their reason for doing an experiment that distinguished them from the structuralists. The functionalists wanted to know how the mind works and what uses the mind has, not simply what contents and structures are involved in mental processes.

Functional psychology changed the new German science by adding influences historically absent from the German intellectual milieu. Specifically, while accepting the underlying Lockean assumptions of the mind inherent in Wundt's formulation, the Americans retained a general commitment to other prevailing aspects of British thought. Most notable was the strong influence of Darwin's theory of evolution. Functional psychology valued the importance of adaptation of both the species and the individual to environmental influences. Adaptation as a survival mechanism was amenable to the American national experience as a pioneering enterprise that saw itself as having transplanted the best of European civilization and left behind the shortcomings of European society in the attempt to tame a wild continent.

American functionalism was a relatively short-lived movement. It introduced to America Wundt's attempt to identify a new science, but in the process of importing structuralism, the functionalists discarded the rigidity of Wundt's system. As a movement within psychology, functionalism prepared the way for the eventual redefinition of psychology in terms of a behavioristic approach that rapidly came to dominate American psychology. On the one hand, functionalism may be viewed as a transitional stage in America between structuralism and behaviorism. On the other, psychology was firmly entrenched in America through the concerted efforts of the functionalists, who successfully conveyed its value for academic as well as applied purposes. Thus, it may be argued that the functionalists were progressives who gave psychology an American imprint that the discipline retains to this day.

BACKGROUND

The Legacy of Nineteenth-Century British Thought

The common language uniting Britain and the United States has forged deep-rooted ties in economic, political, and social spheres throughout the last four centuries. This relationship is reflected in the philosophical bases of science, and for psychology this relationship has implied a reliance on empiricism and the Lockean model of mental process. In a general way, the Lockean model, so central to the development of empirical psychology, also played a significant role in the nurturing of eighteenth-century political thought and had a profound effect on the emergence of the American nation. The social implications of the Lockean model were recognized in the founding ideals of the United States. Jefferson's *Declaration of Independence* justified the actions of the American colonies against Britain by asserting that society is an organic unit that is propelled toward its own betterment. Society itself, according to Jefferson, is composed of men born equal—the *tabula rasa* state proposed by Locke.

As outlined in Chapter 10, the flowering of British science in the nineteenth century confirmed the justification of empiricism. The impact of Darwin's theory, evolution by natural selection, may be best appreciated within the context of the Lockean model: Darwin provided empirical support for the improvement of species through successful adaptation to the environment. Darwin's theory found ready acceptance in the United States because his findings offered a mechanism that explained American progress. The United States was emerging from the nineteenth century as a nation of boundless potential, welcoming Europe's oppressed masses to participate in its opportunities. Thus, America was proving Spencer's interpretations of evolutionary improvement on a vast social scale.

The impact of evolutionary theory was felt far beyond the scope of biology. Galton's study of mental inheritance was one of the initial applications of Darwinism that eventually led to the development of testing as a valuable tool of psychologists. This movement received its initial impetus from British scholars, but the testing movement reached full expansion in the United States and eventually became an important part of functional psychology.

Following Galton's analysis of the inheritance of mental traits and his development of the basis for regression and correlation (see Chapter 10), Karl Pearson (1857–1936) provided the mathematical support for the assessment of the covariation of multiple traits. Pearson started a statistical laboratory at University College in London and, with Galton, founded *Biometrika* in 1901 to publish statistical papers dealing with biological and psychological variables. Also in 1901, Pearson published a theoretical paper on the mathematical possibility of predicting aptitude on the basis of many tests of various mental traits. The statistical implications of Pearson's views were applied to intelligence testing by Charles Spearman (1863–1945), who wrote a paper in 1904 suggesting that intelligence consists of a single general factor and a number of specific factors or traits. Spearman's two-factor theory of mental ability described a common factor of intelligence and a group of specific factors that were related to individual tests. Later workers in Britain, such as Godfrey Thomson and Cyril Burt, became dissatisfied with Spearman's two-factor theory and proposed alternative models, at the same time improving the statistical techniques to provide support for tests of multiple abilities. Finally, an American at the University of Chicago, L. L. Thurstone (1887–1955), used factor analysis as an invaluable aid in interpreting multiple tests, because it provided a means of weighing factors according to the extent that they account for total variability. Factor analysis made possible the development of a composite prediction method for individual abilities.

While Pearson and his followers were examining statistical techniques to predict mental abilities better, Alfred Binet (1857–1911) in France developed the first widely used standardized intelligence test. Asked by the minister of education to devise a method of assessing the intellectual aptitude of schoolchildren, Binet and his coworkers invented specific test items to measure various kinds of intellectual processes. From these efforts came the concept of mental age, an individual index of a child's ability compared to a reference group. A German psychologist, William Stern (1871–1938), later suggested dividing the mental age by the chronological age to calculate an individual's intelligence quotient (IQ). In 1916, a group at Stanford University revised and restandardized the Binet test for American use. In 1917, intelligence testing received significant reinforcement in America when the army adapted it as a selection device for young men drafted for World War I.

The American Character

Before reviewing the development of functional psychology, it is appropriate to describe briefly the American scene at the turn of the twentieth century. The United States entered the twentieth century as a nation just beginning to tap its vast resources and exert its strength in the international community. Nineteenth-century America had been divided by the Civil War, and glaring social inequities based on race and ethnicity were prevalent. However, the United States had remained relatively aloof from the European turmoil of the nineteenth century. By 1900, the United States was a colonial power, having ousted Spain from the Western Hemisphere. Yet even colonial rule was justified as being a missionary effort to bring the benefits of American life to the masses who had been exploited by European imperialism in the former Spanish possessions. Accordingly, there was an exciting and idealistic sense of purpose and righteousness in America, and Americans had an overwhelming faith and confidence in themselves.

That spirit of moral and economic superiority was present in American academia as well. Although American universities existed from the seventeenth century, they were mostly small institutions intended for training clergy and physicians. Until late in the nineteenth century, most Americans in search of quality education went to European institutions. However, American universities began to change their character, moving away from denominational control and toward more liberal studies. President Charles Eliot of Harvard began a radical upgrading of medical education in 1870. Professional and graduate centers of education were established at such institutions as Johns Hopkins University in 1876 and the Columbia School of Political Science in 1880, making it possible for Americans to study at the doctoral level in their own country.

Public support of universities increased dramatically as a result of the Morrill Act of 1862, which provided federal land and support to start state agricultural schools. In those areas of the country without the tradition of private universities, especially in the Midwest, these land grant schools expanded to include comprehensive education in liberal arts and sciences, at both undergraduate and graduate levels.

Psychology was introduced to America at a time coinciding with expansion and revitalization of both the universities and the nation at large. Whereas the new science was received with some skepticism by the conservative academicians of Europe, the American universities reacted with greater acceptance simply because psychology was new. Psychology as an independent discipline benefited from the American atmosphere and gained an identity and a stability that were unmatched in Europe.

EARLY AMERICAN PSYCHOLOGY

Even after American universities improved to the point of competing with European institutions, the character of intellectual pursuit in America was still marked by the earlier emphasis on the applied aspects of human knowledge. American values tended to play down questions of abstract science while glorifying technology. In philosophy, the problems of essence and being, studied in metaphysics, yielded to ethical questions of the concrete standards for human deportment. Issues relating to psychology were considered in such applied fields as medicine and ethics.

Moral Philosophy and Medicine

In colonial America moral values and psychological activity were intertwined with theology. Periodic movements of intense evangelical Christianity have occurred from time to time in American history, one of which was sparked in 1734 when Jonathan Edwards (1703–1758) preached revivalist sermons in Northampton, Massachusetts. Edwards was America's first native philosopher of note, and he inspired a fundamentalist crusade to bring people back to the beauty and purity of God and nature. Edwards had read Locke while he was a student at Yale and chose to study the relationship between God and man through a revision of the deterministic theology of John Calvin. Preaching predestination and faith in God, Edwards urged people to return to the absolute rule of God, who gave everything to humanity, born with nothing. In a similar vein, the New Jersey Quaker John Woolman (1720–1772) tempered his recognition of God's will with humanitarian gestures in proposing the ideal standards of deportment.

Perhaps the American who came closest to being an all-around scholar was Benjamin Franklin (1706–1790), whose appreciation of learning complemented his inventive genius for technological advances. His interest in practical science, reflected in his observations on electricity (see Chapter 7), was not at the expense of theoretical learning. In 1744, he was instrumental in founding the American

Philosophical Society, the first learned society in America. As a scientist, philosopher, and inventor, and later as a statesman, Franklin embodied the American ideals of the eclectic and the useful.

Most colonial physicians gained their knowledge through experience as working practitioners, and the first American medical school—at the University of Pennsylvania—was not started until 1765. A professor there and the most famous physician in revolutionary America was Benjamin Rush (1745–1813). Rush received his medical degree in Edinburgh and carried back to America some of the views of the commonsense Scottish empiricists. However, he is better known as the chief physician of the Revolutionary Army, urging better sanitation and diet to counteract rampant disease among the soldiers. After the war, he made pioneering observations on psychosomatic disorders and psychiatric treatment. He was also a strong proponent of abstinence from alcohol and organized efforts that led to the formation of temperance societies in several regions of the country.

Science and research with no immediate practical application were generally ignored. A tenth of the population of Philadelphia died from an epidemic of yellow fever in 1793, despite Rush's suggestion that the epidemic might have been spreading through vapors given off by decomposing matter. Nobody followed through to investigate Rush's hypothesis until after that plague. The primitive state of medicine may inadvertently have led to the death of George Washington: He was treated for a throat infection by bleedings and purgings, reducing his bodily resistance. Gradually, however, scientific inquiry gained support. In 1780, the American Academy of Arts and Sciences was founded in Boston to promote learning and advance the interest and betterment of society. There was enthusiastic support for applied science and technology, in contrast to basic science. Indeed, the American success story is filled with creative inventions, architectural wonders, and engineering feats. The steamboat, cotton gin, and Erie Canal are only a few of the many accomplishments of Americans who were able to win business and government support for projects with immediate practical applications.

American Pragmatism

Pragmatism is a native American philosophical system. The word *pragmatism* is derived from a Greek root meaning an act or a deed. As a philosophy, pragmatism emphasizes results rather than method. A pragmatic view of science accepts various methodological approaches to knowledge. In ethics, pragmatism stresses the way that the individual makes compromises between desires and reason. Pragmatic philosophy, then, does not contain a comprehensive collection of doctrines or beliefs, but rather consists of a characteristic manner of philosophizing. As the immediate precedent of functional psychology, early pragmatic philosophy created an intellectual atmosphere that studied not so much what a person does as how the person goes about doing it.

William James. As the first person associated with the new empirical science of psychology in America, William James (1842–1910) actually deserves the title of advocate rather than practitioner. Although he introduced experimental psychol-

WILLIAM JAMES (1842–1910).
(Courtesy, New York Public Library.)

ogy to American academia and imported one of Wundt's students specifically to start a laboratory at Harvard University, he remained a philosopher. He appreciated the efforts of others to establish an empirical science of psychology, but was not an empiricist. He excited the interest of his many students in psychology, but James himself did not become committed to the narrow focus of experimental work. His genius was not contained within psychology, and he followed a wide range of interests throughout his long career.

James was born into an eminent and wealthy family that had the resources and motivation to foster scholarly pursuit. His brother, Henry, gained literary fame as a novelist, and William and his four siblings received superb educations in Europe and America. At first, William showed an interest in painting, but his talent was not convincing, so he enrolled at Harvard. While studying biology and medicine there, he was influenced by Louis Agassiz (1807–1873), the Swiss-born naturalist and zoologist. His studies were interrupted by a period of illness, described as a nervous and emotional crisis, and he traveled to Europe to recuperate. While in Germany and France, James read widely in philosophy and psychology, and attended the lectures of some of the eminent continental thinkers. Upon his return to America, he completed his medical degree at Harvard but decided to pursue philosophy. During his long career at Harvard, he became a legend, admired and respected by both students and colleagues. He carried on a voluminous correspondence with such contemporaries as Oliver Wendell Holmes, Henri Bergson, and G. Stanley Hall. Among his prolific writings, his major contributions to psychology were contained in *The Principles of Psychology* (1890), published in two volumes. This work is a comprehensive treatment of psychology and was used as an introductory textbook for many years. It remains a classic work of American psychology.

The pragmatism of William James was based on a firm appreciation of empiricism and can be summarized by the following points:

1. The consequences of theoretical positions form the major criteria for judging differences among the positions. Different philosophical theories may state disparate views, but only their consequences can truly differentiate them. Thus, James accepted empirical tests of the validity of a theory.
2. If a theory asserts a useful, satisfying effect in organizing experience, then the theory should win at least tentative acceptance. This point allows a subjective, utilitarian perspective on individual experience. For example, if a person holds a religious belief that is critical and reassuring for the individual, then the belief is "true" for that person.
3. Experience itself is not reduced to either the elements of consciousness or the mechanical laws of matter. In contrast to Wundt, James argued that experience is not a succession of discrete sensations bound together by associations. Rather, experience is a continuous flow of subjective events.

James believed that mind and body, subjective and objective aspects of experience, are not two different interacting subsystems. Rather, reminiscent of the earlier views of Spinoza, he proposed that mental and physical experiences are different aspects of the same experience. For example, we may read from a book or use the book as a paperweight. We do not have two experiences of the book, but only one experience that is described in two different ways. Thus, James blurred the distinction between mind and body because he believed the distinction was an intellectual artifact used to describe how we experience. Experience itself is a singular entity.

By defining psychology as the "science of mental life" and proposing that experience is a continuous stream of consciousness, James accepted an enlarged scope of psychology compared to the Wundtian model. Because experience must be described in both physical and mental terms, James emphasized a truly physiological psychology that stressed brain functions in accounting for mental experience, or consciousness. In addition, the mind for James was a process that was personal, changing, continuous, and selective. Accordingly, he advocated an empirical approach to the study of experience that focused on the mind in terms of its functions, so that the psychologist must observe the mind in use.

One part of James' psychology, illustrative of his belief that consciousness should be appropriately described in terms of both physical and mental dimensions, concerned his theory of emotions. First formulated by James in 1884, this theory has come down to us as the James–Lange theory of emotions, because a Danish psychologist, Carl Lange (1834–1900), developed a similar interpretation in 1885. James noted that the body responds with certain automatic reflex actions when confronted by emotional stimuli, and these reactions are usually confined to skeletal and visceral levels. When we become aware of these reactions, we then experience the emotion, according to James. For example, if a speeding car races toward you and narrowly misses hitting you, your autonomic nervous system responds automatically and immediately with increased heart rate, rapid breathing, and perspiration, all in preparation for a motor response of fleeing or freezing, if either becomes necessary. Experiencing that sequence of reactions, James asserted, is the emotion of fright. Thus, James argued that we first become aware of the physiological aspects of the experience, and then the psychological aspect comes into focus. Emotions, then, are the result of a sequence of autonomic reactions, not the cause. Of interest in James' view on emotions are the emphases on the two dimensions of the total experience, physical and psychological, and the description of emotions in terms of observable functions of activity.

Boring (1950) suggested three reasons for James' prominent place in the development of American psychology. First, his dynamic personality, clear writing, and effective teaching excited students about psychology. He created an atmosphere that facilitated the growth of psychology in American academia. Second, James offered an alternative to the Wundtian formulation of the new science, represented in America by Titchener. James based his definition of psychology on experience, described as a stream of consciousness and not a collection of sensory elements. Finally, James proposed a distinctive American psychology that was functional in character. Functional psychology was open to practical applications and admitted the data of observable behavior.

Charles Sanders Peirce. About as opposite in temperament to James as one could be, Charles Sanders Peirce (1839–1914) was an important figure in pragmatism because he integrated elements of diverse philosophies to produce an eclectic theory of consciousness. Although he had considerably less influence on functional psychology than James, Peirce's pragmatism nevertheless forms an intellectual basis of note for American psychology, which was recognized by James.

Peirce was the son of a Harvard mathematician and received sound training in mathematics and biology while pursuing his own reading in history and philosophy. After graduating from Harvard, he worked as a scientist for the United States Coast and Geodetic Society until 1879, when he received an appointment as an instructor of logic at Johns Hopkins University. He was unsuccessful as a teacher and left Hopkins after 4 years. Despite the efforts of James to secure a position for him at Harvard, Peirce never again had regular employment. He lived a poor existence supported by occasional jobs reviewing papers, and became increasingly alone and cantankerous. Most of his works were published after his death.

Peirce was influenced by the writings of Kant and Bain, who came from different philosophical traditions. Peirce agreed with Kant that the mind contributes to the organization of experience by relating and unifying sensory information through a priori categories. At the same time, adhering to an empirical position, Peirce believed that only questions able to be subjected to empirical scrutiny have scientific validity. From Bain, Peirce expanded his empiricist assumptions by accepting the basic tenets of British associationism and a definition of individual belief in terms of habits of activity confirmed by the satisfaction of needs.

Like James, Peirce viewed consciousness and mental processes in terms of their practical consequences. Moreover, he defined the higher mental process of judgment in terms of the person seeking the meaning of the consequences of ideas. Any meaningful idea has three mental categories: quality, essence, and its relationship to other ideas. In contrast to James, however, Peirce stressed the logical rather than the psychological consequences of ideas. In so doing, Peirce reflected his belief that the mind is intimately tied to the organizational structure that it imposes on sensory information.

Both James and Peirce contributed to the intellectual atmosphere that readily accepted new formulations of psychology. Elements of their pragmatic views anticipated later systems of American thought. James's empiricism, for example, favored an acceptance of observable behavior as psychological data; Peirce's emphasis on mental organization was consistent with the subsequent development of Gestalt psychology. Pragmatism as a philosophical movement defined the immediate character of American psychology, and functional psychology, in turn, provided the needed transition from the rigid model of Wundt to the varied systems of psychology that began to blossom in America during the 1930s.

Transitional Figures

As mentioned earlier, functional psychology was a diffuse system of psychology, characterized more by an attitude toward psychological study than by coherent theory. Nevertheless, centers of functional psychology did emerge, and these are considered in the next section. First, however, the views of several psychologists should be surveyed. These psychologists were functionalists to the extent that they contributed to the formation of the American approach to psychology. Moreover, they were individualists who expressed rather personalized views of psychology; they did not fit neatly with the more formal expositions of functional psychology.

Hugo Münsterberg. A student of Wundt, Hugo Münsterberg (1863–1916) was recruited from Germany by William James to expand and manage the psychology laboratory at Harvard University. He fulfilled that task, but Münsterberg also possessed a wide vision of psychology's potential for becoming a valuable applied discipline. He gained popular fame for his varied writings on the applications of psychology to social, commercial, and educational issues. On an abstract level, he remained nominally a structural psychologist. However, he generally ignored theory and became part of the American functional spirit.

Münsterberg's life was marked by views that earned him prestige and admiration followed by scorn and ridicule. He was born in the East Prussian port city of Danzig (now Gdańsk, Poland) to a cultured and intellectually oriented family. He received his PhD under Wundt at Leipzig in 1885 and a year later was awarded a medical degree. Wundt had earlier rejected some of his initial research on the will, but Münsterberg continued this work independently and later expanded the rejected paper to a small book, further alienating Wundt when he published it in 1888. Münsterberg was appointed to the faculty of the University of Freiburg in 1887, where he started a laboratory and began to publish papers on time perception, attentional processes, and learning and memory. These papers attracted the attention of psychologists in Germany and America, and William James cited several of them in his *Principles of Psychology*. James met Münsterberg in Paris at the First International Congress of Psychology in 1889. Following that meeting, the two corresponded, and James sent one of his students to Freiburg to work with Münsterberg. In 1892, James secured an offer from Harvard for Münsterberg to become director of the psychology laboratories for 3 years. In addition to expanding the laboratories and directing students, Münsterberg learned English and prepared a German textbook. He returned to Freiburg in 1895 to think over the offer of a permanent professorship at Harvard. In 1897, he returned to Harvard and remained in America for the rest of his life, with the exception of brief visits to Europe and a year as an exchange professor at Berlin.

In addition to his prolific publications in psychology, Münsterberg became a spokesperson for German–American relations. He never sought American citizenship and maintained a fierce nationalism toward Germany. He published a popular book on the American character, culture, and social structures for Germans. In the first years of this century, he was honored by political leaders in both countries, and advocated increased contact between American and German scholars. However, the political winds changed and the good will began to evaporate in the years prior to World War I, when Germany's image among the American public deteriorated. Münsterberg was caught in the middle of the American public outrage over German political and military aggression. In particular, he became a symbol of German arrogance in the American newspapers that earlier liked to quote him on the benefits of German–American cooperation. This vilification no doubt created considerable pressure for Münsterberg and probably contributed to his fatal stroke in 1916, a year before the United States' entry into the war against Germany.

Münsterberg, like James and most psychologists of his era, considered himself a philosopher. Interestingly, Münsterberg condemned pragmatism as a mere updating of the Greek Sophist tradition. He believed psychology would be constraining itself by alignment with pragmatism, which he thought was ultimately too limiting and operational in scope. Rather, he adhered to the idealistic basis of the German model of mental activity and posited a distinction between causal and purposive psychology: Causal psychology is empirically based and examines the relationship between mental events and psychological processes; purposive psychology is the study of the pursuit of goals by activities of the will. Although Münsterberg initially stated that purposive psychology belongs in the metaphysical province of philosophy, he later placed it in psychology proper. Münsterberg's purposive psychology influenced one of his students, Edwin Holt, in his conceptualization of behaviorism (see Chapter 15), and, in turn, Holt influenced Edward Tolman's later expansion of the behavioristic model of psychology.

Almost from its inception, Münsterberg's laboratory at Harvard expanded its research to subject matter beyond the restrictions of the introspective psychology of Wundt and Titchener. He organized divisions of human and infrahuman investigations, and his laboratory soon became one of the more productive centers of experimental psychology. His views of psychological research were broad and eclectic, combining the German tradition of Wundt's structural psychology and Brentano's act psychology, an integration permitted by Münsterberg's conceptualization of causal and purposive psychology.

Münsterberg's applications of psychology may be clearly seen in his writings on a variety of subjects. From his background as a psychologist and a physician, he was interested in psychotherapy and published a review of the area in 1903. He disagreed with Freud on the nature of unconscious motivations, but he valued the interest in psychopathology generated by Freud's emerging theory. His book *On the Witness Stand* (1908) was an initial effort in forensic psychology, and he developed a precursor of the "lie detector" polygraph in his laboratory. His book *Psychology and the Teacher* (1909) paid particular attention to individual differences in learning arising from variability in inherited dispositions, and he suggested several tests to measure student aptitude. In two books, *Vocation and Learning* (1912) and *Psychology and Industrial Efficiency* (1913), Münsterberg described studies of personnel selection and labor management. He even wrote an analysis of film technique in *The Photoplay: A Psychological Study* (1916).

Münsterberg was a remarkable person whose wide intellectual capacity joined easily with the American utilitarian ethic. Although he rejected pragmatism in the abstract, he contributed to the making of functional psychology in practice. Münsterberg's place in the history of American psychology has not been given the emphasis it deserves, probably because of the anti-German attacks on him and his own shortcomings at an interpersonal level. Nevertheless, as William James is credited with popularizing the new science of psychology within academia, Hugo Münsterberg should be credited with popularizing psychology among the masses of people by demonstrating its practical worth.

William McDougall. Freud and William McDougall (1871–1938) have often been compared because of their reliance on inherited instinctual patterns of psychological activity. McDougall also has sometimes been classified as a behaviorist because he emphasized overt, observable behavior as a reflection of psychological activity. However, McDougall was an individualist pursuing his own directions in psychology, which for the most part moved against the grain of the American psychology of his time. More recently, McDougall's psychology has received favorable review for his eclectic combination of instinct and purpose in the analysis of comparative behavior. He formulated most of his important views in his native Britain, but they were better tolerated within the functional spirit of early American psychology.

McDougall received a strong background in the humanities and in medicine at Cambridge and Oxford Universities, followed by a year of physiological study at Göttingen in Germany. After 4 years as an intern at St. Thomas Hospital in England, he joined the Cambridge Anthropological Expedition to New Guinea and Borneo. His studies of primitive societies, published in several volumes, reflected McDougall's early research inclination as a master of detailed observation. He taught briefly at University College in London and then for 16 years at Oxford University. During World War I, he worked as a physician for the British army and studied cases of psychoneurosis, which were to form the basis of his later views on abnormal psychology.

In 1920, he accepted the professorship at Harvard that became vacant when Münsterberg died in 1916. He stayed at Harvard for only 7 years and left for Duke University. His dissatisfaction with Harvard probably resulted from his perception that his views did not receive the admiration and followers that they deserved. Also, he believed that the more moderate climate of North Carolina might benefit his increasing deafness. While at Duke, he chaired the psychology department and fostered an atmosphere of intellectual tolerance for diverse expressions of psychology, including studies of parapsychology.

McDougall's scientific background was in the same nineteenth-century tradition that produced Darwin. Philosophically, he was attuned to the Scottish empiricists as well as the open-ended association views of John Stuart Mill. McDougall was also influenced by James's psychology and dedicated a book to his memory.

McDougall called his psychology ***hormic***, a word derived from the same Greek root as hormone and which means "an impulse." By choosing this label, he stressed that psychological activity has a purpose, or goal, that prods the individual to action, although the person may not have any real understanding or knowledge of the goal itself. The dynamism or propelling force of activity was termed an *instinct* or *urge*. Psychological activity, as opposed to physiological activity, was defined as behavior, and it had seven critical characteristics:

1. Spontaneity of movement
2. Persistence of activity beyond the action of some initiating stimulus
3. Variation in the direction of movement
4. Termination upon the perception of a change in the situation
5. Preparation for new situations
6. The capacity to improve with practice
7. The reflection of the totality of organismic reactions

This restricted definition of behavior excluded reflexive actions, which McDougall believed were physiological responses. McDougall's behavioral views were overshadowed by the wider, less rigorous definition offered by Watson. Nevertheless, for McDougall, behavior arising from inherited instincts provided a mechanism of action that can be modified by experiences, especially in higher animals.

McDougall's formulations stressed the importance of inherited characteristics and of behavior that can be learned and modified by environmental influences. He asserted personal freedom in the variabilities of behaviors in seeking goals, so his psychology was not deterministic. McDougall's views stand in contrast to Watson's complete reliance on environmental determination of behavior. The mind for McDougall has organization and interacts with bodily processes. The individual, then, is portrayed as free to determine her or his own purpose or pathways to personal goals.

One of McDougall's major contributions was his recognition of the social context of human and animal behavior. He emphasized critical social variables that influence interactions within species and how such behaviors are instinctually based and inherited. His *Introduction to Social Psychology* (1908) remained a definitive resource for many years.

McDougall's psychology was somewhat apart from the mainstream of American psychology of his time. Although functional in character, his views did not inspire the imaginations of American psychologists in the way that Watson's were able to do. However, as ethology has gained wider recognition in recent years, it may be argued that McDougall's "behaviorism" had a sounder conceptualization than Watson's. Certainly, the cogent research efforts of ethologists such as Konrad Lorenz and Niko Tinbergen are more consistent with McDougall's views than with any of the other versions of early behaviorism in America.

G. Stanley Hall. Perhaps the most independent of the early American psychologists, Stanley Hall (1844–1924) was instrumental in firmly establishing psychology in the United States through both substantive and practical activities. In addition to his contributions to child psychology and educational issues, he succeeded in securing recognition of psychology as a profession.

Hall's biography may be described as a series of "firsts." His was the first PhD granted from Harvard's philosophy department (1878), and he became the first American to work in Wundt's psychology laboratory in Leipzig (1879). He started the first legitimate psychology research laboratory in America (1883) at Johns Hopkins University. In 1887, he founded the first English language journal devoted exclusively to psychology, *The American Journal of Psychology*. The following year he became the first president of Clark University in Worcester, Massachusetts, and in 1892 organized the American Psychological Association, becoming its first president. He was involved in the founding of other journals: *Pedagogical Seminary* (1891), after 1927 known as the *Journal of Genetic Psychology*; the *Journal of Religious Psychology* (1904–1914); the *Journal of Race Development* (1910), later known as the *Journal of International Relations* and still later known as *Foreign Affairs*; and the *Journal of Applied Psychology*, in which Hall ended up investing $8,000 of his own money.

Hall was born on a farm near Boston and received his bachelor's degree in 1867 from Williams College in western Massachusetts. He then went to Union Theological Seminary in New York City to prepare for a career as a clergyman. Boring (1950) recounted the story of Hall's preaching a trial sermon before a critic on the faculty. At the end of the sermon the faculty member, forsaking any criticism, simply began to pray for the salvation of Hall's soul. Probably taking a hint, Hall went to Germany for 3 years, where he studied philosophy and also attended Du Bois–Reymond's lectures on physiology. Returning to New York in 1871, he completed his divinity degree and served briefly at a country church. He then secured a position at Antioch College near Dayton, Ohio, and taught a variety of courses. Impressed by Wundt's *Physiological Psychology*, Hall set out again for Germany to learn from Wundt. However, President Eliot of Harvard offered him a minor teaching post in English, which also allowed him to work with William James. He received his doctorate in 1878 for a dissertation on muscular perception. From then to 1880 Hall lived in Germany, where he worked for Wundt during the first year of the Leipzig laboratory.

In 1881, Hall joined the newly founded Johns Hopkins University, devoted to graduate education, where he worked with young people who later went on to positions of note within psychology, among them John Dewey, James McKeen Cattell, and

Edmund Clark Sanford. In 1888, Hall was named to the presidency of Clark University. He brought along Sanford to start the psychology laboratory while he began a department of educational psychology. Clark's psychology department soon gained a fine reputation and in 1909 was the site of Freud's lectures at Hall's invitation. In the year of his death, Hall was elected to a second term as president of the American Psychological Association; the only other person to be so honored was William James.

Hall's many accomplishments secured a firm foundation for psychology in America. However, like James, Hall did not have the temperament for laboratory work. Rather, he created an intellectual atmosphere to support those who were more empirically inclined. Nevertheless, Hall did contribute to the emerging body of psychological knowledge. Specifically, he was convinced of the importance of genetics and evolution for psychology, which was reflected in his writings and his support of the study of developmental psychology. In addition, Hall pioneered survey techniques, which have remained a fixture of social science research.

These three early psychologists, Münsterberg, McDougall, and Hall, were independent thinkers. They did not start systems of psychology, nor develop coherent theoretical frameworks, nor leave behind loyal followers. However, they made psychology functional and left it firmly entrenched in America.

FUNCTIONAL PSYCHOLOGY

We now come to the more formal statements of functional psychology. As stated earlier, functional psychology was more an attitude toward the results of psychological investigation than a comprehensive system. However, centers of functional psychology developed at the University of Chicago and at Columbia University. No substantial difference discriminated these two centers of functionalism. Indeed, the great American philosopher–psychologist John Dewey was associated with both universities. Rather, both Chicago and Columbia served as focal places for the spread of the new science in America, and both imposed a functional identification on psychology.

Chicago Functionalism

At Chicago, psychology was easily related to other disciplines. Applied directions in education were especially paramount, and research on issues of psychological and biological importance served as a precursor to the subsequent emergence of behavioral psychology.

John Dewey. John Dewey (1859–1952) initiated functionalism at the University of Chicago, and his long career was characterized by a commitment to social change. He fully appreciated the democratic implications of Darwin's theory, and saw education as the key to individual improvement and the betterment of society. Accordingly, rather than dedicating himself to enhancing the field of psychology per se, Dewey used psychology as a means to his social vision.

After receiving his degree in 1884 at Hopkins for a dissertation on Kant's psychology, Dewey spent the next 20 years in the Midwest, first at Michigan and then at Chicago, before moving on to Columbia University in 1904. Still a young man, Dewey published the first textbook, *Psychology* (1886), on the new science in America. Although this work defined psychology in functional terms, Dewey was very much the philosopher, describing sensation, for example, as elementary consciousness arising as a response of the soul. Dewey's major contribution to psychology was contained in a famous paper published while he was at Chicago, "The Reflex Arc Concept in Psychology" (1896). Anticipating the later Gestalt interpretation of behavioral activities, Dewey argued against an elementaristic analysis of reflexive responses, in contrast to the reflexology that shortly developed through Pavlovian and Watsonian behaviorism. Dewey emphasized the totality of movement, contending that coordination is more than the sum of reflexes. Dewey rejected the view that reflexes are discrete series of stimulus actions followed by responses and separated by intervening sensations. Reflexes are smooth and orderly sequences of coordinated movements that are indivisible.

By the time Dewey moved to Columbia, his views had increasingly evolved toward education and social philosophy. His major contribution, from his time at Chicago, consisted of leading a group of young scholars convinced of the utility of psychology and advocating the position that American psychology and functional psychology are synonymous.

James Angell. The organizer of Chicago functionalism, James Angell (1869–1949) went to Chicago in 1894 and stayed until 1920. Born in Vermont, Angell was the grandson of a president of Brown University and the son of the president of the University of Vermont and later the University of Michigan; he himself became the president of Yale in 1921. He did his undergraduate work at Michigan while Dewey was there, and received a master's degree at Harvard under William James in 1892. He went to Halle, Germany, for doctoral work and completed all requirements, but left before revising his dissertation to accept a job offer from the University of Minnesota. Thus, he never received his doctorate, but his new position allowed him to get married, and he effectively compensated for the lack of a doctorate by collecting more than 20 honorary degrees during his long and distinguished career.

His presidential address to the American Psychological Association in 1906, published the following year in the *Psychological Review*, was titled "The Province of Functional Psychology," and it contained a very clear statement of the agenda of functional psychology. Essentially, Angell defined the core of functional psychology as the acceptance of a biological approach to determine how the mind works in adjustments of the psychophysical person to the environment. This definition placed functional psychology in line with British natural science and Darwinism. In contrast to Wundt, Angell stated that consciousness progressively improves the adaptive activities of a person, and that attentional processes are the center of consciousness. Elaborating, Angell described three areas included in functional psychology. First, functional psychology studies mental operations, as opposed to the mental elements of a structural psychology. Second, functional psychology's emphasis on the adaptive activities of the mind means that the mind is viewed in a mediational role between the person's needs and the environment. Because consciousness habituates to environmental events after successful accommodation, according to Angell, novel stimuli elicit attentional fluctuations in consciousness and assume a critical role. Third, functional psychology assumes a psychophysical, mind–body interaction; hence, traditional psychophysics would continue to occupy an important place in research.

Under Angell's leadership, functional psychology at Chicago flourished and research papers on human as well as infrahuman levels were widely published. Perhaps Angell's most famous student was the founder of American behaviorism, John B. Watson, whose dissertation was titled *Animal Education: The Psychical Development of the White Rat* (1903). Although Angell subsequently rejected Watson's behaviorism as philosophically absurd and psychologically pernicious, Watson's views were nevertheless a logical consequence of some of the basic goals of functional psychology.

Harvey Carr. The major spokesperson of Chicago functionalism after Angell, Harvey Carr (1873–1954) received his doctorate there in 1905. He became department chair in 1919 and during the next 19 years presided over the awarding of 150 doctorates. By the time Carr wrote his influential textbook, *Psychology*, in 1925, the development of functional psychology was essentially complete. Moreover, the major reason for functional psychology—its attack on Wundt and Titchener and their structural psychology—no longer existed as a viable force in psychology. Although functional psychology continued as a nominal system, it was steadily absorbed into American behaviorism during Carr's tenure at Chicago.

Carr defined psychology as the science of mental processes, and emphasized motor responses, adaptive activities, and motivation. Carr recognized both subjective, introspective methods and objective measures in psychology. However, with his background in animal psychology, he tended toward the objective at the expense of the subjective methods of research. In the experiments done at Chicago, the widespread use of objective measurements of psychological activities paved the way for the subsequent research approach that emphasized overt, observable behavior as the primary source of psychological data. Accordingly, Carr's place in functional psychology was to summarize the basic principles commonly shared in this movement: First, mental processes are adaptive and have purpose. Second, mental activity is elicited by environmental stimuli. Third, motivation always affects mental processes and modifies stimulus influences. Fourth, behavioral responses have consequences. Finally, all mental activity is continuous and coordinated.

Columbia Functionalism

Psychology at Columbia University had a broadly based functional character with varied applications. The three psychologists who will be considered illustrate the diversity of approaches to functional psychology at Columbia.

James McKeen Cattell. James McKeen Cattell (1860–1944) was probably second only to Hall in his efforts to establish a sense of professionalism among psychologists. Cattell received his undergraduate education at Lafayette College in Easton, Pennsylvania, and then went to Germany, where he studied under Lotze and Wundt. After returning to America for a year of study at Hopkins, Cattell acquired a definite commitment to psychology. He went back to Germany, boldly told Wundt that he needed an assistant—namely, himself—and worked productively for 3 years, receiving his doctorate in 1886. While in Wundt's laboratory, Cattell became fascinated with reaction time experiments, and studied individual differences in reaction times, an unusual topic for one of Wundt's students. After a year of teaching in the United States, Cattell spent 1888 lecturing at Cambridge University. While in England, he met Sir Francis Galton, with whom he shared an interest in individual variability. From 1888 to 1891, Cattell was a professor of psychology at the University of Pennsylvania, and from 1891 to 1917, he held a similar position at Columbia. He started psychology laboratories at both institutions. Cattell, along with many prominent Americans, including Secretary of State William Jennings Bryan, was ardently opposed to America's entry into World War I. Cattell was fired from Columbia University for his pacifistic position, and devoted the remainder of his life to his interests in psychological testing and his many editorial duties. In 1894, he cofounded the *Psychological Review* with James Baldwin (1861–1934), and in 1900 started *Popular Science* (later *Scientific Monthly*). Cattell edited *American Men of Science* for 32 years and served at various times as editor of *Science, School and Society* and *American Naturalist*.

By the 1890s, Cattell's interest in individual differences led him to promote mental testing. In 1892, he published a monograph, *On the Perception of Small Differences*, in which he introduced detailed statistical analyses of errors in judgments made by subjects in traditional psychophysical experiments. That study was followed by research along the direction taken by Galton. In 1896, Cattell published a report of the physical and mental measures of students at Columbia, followed by an evaluative survey of eminent scientists. He started his own company, The Psychological Corporation, to market psychological expertise and measurement instruments to the public.

Cattell's lifelong interest in individual differences was functional in both theoretical terms and applications. He was concerned with the measurement of human capacity, which he viewed in an evolutionary sense, as did Galton. As a leading American psychologist, Cattell influenced many students. His advocacy of the use of statistics and testing bolstered an entire applied specialization within psychology.

Edward Lee Thorndike. A biographical sketch of Edward Lee Thorndike (1874–1949) is given in Chapter 15 as a predecessor of American behaviorism. Indeed, Thorndike's earlier work on animal learning reflects his appropriate classification in the behaviorist tradition that grew out of the functional spirit of American psychology. He received his doctorate under Cattell in 1898, and his subsequent association with Columbia Teacher's College influenced his later interest in human intelligence and testing.

Thorndike published two works outlining applications of learning and testing principles: *Educational Psychology* (1903) and *Introduction to the Theory of Mental and Social Measurement* (1904). Both texts became necessary reading for a generation of students of psychology and the social sciences. Thorndike described intelligence through a somewhat elementaristic approach by stressing that intelligence is composed of a number of abilities. Although Thorndike's views on association processes, presented in Chapter 15, earned him greater fame in behavioristic psychology, his capacity to use his research reflected an applied direction, entirely consistent with American functionalism.

Robert S. Woodworth. After receiving his doctorate under Cattell in 1899, Robert S. Woodworth (1869–1962) remained at Columbia for his entire career, with the exception of a postdoctoral year studying with the British neurophysiologist Charles Sherrington. Woodworth's first major work, *Dynamic Psychology* (1918), was an eclectic combination of prevailing views on psychology. Among his other works, *Contemporary Schools of Psychology* (1931) and *Dynamics of Behavior* (1958) offer careful functional perspectives on psychology. His *Experimental Psychology* (1938), revised in 1954 with Harold Schlosberg, was for many years the dominant textbook in university-level laboratory courses in psychology.

Woodworth's "dynamic" psychology focused on motivation. His views were not dynamic in terms of any essential deviation from the Lockean model of empiricism. Rather, Woodworth followed a fairly accepted interpretation of psychological processes, consistent with Chicago functionalism and Thorndike's views, but stressed individual motivation and underlying physiological correlates as central to adaptation. He used the term *mechanism* to describe the psychological act of adaptation, similar to the position of Carr. Mechanisms are elicited by drives, both internal and external in origin. For Woodworth, the entire repertoire of psychological activity gains a coherence and unity because of the individual's sense of purpose.

Women in Early American Psychology

An additional topic in this chapter on the history of early American psychology that deserves special mention concerns the role of women. Although women have contributed to psychology throughout its long history, because of psychology's particular success in the United States it is appropriate to underscore the role of women in the founding of American psychology. At the same time, it is important to note that psychology has not been immune to the biases and prejudices that have historically plagued all disciplines and have resulted in limited opportunities for women at all levels of intellectual endeavor. Although many women were barred from making contributions to the study of psychology, it may be argued that psychology has a record of access for women when compared to other sciences. Perhaps this relative openness reflected psychology's newness as a discipline emerging in the twentieth century at a time when women were making significant strides toward participation in the universities, the political arena, and the marketplace. Nevertheless, success for women in all fields, including psychology, often meant great personal sacrifices that included leaving the security of home and family to obtain a competitive education, remaining unmarried, becoming financially self-supporting, and continually proving themselves under male-dominated review.

The three women briefly described here were successful early American psychologists. Each reflected the particular American spirit in psychology, which is functional; each provided significant research in fostering the development of psychology; and each influenced many students.

Mary Whiton Calkins. After graduating from Smith College and touring Europe for a year during which she studied at the University of Leipzig, Mary Whiton Calkins (1863–1930) started a 40-year affiliation with Wellesley College, where she began as a tutor of ancient Greek. Recognizing her intelligence and interest in the new science, Wellesley officials asked her to develop a course in experimental psychology and urged her to pursue her own education in psychology. She finished the doctoral degree requirements at Harvard under William James and Hugo Münsterberg, but was not awarded the degree because Harvard University itself was not at that time coeducational, and Radcliffe, the women's college of Harvard, did not grant doctoral degrees. Thus, despite the awarding of the degree from the Department of Philosophy and Psychology, the Harvard Corporation would not confer the PhD. In 1896 Calkins published a paper in *Psychological Review*, in which she reported a method for presenting pairs of verbal items having no existing meaningful relation. She used this technique to vary the influence of the major determinants of memory—primacy, frequency, recency, and vividness. In extending the research of Ebbinghaus, Calkins also provided data supporting the secondary laws of association, originally proposed by the Scottish philosopher Thomas Reid.

Calkins was a faculty member of Wellesley College for most of her career, and established a laboratory there in 1891. In 1909 she published an influential introductory psychology textbook, *A First Book in Psychology*. In 1905 she was elected the first woman president of the American Psychological Association and in 1918 was the first woman to serve as president of the American Philosophical Society.

After about 10 years of amazing laboratory-based productivity, Calkins' attention shifted to more theoretical and philosophical concerns. She is remembered chiefly for her contributions toward a psychology of the self. Calkins' self-psychology emphasized the essential unity and coherence of consciousness, very dependent on both interpersonal and environmental interactions. This perspective is very interesting because for the remainder of her career she consistently provided an alternative to the emergence and eventual domination of behaviorism. In a very real sense Calkins reflected the wide definition of psychology from two of her Harvard mentors, Hugo Münsterberg and William James, that psychology can accommodate investigations that are at several levels, from mental elements to the unity of conscious experience. Her career represents the full gamut of eclectic concerns reflecting the functional spirit of American psychology.

Christine Ladd-Franklin. Born and raised in New England, Christine Ladd (1847–1930) graduated from Vassar in 1869 with a strong background in physics and mathematics. She taught in secondary schools in several locations and began to submit solutions and papers to various popular and scholarly journals. When Johns Hopkins University opened in 1876, Ladd applied and, after overcoming resistance because of her sex, was admitted in 1878, where she worked at first with the mathematician, James J. Sylvester. In 1879, as she began working with Charles Sanders Peirce and was teaching part-time at Hopkins, Ladd's interest turned to symbolic logic and experimental psychology. After completing all of the requirements for a doctorate at Hopkins, she married a fellow graduate student and instructor, Fabian Franklin. Because Hopkins did not grant degrees to women, it was not until 1926 that the university granted her the PhD. In 1891–92, Franklin's sabbatical leave from Hopkins permitted a trip to Europe, where Ladd-Franklin studied at Göttingen and Berlin universities. In 1895, Franklin left Hopkins to embark on an editorial career, first in Baltimore, then in New York. After her move to New York, Ladd-Franklin maintained a type of unpaid courtesy appointment at Columbia University as an academic base for her many publications.

Best remembered for her theory of color vision, Ladd-Franklin attempted to reconcile the trichromatic theory of Helmholtz with the tetrachromatic theory of Hering and Georg Elias Müller. Her solution, building on the finding that yellow–blue cones develop before red–green sensitivity, proposed that from white (gray) sensitivity, blue and yellow emerge, and from yellow sensitivity, green and red emerge. Color vision, then, is a product of these stages, which she also related to other visual processes, such as afterimages, as well as to pathology, such as color blindness.

Ladd-Franklin was recognized as an important psychologist in her lifetime. She received an honorary LLD from Vassar in 1887, and she consistently advocated women's rights. She was active in women's organizations, including a forerunner of the American Association of University Women.

Margaret Floy Washburn. Like Ladd-Franklin, Margaret Floy Washburn (1871–1939) was also a graduate of Vassar and became the first woman to receive a doctorate in psychology in the United States. She completed her training with Titchener at Cornell University in 1894, and Titchener sent her thesis results to Wundt for publication in *Philosophische Studien* in 1895. She also translated some of Wundt's writings into English. In 1903, she joined the faculty of Vassar College, where she remained until her death. A pioneer in animal psychology, she wrote *The Animal Mind* in 1908, which attempted to look at conscious states reflected in observable behavior. She returned to this theme in her 1916 work *Movement and Mental Imagery*, which proposed a reconciliation between behaviorism and introspection.

Widely recognized during her lifetime for her contributions as a scholar and academic, in 1921 she was elected president of the American Psychological Association. She was the first woman psychologist and second woman scientist to be elected to the National Academy of Sciences (1932).

These psychologists offered significant contributions to the development of American psychology and provided admirable role models for future generations of women psychologists.

IMPACT

Because of the unsystematic nature of functional psychology, it is difficult to appreciate this movement without concentrating on individual functionalists. The psychologists surveyed are representative, but brief mention of other leaders of the movement is also appropriate. George Trumbull Ladd (1842–1921) stressed the adaptive value of the mind and argued for the necessity of an active self-concept. Edward Wheeler Scripture (1864–1945) was a detailed methodologist who studied speech patterns and phonetics. We have already mentioned James Baldwin in connection with the start of the *Psychological Review*; he also did much to integrate Darwinism and functional psychology, and he founded Princeton's psychology laboratories. Joseph Jastrow (1863–1944), a student of Peirce, went on to work in psychophysical research and became a popular writer on psychology. Edmund Clark Sanford (1859–1924) founded the psychology laboratory at Clark University and wrote an early textbook on experimental psychology. Finally, Edmund Burke Delabarre (1863–1945), a student of James and Münsterberg, chaired the psychology department at Brown University and investigated visual perception. He took over the Harvard laboratories during Münsterberg's occasional trips to Europe. Collectively, their views were varied, but they were people of talent who valued psychology and placed the discipline on a firm foundation in America.

This diversity of functional psychology, reflected by the lack of systematic substance, dictated functionalism's disintegration. The functionalists were arguing against structural psychology, represented in America by Titchener, just as the Gestaltists were arguing with Wundt in Germany. This context should be recalled, because in many respects functional psychology was defined in terms of structural psychology; that is, Wundt's system served as the reference, albeit negative, for functional psychology. In contrast to the Gestalt development, however, functional psychology did not grow to the point of offering a comprehensive alternative model of psychology. As structural psychology began to wither, so, too, did functional psychology. It served its purpose by providing a transition from structural psychology to behaviorism.

CHAPTER SUMMARY

Functional psychology was less a system than an attitude that valued the utility of psychological inquiry. Assuming a philosophical underpinning from the pragmatism of William James and Charles Sanders Peirce, functional psychology fit well into the pioneering spirit of America. From its beginning, functional psychology had a clear emphasis on applying psychology to individual and social improvement, as was evident from the works of Münsterberg, McDougall, and Hall. The tradition of British natural science and evolutionary theory was integrated into psychology in the views on adaptation championed by the Chicago functionalists, such as Dewey, Angell, and Carr. Mental testing and the study of human capacity constituted important areas of investigation among the Columbia functionalists, represented by Cattell, Thorndike, and Woodworth. Although its reaction to structural psychology kept functional psychology from developing a systematic alternative model of psychological inquiry, this phase of American psychology resulted in two critical benefits. First, functionalism firmly entrenched the new science of psychology in America and imposed on it a particular American orientation toward applied psychology. Second, functional psychology provided a necessary transition from the restricted context of structural psychology to more viable models of psychology, permitting the science to progress.

BIBLIOGRAPHY

Primary Sources

ANGELL, J. R. (1907). The province of functional psychology. *Psychological Review, 14,* 61–91.

CALKINS, M. W. (1896). Association. *Psychological Review, 3,* 32–49.

CALKINS, M. W. (1900). Psychology as science of selves. *Philosophical Review, 9,* 490–501.

CALKINS, M. W. (1909, 1914). *A first book in psychology*. New York: Macmillan.

CALKINS, M. W. (1961). Mary Whiton Calkins. In C. Murchison (Ed.), *A history of psychology in autobiography*, Vol. 1. New York: Russell & Russell, 31–62.

CARR, H. (1925). *Psychology*. New York: Longmans Green.

CARR, H. (1930). Functionalism. In C. Murchison (Ed.), *Psychologies of 1930*. Worcester, MA: Clark University Press.

CATTELL, J. McK. (1904). The conceptions and methods of psychology. *Popular Science Monthly, 46,* 176–186.

CATTELL, J. McK. (1943). The founding of the association and of the Hopkins and Clark laboratories. *Psychological Review, 50,* 61–64.

DEWEY, J. (1886). *Psychology*. New York: Harper.

DEWEY, J. (1896). The reflex arc concept in psychology. *Psychological Review, 3,* 357–370.

GALTON, F. (1889). *Natural inheritance*. London: Macmillan.

HALL, G. S. (1917). *The life and confessions of a psychologist*. Garden City, NY: Doubleday.

JAMES, W. (1890). *The principles of psychology*. New York: Holt.

JAMES, W. (1902). *Varieties of religious experience*. New York: Longmans Green.

JAMES, W. (1907). *Pragmatism*. New York: Longmans Green.

JAMES, W. (1985). Habit (1892). *Occupational Therapy in Mental Health, 5,* 55–67.

LADD-FRANKLIN, C. (1911, 1924). The nature of the colour sensation: A new chapter on the subject. In H. Helmholtz (Ed.), *Physiological optics*, 3rd ed. Rochester, NY: Optical Society of America, 455–468.

LADD-FRANKLIN, C. (1929). *Colour and colour theories*. New York: Harcourt Brace Jovanovich.

MCDOUGALL, W. (1908). *Introduction to social psychology*. London: Methuen.

MÜNSTERBERG, H. (1903). *Psychotherapy*. New York: Moffat Yard.

MÜNSTERBERG, H. (1904). *The Americans* (E. B. Holt, Trans.). New York: McClure Philips.

MÜNSTERBERG, H. (1908). *On the witness stand.* New York: Doubleday.

MÜNSTERBERG, H. (1909). *Psychology and the teacher*. New York: Appleton.

MÜNSTERBERG, H. (1912). *Vocation and learning.* St. Louis: The People's University.

MÜNSTERBERG, H. (1913). *Psychology and industrial efficiency*. Boston: Houghton Mifflin.

MÜNSTERBERG, H. (1916). *Psychology, general and applied*. New York: Appleton.

MÜNSTERBERG, H. (1916). *The photoplay: A psychological study*. New York: Appleton.

PEARSON, K. (1901). On lines and planes of closest fit to systems of points in space. *Philosophical Magazine, 6,* 559–572.

PEIRCE, C. S. (1962). *The collected papers of Charles Sanders Peirce* (C. Hartshorne, P. Weiss, & A. Burks, Eds.). Cambridge: Harvard University Press.

SPEARMAN, C. (1904). General intelligence, objectively determined and measured. *American Journal of Psychology, 15,* 201–293.

THORNDIKE, E. L. (1931). *Human learning*. New York: Appleton.

THURSTONE, L. L. (1935). *Vectors of the mind*. Chicago: University of Chicago Press.

WASHBURN, M. F. (1908). *The animal mind*. New York: Macmillan.

WASHBURN, M. F. (1961). Margaret Floy Washburn. In C. Murchison (Ed.), *A history of psychology in autobiography*, Vol. II. New York: Russell & Russell, 333–358.

WOODWORTH, R. S. (1918). *Dynamic psychology*. New York: Columbia University Press.

WOODWORTH, R. S. (1931, 1948). *Contemporary schools of psychology* (Rev. ed.). New York: Ronald Press.

WOODWORTH, R. S., & SCHLOSBERG, H. (1954). *Experimental psychology* (Rev. ed.). New York: Holt, Rinehart & Winston.

Studies

BENDY, M. (1974). Psychiatric antecedents of psychological testing (before Binet). *Journal of the History of the Behavioral Sciences, 10,* 180–194.

BORING, E. G. (1950). *A history of experimental psychology*, 2nd ed. Englewood Cliffs, NJ: Prentice Hall.

BRENNAN, J. F. (1975). Edmund Burke Delabarre and the petroglyphs of Southeastern New England. *Journal of the History of the Behavioral Sciences, 11,* 107–122.

BURNHAM, W. H. (1925). The man, G. Stanley Hall. *Psychological Review, 32,* 89–102.

CADWALLADER, T. C. (1974). Charles S. Peirce (1839–1914): The first American experimental psychologist. *Journal of the History of the Behavioral Sciences, 10,* 291–298.

CADWALLADER, T. C., & CADWALLADER, J. V. (1990). Christine Ladd-Franklin (1847–1930). In A. N. O'Connell and N. F. Russo (Eds.), *Women in psychology: A bio-bibliographic sourcebook*. New York: Greenwood Press, 220–229.

CAMFIELD, T. M. (1973). The professionalization of American psychology, 1870–1917. *Journal of the History of the Behavioral Sciences, 9,* 66–75.

CARLSON, E. T., & SIMPSON, M. M. (1970). Perkinism vs. mesmerism. *Journal of the History of the Behavioral Sciences, 6,* 16–24.

FISHER, S. C. (1925). The psychological and educational work of Granville Stanley Hall. *American Journal of Psychology, 36,* 1–52.

FULCHER, J. R. (1973). Puritans and the passions: The faculty psychology in American puritanism. *Journal of the History of the Behavioral Sciences, 9,* 123–139.

FURUMOTO, L. (1979). Mary Whiton Calkins (1863–1930). Fourteenth President of the American Psychological Association. *Journal of the History of the Behavioral Sciences, 15,* 346–356.

FURUMOTO, L. (1990). Mary Whiton Calkins (1863–1930). In A. N. O'Connell and N. F. Russo (Eds.), *Women in psychology: a bio-bibliographic sourcebook.* New York: Greenwood Press, 57–65.

GILLHAM, N. W. (2001). *A life of Sir Francis Galton from African exploration to the birth of eugenics.* New York: Oxford University Press.

GUBER, C. (1972). Academic freedom at Columbia University, 1917–1918: The case of James McKeen Cattell. *American Association of University Professors Bulletin, 58,* 297–305.

HARRISON, F. (1963). Functionalism and its historical significance. *Genetic Psychology Monographs, 68,* 387–423.

HENLE, M., & SULLIVAN, J. (1974). Seven psychologies revisited. *Journal of the History of the Behavioral Sciences, 10,* 40–46.

HEIDBREDER, E. (1972). Mary Whiton Calkins: A discussion. *Journal of the History of the Behavioral Sciences, 8,* 56–68.

JONCICH, G. (1968). *The sane positivist: A biography of Edward L. Thorndike.* Middletown, CT: Wesleyan University Press.

KLOPPER, W. G. (1973). The short history of projective techniques. *Journal of the History of the Behavioral Sciences, 9,* 60–65.

KRANTZ, D. L., HALL, R., & ALLEN, D. (1969). William McDougall and the problems of purpose. *Journal of the History of the Behavioral Sciences, 5,* 25–38.

MCCURDY, H. C. (1968). William McDougall. In B. Wolman (Ed.), *Historical roots of contemporary psychology.* New York: Harper & Row, 4–47.

MCKINNEY, F. (1978). Functionalism at Chicago—Memories of a graduate student: 1929–1931. *Journal of the History of the Behavioral Sciences, 14,* 142–148.

MILLS, E. S. (1974). George Trumbull Ladd: The great textbook writer. *Journal of the History of the Behavioral Sciences, 10,* 299–303.

MOSKOWITZ, M. J. (1977). Hugo Münsterberg: A study in the history of applied psychology. *American Psychologist, 32,* 824–842.

MUELLER, R. H. (1976). A chapter in the history of the relationship between psychology and sociology in America: James Mark Baldwin. *Journal of the History of the Behavioral Sciences, 12,* 240–253.

MURPHY, G. (1971). William James and the will. *Journal of the History of the Behavioral Sciences, 7,* 249–260.

NANCE, R. D. (1970). G. Stanley Hall and John B. Watson as child psychologists. *Journal of the History of the Behavioral Sciences, 6,* 303–316.

NOEL, P. S., & CARLSON, E. T. (1973). The faculty psychology of Benjamin Rush. *Journal of the History of the Behavioral Sciences, 9,* 369–377.

PASTORE, N. (1977). William James: A contradiction. *Journal of the History of the Behavioral Sciences, 13,* 126–130.

RAPHELSEN, A. C. (1973). The pre-Chicago association of early functionalists. *Journal of the History of the Behavioral Sciences, 9,* 115–122.

ROBACK, A. (1964). *A history of American psychology* (Rev. ed.). New York: Collier.

RUCKMICK, C. (1912). The history and status of psychology in America. *American Journal of Psychology, 23,* 517–531.

RYAN, T. A. (1982). Psychology at Cornell after Titchener: Madison Bentley to Robert MacLeod, 1928–1948. *Journal of the History of the Behavioral Sciences, 18,* 347–369.

SAMELSON, F. (1977). World War I intelligence testing and the development of psychology. *Journal of the History of the Behavioral Sciences, 13,* 274–282.

SCARBOROUGH, E. Margaret Floy Washburn (1871–1939). In A. N. O'Connell and N. F. Russo (Eds.), *Women in psychology: A bio-bibliographic sourcebook*. New York: Greenwood Press, 342–349.

SCHNEIDER, W. H. (1992). After Binet: French intelligence testing, 1900–1950. *Journal of the History of the Behavioral Sciences, 28,* 111–132.

SOKAL, M. M. (1981). The origins of the Psychological Corporation. *Journal of the History of the Behavioral Sciences, 17,* 54–67.

SOKAL, M. M. (1990). G. Stanley Hall and the institutional character of psychology at Clark University (1889–1920). *Journal of the History of the Behavioral Sciences, 26,* 114–124.

WALLIN, J. E. (1968). A tribute to G. Stanley Hall. *Journal of Genetic Psychology, 113,* 149–153.

13

The Gestalt Movement

The German word *Gestalt* cannot be rendered into a single English word. It describes a configuration or form that is unified. A Gestalt may refer to a figure or object that is different from the sum of its parts. Any attempt to explain the figure by analyzing its parts results in the loss of the figure's Gestalt. For example, a square has a unity and an identity that cannot be fully appreciated by its description as four straight lines connected by right angles. Accordingly, Gestalt expresses the fundamental premise of a system of psychology that conceptualizes psychological events as organized, unified, and coherent phenomena. This view stresses the integrity of a clear psychological level of human activity, which loses its identity if analyzed into preconceived components. Gestalt psychology is characteristically antireductionistic. For example, if learning is considered as a psychological activity, then according to Gestalt psychologists, it cannot be reduced to the physiological mechanisms of conditioning. Gestaltists argue that the very attempt to reduce the psychological event to its physiological components results in the loss of the psychological event. The Gestalt is removed, so that the psychological event of learning is not explained, and only a physiological mechanism is described.

Gestalt psychology was a German movement that directly challenged Wundt's structural psychology. The Gestaltists inherited the tradition of the act psychology of Brentano and Stumpf, as well as the Würzburg school, which attempted to devise an alternative to the model of psychology proposed by the reductionistic and analytic natural science model of Wundt. The Gestalt movement was more consistent than Wundt's system with the major theme of mental activity in German philosophy, following in the tradition of Kant; that is, underlying Gestalt psychology was the nativistic proposition that the organization of mental activity predisposes the individual to interact with the environment in characteristic ways. Accordingly, the goals of Gestalt psychology were to investigate the organization of mental activity and to determine the exact nature of person–environmental interactions.

By 1930, the Gestalt movement had largely succeeded in replacing the Wundtian model in German psychology. However, the movement's success was short-lived because of the advent of Hitlerism, with its accompanying intellectual sterility and physical barbarism. The leaders of the movement fled to America, but there Gestalt psychology never enjoyed the dominance it had achieved in Germany. Essentially, the initial views of Gestalt psychology were contrasted with Wundt's psychology. Those arguments were out of touch with American psychology because Wundt's system was largely irrelevant by the 1930s. American psychology had evolved through the functional period and by the 1930s was dominated by behaviorism. Accordingly, the framework of Gestalt psychology was not in step with American developments. Nevertheless, the attractiveness of Gestalt psychology influenced many behaviorists, as reviewed in Chapter 15, and the movement played an important role in the expansion of the behavioristic model of psychology in America.

BACKGROUND IN GERMANY

Before the specific views of the Gestalt psychologists are examined, two immediate precedents should be reviewed. Both contributed to the intellectual climate of German psychology as well as to the acceptance and success of Gestalt psychology.

The Würzburg Legacy

As discussed in Chapter 11, the German intellectual atmosphere may be broadly described as containing two schools of psychology. One was the Wundtian natural science model of psychology, which revolved around the study of immediate experience through controlled introspection. By virtue of his strong personality and tremendous productivity, Wundt introduced a unified system of structural psychology that restricted the new science to a view of mental contents as wholly dependent on sensory input. The other school consisted of a loose collection of writers who, although they held various views, shared a dissatisfaction with Wundt's model. Brentano and Stumpf, as we have seen, were dominant figures in the attempt to remove psychology from the limitations of Wundt's rigid formulation.

As a development of the anti-Wundt movement in Germany, the Würzburg school under Külpe attempted to define mental activity in terms of nonsensory consciousness. The "imageless thought" controversy served as a catalytic agent for various trends that were moving toward a viewpoint involving more self-initiated activity in mental processes. Indeed, Külpe viewed the mind as predisposed to the ordering of environmental events along dimensions of quality, intensity, time, and space, restoring to German psychology the mental categories of Kant. These predispositions, together with the recognition of the existence of mental contents having nonsensory origin, radically challenged the assumptions underlying mental processes, such as those of Wundt's psychology. The Würzburg psychologists asserted that the mind has characteristic sets, or determining tendencies, that result in patterns of perception. Depending on the organism's set at a given time, associations may change from one pattern or sequence to another. This type of mental activity, then, is dependent on the mind's organization.

As noted in Chapter 11, the Würzburg school enjoyed a relatively brief period of productivity but did not devise a comprehensive, alternative model of psychology to compete with the formulations of Wundt. Nevertheless, their careful observations, leading to their advocacy of nonsensory consciousness, presented a serious challenge to Wundt's model. It was the Gestaltists who followed the conclusions of the Würzburg school to a systematic position in opposition to Wundt.

German Phenomenology

As the word indicates, phenomenology is the study of phenomena. A phenomenon is literally that which appears. However, in the context of phenomenology, phenomena are taken as events studied for themselves, without concern for underlying causality or inferences. For psychology, a phenomenological approach characteristically emphasizes experiences as perceived by an individual. It stands in sharp contrast to any form of analysis that breaks a psychological event into elements or reduces an event to other levels of explanations.

We have already seen examples of phenomenological approaches. The empirical strategies of many eighteenth- and nineteenth-century physiologists typically contained keen observations, as opposed to tight experimental controls. The sensory investigations of Purkinje stand out as a clear example of this type of phenomenological study in physiology. Certainly, the act psychology of Brentano and Stumpf was more amenable to the observational foundation of phenomenology than to an experimental method of controls imposed over psychological variables. Moreover, the scientific positions of Dilthey and Bergson were more conducive to the descriptive data of phenomenology than to the causal inferences of experimentation. Accordingly, phenomenology is a traditional method of empiricism, but its variations have differed according to the particular assumptions attached to the subject matter under investigation, especially in psychology.

The modern expression of phenomenology came from a student of Brentano, Edmund Husserl (1859–1938). Husserl's application of phenomenology to psychological issues called for a pure science of consciousness; he advocated detailed and sophisticated description of experienced mental activity. Husserl developed a method of observation that elaborated all levels of the modes by which phenomena could appear in consciousness. Yet Husserl's method was not analytical and was opposed to reduction. Thus Husserl's phenomenology left a purely psychological level of investigation intact. Husserl and the Gestalt psychologists had different views of the content of psychology. It was left to later thinkers to explore fully Husserl's brand of phenomenology for psychology, which is examined in Chapter 16. However, Husserl's phenomenology and the Gestalt movement were both products of the same intellectual forces in Germany at the beginning of the twentieth century. While pursuing different implications, both the Gestalt psychologists and Husserl were dubious of the analytical character of the controlled laboratory method, and they searched for alternative formulations of psychology that recognized the inherent organization and activity of the mind.

THE FOUNDING OF GESTALT PSYCHOLOGY

As stated earlier, Gestalt psychologists emphasized the organization and unity of their data, defined in terms of phenomena. The wholeness and unity of experiences were examined in terms of forms. In contrast to the study of immediate experience of Wundtian psychology, the Gestaltists' study of phenomena was purposefully defined as mediated experience. Gestalt psychology included the investigation of objects and their meaning; it valued mediated perceived thoughts over sensory events. The individual was viewed as actively interacting with the environment, within a dynamic field or system of interactions. Whereas not completely rejecting analytic methodologies as the pure phenomenologists did, the Gestalt psychologists asserted the freedom to use varieties of methodologies that did not interfere with the integrity of phenomena.

Gestalt psychology was originated and nurtured by the writings of three persons: Max Wertheimer, Wolfgang Köhler, and Kurt Koffka. All were educated in the exciting intellectual atmosphere of early twentieth-century Germany, and all later fled Nazi persecution and immigrated to America.

Max Wertheimer

Born in Prague and educated at Charles University there, Max Wertheimer (1880–1943) studied with Stumpf in Berlin for several years before joining Külpe at Würzburg, where he received his doctorate in 1904. While traveling on a summer holiday in 1910, Wertheimer came upon an idea for an experiment on apparent movement. According to the story recounted by Boring (1950, p. 595), Wertheimer left the train at Frankfurt am Main and bought a toy stroboscope, a primitive device for showing moving pictures, to try to determine the optimal conditions for the phenomena to occur. We encounter this type of illusion daily in motion pictures or through special lighting effects that simulate motion through the appropriate, successive presentation of static stimuli. Wertheimer stayed on to become associated with the Psychological Institute of Frankfurt and gained access to a tachistoscope at the university to investigate the illusion more thoroughly. There he met Koffka and Köhler, who served as subjects for his experiments. He named this illusion the "phi phenomenon," and in 1912 he published "Experimental Studies of the Perception of Movement." His findings marked the formal beginnings of Gestalt psychology. The major implication of Wertheimer's research was that the phi phenomenon cannot be reduced to the stimulus elements presented to the subjects, as Wundt's system would predict. The subjective experience of motion is the result of a dynamic interaction between an observer and the stimuli.

Wertheimer was the scholarly guiding force among the early founders of Gestalt psychology. After working in military research during World War I, he lectured at various universities before accepting a professorship at Frankfurt in 1929. In 1933, he fled Germany for the United States and joined the faculty of the New School for Social Research (now the New School University) in New York City, where he stayed until his death. Through his teaching and personal meetings with American psychologists, he attempted to extend the scope of Gestalt principles beyond perceptual problems to thought processes. His final ideas on cognitive psychology within a Gestalt perspective were contained in *Productive Thinking*, published posthumously in 1945. In it he suggested guidelines to facilitate the development of creative strategies in problem solving. Despite the difficulties of moving to a new country and having to master a foreign language, he served as an inspirational force to the Gestalt movement in the United States.

Wolfgang Köhler

Perhaps the most systematic of the early Gestaltists, Wolfgang Köhler's (1887–1967) many publications gave Gestalt psychology definitive form. Born in Reval near the Baltic Sea area of East Prussia, Köhler attended various universities before receiving his doctorate under Stumpf's direction at Berlin in 1909. After working with Wertheimer in Frankfurt, Köhler went to the Canary Islands in 1913 to study chimpanzees. As a German national caught on the other side of Allied lines at the outbreak of World War I, he remained there until the end of the hostilities. In 1917, he published his *Intellegenzprüfungen an Menschenaffen*, based on his work during internment; it was translated into English in 1925 as *The Mentality of Apes*. After his experiences with Wertheimer and Koffka at Frankfurt, Köhler was able to offer an innovative approach to his studies of discrimination learning and problem solving. He applied a Gestalt interpretation to the acquisition of relationships between stimuli, as opposed to learning the absolute value of stimulus dimensions. Further, he found that the chimps used insightful strategies in solving puzzles rather than relying only on trial-and-error learning. Köhler's most intelligent subject, Sultan, was able to master various tasks to secure food reward, easily switching among problem-solving strategies. The rapidity of solutions impressed Köhler as evidence of insightful learning. Köhler's book is important in the history of the movement because he demonstrated specific instances of many of the principles of mental organization underlying Gestalt psychology.

Köhler returned to Germany in 1920 and taught at Göttingen for a year before being named to succeed Stumpf at Berlin in 1922. This prestigious appointment was largely the result of a scholarly and erudite work published in 1920, *Die Physischen Gestalten in Ruhe und im stationärem Zustand* (*Static and Stationary Physical Gestalts*). From 1934 to 1935, Köhler lectured at Harvard, and in 1935 finally left

Germany to join the faculty of Swarthmore College, where he stayed until his retirement. He adapted to America better than did Wertheimer, becoming the major spokesman for the Gestalt movement, and continued to edit the primary Gestalt journal, *Psychologische Forschung* (*Psychological Research*). This journal, started in Germany by the founders of the Gestalt movement, published 22 volumes before it was suspended in 1938. Finally, in 1959, Köhler was elected president of the American Psychological Association, a fitting testimonial to his life of creative research and scholarly writing.

Kurt Koffka

Like Köhler, Kurt Koffka (1886–1941) received his doctorate from Stumpf in 1909 at the university in Berlin, the city of Koffka's birth. After working with Wertheimer and Köhler in Frankfurt, Koffka joined the University of Giessen, near Frankfurt, where he remained until 1924. He traveled in the United States while a visiting professor at several American universities, and secured a faculty position at Smith College in 1927, remaining there until his death.

It was Koffka who introduced Gestalt psychology to wide American audiences through his publication in 1922 of "Perception: An Introduction to the Gestalt-Theorie" in the *Psychological Bulletin*. In addition, Koffka published a book on developmental child psychology, *The Growth of the Mind* (1921), which was widely acclaimed in America as well as in Germany. However, his major aim—to write the definitive work on the Gestalt movement—was not fulfilled by his *Principles of Gestalt Psychology* (1935), which is now remembered chiefly as a very difficult book. Koffka was perhaps the most prolific writer of the three founders, yet lacked the inspirational quality of Wertheimer and the thoughtful, systematic capacity of Köhler. However, he did succeed in bringing the teachings of the Gestalt movement to a great number of psychologists, especially in the United States.

BASIC PRINCIPLES OF GESTALT PSYCHOLOGY

Consistent with the original work of Wertheimer on the phi phenomenon, the principles of Gestalt psychology grew out of research on sensory and perceptual processes. Much of the terminology and many of the examples illustrating Gestalt principles were derived from such studies. Only later were these principles extended to other psychological activities. This extension was especially appropriate in the applications of Gestalt principles to learning processes, when the focus of the Gestalt movement shifted to America. The Gestalt movement provided an alternative to Thorndike's trial-and-error learning as the behavioristic model was expanded.

In Gestalt psychology the focus of person–environment interactions is termed the *perceptual field*. The major characteristic of any perceptual field is organization, which has a natural tendency to be structured in terms of figure and ground. Thus, seeing the salient features of shapes and forms on a background within a perceptual field is a spontaneous innate activity and not an acquired skill; we are predisposed

FIGURE 13–1 Circle *a* is complete and shows good Gestalt; line *b* is incomplete but is perceived as a circle because of closure.

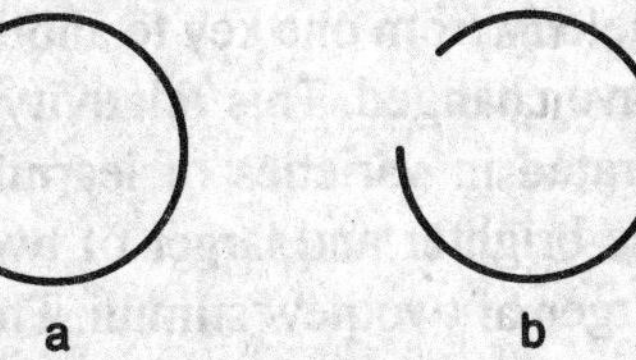

to perceive in such a manner. A good figure is complete, tending toward symmetry, balance, and proportion. For example, a circle has a perfect Gestalt in terms of completeness. Incomplete figures tend to be perceived as complete, and this organizational characteristic is called closure; for example, a curved line with ends not quite touching will still be perceived as a circle because of the compelling tendency for closure. The two drawings in Figure 13–1 compare a complete Gestalt with a form illustrating closure. Similarly, in Figure 13–2 the importance of the relative saliency of context is illustrated; center circles in both illustrations are of equal size, but the surrounding circles make them appear to be of different sizes.

Organization leading to meaning, then, is the key to our perceptual structure, according to the Gestaltists. Other principles of organization in addition to closure include proximity and similarity. Organized figures are stable and tend to retain their stability as structural wholes despite changes in the stimulus characteristics; the Gestaltists called this *object constancy.* For example, an actor viewed on a television program is still perceived as a man, although the figure itself may be only a few inches tall.

The critical dimension in comparisons among figures or objects in the environment, according to the Gestaltists, is the relationships between parts of a figure, not the characteristics of the parts. If particular aspects of stimuli change, but not the relationships, the perception remains the same. Accordingly, the Gestaltists were able to accommodate Stumpf's argument with Wundt that the transposition of a

FIGURE 13–2 Because of the relationship of contexts, the center circle in *a* is perceived as larger than the center circle in *b*. Both center circles are the same size.

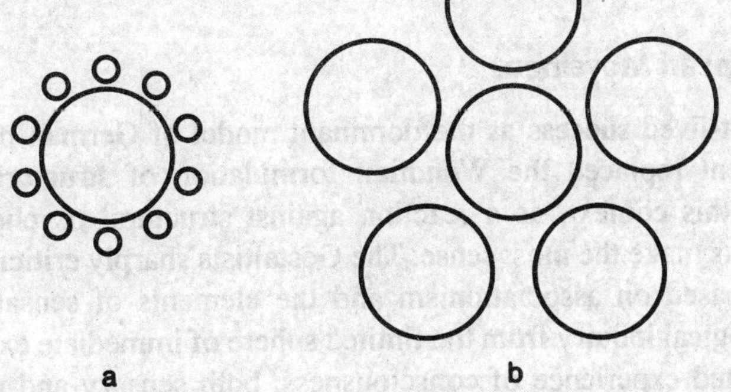

melody from one key to another retains the melody although the elements (the notes) have changed. This relativity as the basis for acquiring discriminations was demonstrated in varieties of learning situations. For example, rats trained to respond to the brighter and larger of two stimulus objects continued to select the brighter and larger of two new stimuli. The recognition of the salient relationships and the transfer of that knowledge from one learning situation to another was called *transposition*, and it was repeatedly demonstrated with varying species tested under many stimulus parameters.

Perhaps the weakest, and certainly the most elusive, part of Gestalt theory concerns the explanation of underlying brain activity mediating perceptual processes, which Gestaltists called *isomorphism*. The Soviet physiologist Ivan Pavlov, discussed in Chapter 15, criticized the Gestalt movement in a 1935 paper in which he pointed to the idealistic basis of Gestalt psychology as being mentalistic and without a mechanical, physiological foundation. The Gestalt attempt to define the physiological basis of perceptual processes was a deliberate attempt to get away from the mechanical notion of cortical excitation corresponding to stimulus action and associated sensory processing. Rather, the Gestaltists argued for what they called "brain experience," which involved the assertion that a perceptual field has an underlying excitatory brain field that corresponds to the perceptual field in order, but not necessarily in exact form. Hence, their choice of the term *isomorphic*, derived from Greek *iso* (similar) and *morphic* (shape). Isomorphic representation is defined as parallel processes between perceptual and physiological levels. The perceptual experience and the brain experience do not correspond on a one-to-one basis, but rather correspond in terms of relations. Accordingly, as Köhler described isomorphism, this principle relates the perceptual field to the brain field. The former is elicited by stimulus activity, whereas the latter consists of electrochemical activity. When Gestalt principles were formulated, isomorphism ran contrary to the prevailing concepts contained in Pavlovian and Sherringtonian neurophysiology. However, although the Gestalt notion of isomorphism remained rather vague, subsequent developments viewing cortical activity as a cybernetic system have made the hypothesis of isomorphism somewhat more tenable, at least as an analog, as we shall see in Chapter 15.

IMPLICATIONS OF GESTALT PSYCHOLOGY

As a European Movement

In its short-lived success as the dominant model of German psychology, the Gestalt movement replaced the Wundtian formulation of structural psychology. Indeed, it is in this context, as a reaction against structural psychology, that the Gestalt arguments make the most sense. The Gestaltists sharply criticized any model of psychology based on associationism and the elements of sensation. They expanded psychological inquiry from the limited sphere of immediate experience to include the mediated experience of consciousness, both sensory and nonsensory. In this sense, Gestalt psychology was an act psychology in the tradition of Brentano and Stumpf. In contrast to Wundt, Gestalt psychology studied the *how* of mental processing, not the *what*.

By admitting phenomenology as a methodological approach, the Gestalt movement expanded the empirical basis of psychology. Its practitioners demonstrated how a psychology defined in terms of the higher mental processes of mediated experience could nevertheless retain a scientific, empirical framework. In moving beyond their initial investigations of perceptual processes to a comprehensive psychology, the Gestaltists emphasized that consciousness and behavior should not be viewed separately but must be considered together. Thus the Gestalt movement in Europe showed every promise of integrating the positive developments of European psychology to consider all of the complexities of human activity.

Unfortunately, that promise was not fulfilled, as European intellectual life disintegrated. What remained of the Gestalt movement in Germany after its leaders fled to America quickly became distorted in the propaganda of Hitlerism. When European intellectual life revived after World War II, too much time had elapsed and other models of psychology appeared. During the interim, the Gestalt movement had been absorbed into American neobehaviorism and was no longer recognizable as a separate system of psychology.

As an American Movement

When the chief Gestaltists left Germany and came to America, the prevailing system in the United States was not structural psychology but behaviorism. Moreover, American behaviorism had evolved out of the functional period of early American psychology with its characteristic utilitarian flavor, in contrast to the European orientation of defining the mind per se and caring less about its functions. Accordingly, the Gestalt movement was not in tune with ongoing developments on the American psychological scene.

Although unable to compete with behaviorism, the Gestalt movement played a major role in redefining behaviorism. A major behaviorist who showed an early inclination toward Gestalt theory was Edward C. Tolman, whose work is considered in more detail in Chapter 15. A few of the research areas Tolman investigated were provoked by Gestalt thoughts. Tolman did important experiments on latent learning and showed that learned responses can be acquired without apparent manifestation in observed performance. This learning versus performance distinction is not readily explained by stimulus–response (S–R) reductionistic theory but is predicted from the Gestalt view of organized fields of behavior or, as referred to by Tolman, cognitive learning. Similarly, transpositional learning in the acquisition of successive discrimination problems has been found in various species and could not be easily accommodated by initial S–R formulations.

Perhaps the most important Gestalt studies of learning processes come from those investigations, beginning with Köhler's reports on higher learning processes in chimpanzees, dealing with problem solving and insight. The traditional S–R models of learning were based on logical deductions (for example, refinement of responses through trial-and-error elimination of choices) or associations (such as principles of conditioning). The Gestaltists' demonstrations of insight, rapid solutions, and creativity opened the consideration of learning processes in psychology to a broader spectrum of possibilities. Entire areas of research were proposed, ranging from remote retrieval of memory traces to the study of understanding. As mentioned earlier, Wertheimer's *Productive Thinking* offered a refreshing perspective on the facilitation of potential strategies of problem solving.

Field Theory

One application derived from Gestalt theory involved a view of social activities and personality dynamics, termed *field theory*, which received articulate expression in the work of Kurt Lewin (1890–1947). Lewin's views were a product of the active model of the mind prevalent in German philosophy, and in some respects certain parallels can be seen between Freud and Lewin in their formulations within the German tradition. However, Lewin was most directly influenced by the specific principles of the Gestalt movement, and although much of his work was done independently, he contributed heavily to applications of Gestalt principles that are prevalent to this day.

Lewin received his doctorate at Berlin in 1914, where he studied mathematics and physics as well as psychology. After military service during World War I, he returned to Berlin and became involved with the Gestalt group led by Köhler. He quickly gained international fame and spent several years as a visiting professor at Stanford and Cornell. He immigrated to the United States permanently in 1935 and spent the next 9 years at the University of Iowa, where he made innovative studies of childhood socialization. In 1944, he went to the Massachusetts Institute of Technology to lead a research center devoted to group dynamics, which continued working after his death.

In the Gestalt tradition, Lewin argued that personality should be viewed in the context of a dynamic field of individual–environmental interactions. Lewin taught that the restriction of psychological description to group averages or statistical summaries loses sight of the individual. If all of the general laws of behavior were known, the psychologist would still need to appreciate the specific individual's interactions with the environment to make any meaningful predictions. Lewin's model of the interactive field of an individual was based on his notion of hodological space, which was defined as a geometric system emphasizing (1) movement along psychologically directed pathways, (2) the dynamics of person–environment interactions, and (3) the person's behavior at environmental obstacles or barriers. Moreover, the person was viewed in terms of an individual life space, containing not only the predominance of the present hodological space with psychologically directed pathways of movement, but also representations of past experiences and future expectations.

The dynamics of the life space are governed by motivational constructs consisting of several components. Individual needs may arise through physiological conditions, a desired environmental object, or an internalized goal. Such needs produce tensions, or emotional states, that must be reduced. Objects in the environment related to needs have values of attraction or repulsion, and these values are termed *valences*. For example, an apple may have a positive valence for a hungry child, but if the child has just finished eating ten green apples, the apple may assume a negative valence. The directed action toward or away from an object is termed a *vector*; two opposing vectors define a *conflict*. Finally, there are barriers in the environment that may come from other objects, people, or a moral code, and thwart activity. Putting these constructs together as diagrammed in Figure 13–3, let us suppose that C represents a hungry child who wants an apple (A) available in the environment, but the child is told by a parent not to eat before dinner, imposing a barrier. The apple has a positive valence, and the barrier has a negative valence and prevents movement. The hodological space produces a vector of action (arrow) toward the apple until the barrier is confronted and conflict arises. The conflict will be resolved only when the positive or negative valence of one object exceeds the other and results in motion either further toward or away from the apple. Thus, Lewin's model is a motivational system seeking an equilibrium of forces within the life space.

This survey of parts of Lewin's field theory reflects an interesting application of Gestalt theory to personality and social behaviors. Lewin's views fascinated many psychologists because of the complex types of behavior that can be considered in the context of life space. As the behavioristic model of psychology expanded to include cognitive variables, Lewin's teachings were readily adopted to develop a comprehensive theory of behaviorism.

FIGURE 13–3 Diagrammatic representation of the life space of a child (C) attracted to an apple (A) having a positive valence. Prohibitions by the parent set up a barrier with a negative valence, which thwarts the vector of approaching movement.

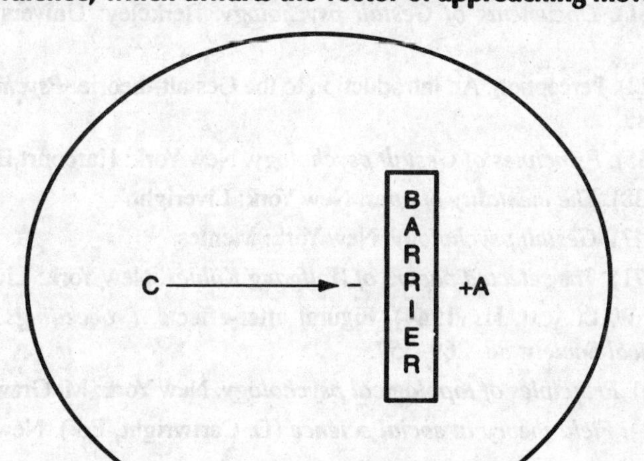

CHAPTER SUMMARY

Gestalt psychology originated as a German intellectual movement heavily influenced by the precedents of the Würzburg school and phenomenological approaches to science. The early Gestaltists directly challenged Wundt's structural psychology and were largely successful in pursuing the traditions of Brentano and Stumpf. Originating in Wertheimer's research on apparent movement, or the phi phenomenon, the Gestalt principles were founded on the assumption of the inherent organization of person–environment interactions. The writings of Köhler and Koffka expanded the perceptual basis to formulate a comprehensive system of psychology especially amenable to higher thought processes of insight, understanding, and productive thinking. When the movement was threatened with destruction by the intellectual sterility of Nazi tyranny, the leaders fled to America. Unfortunately, the Gestalt movement was out of tune with the prevailing behavioristic character of American psychology. However, the Gestaltists assumed an important role in broadening the basis of behaviorism to foster a complete view of learning processes. One application of Gestalt views, contained in Lewin's field theory, met with success in providing an empirical model of personality and social activities. The Gestalt movement, although it did not retain a separate identity, contributed greatly to the reformulation of psychology.

BIBLIOGRAPHY

Primary Sources

HELSON, H. (1925, 1927). The psychology of Gestalt. *American Journal of Psychology, 36,* 342–370, 494–526; *27,* 25–62, 189–223.

HENLE, M. (1961). *Documents of Gestalt psychology*. Berkeley: University of California Press.

KOFFKA, K. (1922). Perception: An introduction to the Gestalt-theorie. *Psychological Bulletin, 19,* 531–585.

KOFFKA, K. (1935). *Principles of Gestalt psychology*. New York: Harcourt Brace Jovanovich.

KÖHLER, W. (1938). *The mentality of apes*. New York: Liveright.

KÖHLER, W. (1947). *Gestalt psychology*. New York: Mentes.

KÖHLER, W. (1971). *The selected papers of Wolfgang Köhler*. New York: Liveright.

KÖHLER, W., & WALLACH, H. (1944). Figural after-effects. *Proceedings of the American Philosophical Society, 88,* 269–357.

LEWIN, K. (1936). *Principles of topological psychology*. New York: McGraw-Hill.

LEWIN, K. (1951). *Field theory in social science* (D. Cartwright, Ed.). New York: Harper & Row.

LEWIN, K., LIPPITT, R., & WHITE, R. (1939). Patterns of aggressive behavior in experimentally created social climates. *Journal of Social Psychology, 10,* 271–299.

WERTHEIMER, M. (1950). Gestalt theory. In W. D. Ellis (Ed.), *A source book of Gestalt psychology*. London: Routledge and Kegan Paul.

WERTHEIMER, M. (1959). *Productive thinking* (Enlarged ed.). New York: Harper & Row.

ZEIGARNIK, B. (1927). Uber das Behalten von erledigten und unerledigten Handlungen. *Psychologische Forschung, 9,* 1–85.

Studies

BORING, E. G. (1950). *A history of experimental psychology*, 2nd ed. Englewood Cliffs, NJ: Prentice Hall.

DOLEZAL, H. (1975). Psychological phenomenology face to face with the persistent problems of psychology. *Journal of the History of the Behavioral Sciences, 11,* 223–234.

GIBSON, J. J. (1971). The legacies of Koffka's Principles. *Journal of the History of the Behavioral Sciences, 7,* 3–9.

HARROWER-ERICKSEN, M. (1942). Kurt Koffka: 1886–1941. *American Journal of Psychology, 55,* 278–281.

HEIDER, F. (1970). Gestalt theory: Early history and reminiscences. *Journal of the History of the Behavioral Sciences, 6,* 131–139.

HENLE, M. (1978). Gestalt psychology and Gestalt therapy. *Journal of the History of the Behavioral Sciences, 14,* 23–32.

HENLE, M. (1978). Kurt Lewin as metatheorist. *Journal of the History of the Behavioral Sciences, 14,* 233–237.

HENLE, M. (1984). Robert M. Ogden and Gestalt psychology in America. *Journal of the History of the Behavioral Sciences, 20,* 9–19.

HOCHBERG, J. (1974). Organization and the Gestalt tradition. In E. Carterette and M. Friedman (Eds.), *Handbook of perception, Vol. 1: Historical and philosophical roots of perception*. New York: Academic Press.

LINDENFIELD, D. (1978). Oswald Külpe and the Würzburg school. *Journal of the History of the Behavioral Sciences, 14,* 132–141.

MACLEOD, R. B. (1964). Phenomenology: A challenge to experimental psychology. In T. Wann (Ed.), *Behaviorism and phenomenology: Contrasting bases for modern psychology*. Chicago: University of Chicago Press, 47–78.

MORROW, A. (1969). *Kurt Lewin*. New York: Basic Books.

WERTHEIMER, M., KING, D. B., PECKLER, M. A., RANEY, S., & SCHAEF, R. W. (1992). Carl Jung and Max Wertheimer on a priority issue. *Journal of the History of the Behavioral Sciences, 28,* 45–56.

14

Psychoanalysis

The psychoanalytic movement's place in contemporary psychology is both unique and paradoxical. On the one hand, psychoanalysis is probably the most widely known, although perhaps not universally understood, of the systems of psychology. Its founder, Sigmund Freud, is certainly one of the most famous persons of the last century. On the other hand, the psychoanalytic movement has little in common with the other expressions of psychology. Psychoanalysis is most clearly aligned with the German tradition of the mind as an active, dynamic, and self-generating entity. Freud was trained in science, yet his system shows little appreciation for systematic empiricism. As a physician, Freud used his keen powers of observation to build his system within a medical framework, basing his theory on individual case studies. He did not depart from his understanding of nineteenth-century science in the effort to organize his observations. He did not attempt to test his hypotheses rigorously by independent verification. As he himself testified, he *was* psychoanalysis, and he did not tolerate dissension from his orthodox views. Nevertheless, Freud had a tremendous impact on twentieth-century psychology. Perhaps more importantly, the influence of psychoanalysis on Western thought, as reflected in literature, philosophy, and art, significantly exceeds the impact of any other system of psychology.

BACKGROUND

The Active Mind

In our consideration of the philosophical precedents of modern psychology in Germany during the seventeenth, eighteenth, and nineteenth centuries, we saw that the tradition of Leibniz and Kant clearly emphasized mental activity. In contrast to British empiricism, which viewed the mind as passive, or the French sensationalistic view of the mind as an unnecessary construct, the German tradition held that the mind itself generates and structures human experience in characteristic ways. Whether through Leibniz's monadology or Kant's categories, the psychology of the individual could be understood only by examining the dynamic, inherent activity of the mind.

As psychology emerged as an independent discipline in the latter part of the nineteenth century under Wundt's tutelage, the British model of mental passivity served as the guiding force. Wundt's empiricistic formulation was at odds with German philosophical precedents, recognized by both Stumpf and Brentano. Act psychology and the psychology of nonsensory consciousness represented by the Würzburg School were closer to the German philosophical assumptions of mental activity than to Wundt's structural psychology. The Gestalt movement encompassed these alternatives to Wundt's psychology in Germany. Eventually Wundt's system was replaced by Gestalt psychology, making the dominant psychology in Germany prior to World War II one based on a model of the mind that admitted inherent organizational activity.

However, the assumptions underlying mental activity in Gestalt psychology were highly qualified. The Gestalt construct for mind involved the organization of perception, based on the principle of isomorphism, which resulted in a predisposition toward patterns of person–environmental interactions. The emphasis on organization meant that the manner of mental processes, not their content, was inherently structured. In other words, people were not born with specific ideas, energies, or other content in the mind; rather, they inherited the organizational structure to acquire mental contents in characteristic ways. Accordingly, the Gestalt movement, while rejecting the rigidity of Wundt's empiricistic views, did not reject empiricism as such. Instead, the Gestaltists advocated a compromise between the empiricist basis of British philosophy and the German model of activity. They opened psychological investigation to the study of complex problem solving and perceptual processes.

Consistent with the Gestalt position, psychoanalysis was firmly grounded in an active model of mental processes, but it shared little of the Gestalt commitment to empiricism. Freud's views on personality were consistent not only with the activities of mental processing suggested by Leibniz and Kant, but also with the nineteenth-century belief in conscious and unconscious levels of mental activity. In accepting the teachings of such philosophers as von Hartmann and Schopenhauer, Freud developed motivational principles that depended on energy forces beyond the level of self-awareness. Moreover, for Freud, the development of personality was determined by individual, unconscious adaptation to these forces. The details of personality development as formulated by Freud are outlined below; however, it is also important to recognize the context of Freud's thinking. Psychoanalysis carried the implication of mental activity further than any other system of psychology. As the major representative of an extreme reliance on mental activity to account for personality, psychoanalysis is set apart from other movements in contemporary psychology. In addition, psychoanalysis did not emerge from academic research, as did the other systems; rather, it was a product of the applied consequences of clinical practice.

The Treatment of Mental Illness

Apart from his fame as the founder of the psychoanalytic movement in modern psychology, Freud is also remembered for his pioneering efforts in upgrading the treatment of mental and behavioral abnormalities. He was instrumental in psychiatry's being recognized as a branch of medicine that specifically deals with psychopathology. Before Freud's attempt to devise effective methods of treating the mentally ill, people who deviated from socially acceptable norms were usually treated as if they were criminals or demonically possessed. Although shocking scandals in the contemporary treatment of mental deviancy appear occasionally, not so long ago such abuses were often the rule rather than the exception.

The treatment of the mentally ill is not a pleasant chapter in Western civilization. It has already been pointed out that abnormal behavior has often been mixed up with criminal behavior as well as with heresy and treason. Even during the enlightened period of the European Renaissance, the tortures and cruelties of the Inquisition were readily adapted to treat what we now consider mental illness. Witchcraft continued to offer a reasonable explanation of such behavior until relatively recent times. Prisons were established to house criminals, paupers, and the insane without differentiation. Mental illness was viewed as governed by obscure or evil forces, and the mentally ill were looked upon as crazed by such bizarre influences as moon rays. Lunatics, or "moonstruck" persons, were appropriately kept in lunatic asylums. As recently as the latter part of the nineteenth century and the beginning of the twentieth century, the institution for the insane in Utica, New York, which was progressive by the standards of the time, was called the Utica Lunatic Asylum. The name reflected the prevailing attitude toward mental illness.

Reforms in the treatment of the institutionalized insane were slowly introduced during the nineteenth century. In 1794, Philippe Pinel (1745–1826) was appointed the chief of hospitals for the insane in Paris, and managed to improve both the attitude toward and the treatment of the institutionalized insane. In the United States, Dorothea Dix (1802–1887) accomplished the most noticeable reforms in the treatment of the mentally ill. Beginning in 1841, Dix led a campaign to improve the condition of indigent, mentally ill persons kept in jails and in poorhouses. However, these reforms succeeded in improving only the physical surroundings and maintenance conditions of the mentally ill; legitimate treatment was minimal. Efforts to develop comprehensive treatment were plagued by various quacks. A pseudoscience developed by Mesmer dealt with the "animal spirits" underlying mental illness. Similarly, the phrenology of Gall and Spurzheim advocated a physical explanation based on skull contours and localization of brain functions.

Gradually attempts were made to develop legitimate and effective techniques to treat emotional and behavioral abnormalities. One of the more productive investigations involved hypnotism and was pioneered by a French physician, Jean Martin Charcot (1825–1893). He gained widespread fame in Europe, and Freud studied under him, as did many other physicians and physiologists. Charcot treated hysterical patients with symptoms ranging from hyperemotionality to physical conversions of underlying emotional disturbances. He used hypnotism as a tool to explore underlying emotional problems that the patient could not confront when conscious. Another French physician in Nancy, Hippolyte Bernheim (1837–1919), developed a sophisticated analysis of hypnotism as a form of treatment, using underlying suggestibility to modify the intentions of the patient. Finally, Pierre Janet (1859–1947), a student of Charcot, used hypnotism to resolve the forces of emotional conflict, which he believed were basic to hysterical symptoms. However, it was Freud who went beyond the techniques of hypnotism to develop a comprehensive theory of psychopathology, from which systematic treatments evolved.

SIGMUND FREUD

Biography

Because psychoanalysis is identified to such a great extent with Freud himself, it is worthwhile to outline the major points in Freud's distinguished life. Sigmund Freud (1856–1939) was born May 6, 1856, in Freiberg, Moravia, at that time a northern province of the Austro-Hungarian Empire and today part of the Czech Republic.

SIGMUND FREUD (1856–1939).
(Courtesy, The Bettmann Archive, Inc.)

Freud was the eldest of eight children, and his father was a relatively poor and not very successful wool merchant. When his business failed, Freud's father moved with his wife and children first to Leipzig and then to Vienna when Sigmund was 4 years old. Freud remained in Vienna for most of the rest of his life. His precocious genius was recognized by his family, and he was allowed many concessions and favors not permitted his siblings. For example, young Freud was provided with better lighting to read in the evening, and when he was studying, noise in the house was kept to a minimum so he would not be disturbed.

Freud's interests were varied and intense, and he showed an early inclination and aptitude for various intellectual pursuits. Unfortunately, Freud was a victim of nineteenth-century anti-Semitism, which was obvious and severe in central and Eastern Europe. Specifically, his Jewish birth precluded certain career opportunities, most notably an academic career in university research. Indeed, medicine and law were the only professions open to Vienna Jews. Freud's early reading of Darwin intrigued and impressed him to the point that a career in science was most appealing. However, the closest path that he could follow for training as a researcher was an education in medicine. Freud entered the University of Vienna in 1873 at age 17. Because of his interests in a variety of fields and specific research projects, it took him 8 years to complete the medical coursework that normally required 6 years. In 1881, he received his doctorate in medicine. While at the university, Freud was part of an investigation of the precise structure of the testes of eels, which involved his dissecting over 400 eels. He later moved on to physiology and neuroanatomy and conducted experiments examining the spinal cord of fish. While at Vienna, Freud also took courses with Franz Brentano, which formed his only formal introduction to nineteenth-century psychology.

After a 4-year engagement, Freud married Martha Bernays in 1886. Recognizing that a scientific career would not provide adequate support, because anti-Semitism worked against the advancement of Jews in academia, Freud reluctantly decided to begin a private practice. Although he and his wife were very poor in the early years of their marriage, he was able to support her and his growing family, which eventually included six children. Freud's early years in private practice were very difficult, requiring long hours for a meager financial reward at work that basically did not challenge him.

During his hospital training, Freud had worked with patients with anatomical and organic problems of the nervous system. Shortly after starting private practice, he became friendly with Josef Breuer (1842–1925), a general practitioner who had acquired some local fame for his respiration studies. Breuer's friendship provided needed stimulation for Freud, and they began to collaborate on several patients with nervous disorders, most notably the famous case of Anna O., an intelligent young woman with severe, diffuse hysterical symptoms. In using hypnosis to treat Anna O., Breuer noticed that certain specific experiences emerged under hypnosis that the patient could not recall while conscious. Her symptoms seemed to be relieved after talking about these experiences under hypnosis. Breuer treated Anna O. daily for over a year, and became convinced that the "talking cure," or "catharsis," involving discussion of unpleasant and repulsive memories revealed under hypnosis, was an effective means of alleviating her symptoms. Unfortunately, Breuer's wife became jealous of the relationship; what would later be called the positive transference of emotional feelings to the therapist at characteristic stages of therapy looked suspicious to her. As a result, Breuer terminated his treatment of Anna O.

In 1885, Freud received a modest grant that allowed him to go to Paris to study with Charcot for 4 and a half months. During that time he not only observed Charcot's method of hypnosis but also attended his lectures, learning of Charcot's views on the importance of unresolved sexual problems in the underlying causality of hysteria. When he returned to Vienna, Freud gave a report of his work with Charcot to the medical society, but its cool reception left Freud with resentment that affected his future interactions with the entrenched medical establishment.

Freud continued his work with Breuer on hypnosis and catharsis, but gradually abandoned the former in favor of the latter. Specifically, he rejected hypnosis as a treatment with general applicability for three reasons. First, not everyone can be hypnotized; hence, its usefulness is limited to a select group. Second, some patients refused to believe what they revealed under hypnosis, prompting Freud to conclude that the patient must be aware during the step-by-step process of discovering memories hidden from accessible consciousness. Third, when one set of symptoms was alleviated under hypnotic suggestibility, new symptoms often emerged. Freud and Breuer were moving in separate directions, and Freud's increasing emphasis on the primacy of sexuality as the key to psychoneurosis contributed to their break. Nevertheless, in 1895 they published *Studies on Hysteria*, often cited as the first work of the psychoanalytic movement, although it sold only 626 copies during the following 13 years.

Freud came to rely on catharsis as a form of treatment. Catharsis involves encouraging patients to speak of anything that comes to mind, regardless of how discomforting or embarrassing it might be. This "free association" took place in a relaxed atmosphere, usually achieved by having the patient recline on a couch. Freud reasoned that free association, like hypnosis, would allow hidden thoughts and memories to be manifested in consciousness. However, in contrast to hypnosis, the patient would be aware of these emerging recollections. Also ongoing during the course of free association is the process of transference, involving emotionally laden experiences that allow the patient to relive earlier, repressed episodes. Because the psychoanalyst is part of the transference process and is often the object of the emotions, Freud recognized transference as a powerful tool to assist the patient in resolving sources of anxiety.

In 1897, Freud began a self-analysis of his dreams, which evolved into another technique important to the psychoanalytic movement. In the analysis of dreams, Freud distinguished between the manifest content (the actual depictions of a dream) and the latent content, which represented the symbolic world of the patient. In 1900, he published his first major work, *The Interpretation of Dreams*. Although it sold only 600 copies in 8 years, it later went through eight editions in his lifetime. In 1901, he published *The Psychopathology of Everyday Life*, the book in which his theory began to take shape. Freud argued that the psychology of all people, not just those with neurotic symptoms, could be understood in terms of unconscious forces in need of resolution.

When his writings began to win him a reputation as a pioneer in psychiatry, Freud attracted admiring followers, among them Alfred Adler and Carl Jung. In 1909, G. Stanley Hall, president of Clark University, invited him to the United States to give a lecture series as part of that institution's twentieth anniversary. The lectures were published in the *American Journal of Psychology* and later in book form, serving as an appropriate introduction to psychoanalytic thought for American audiences.

Because psychoanalysis was perceived as radical by the medical establishment, early believers formed their own associations and founded journals to disseminate their views. However, Freud's demand for strict loyalty to his interpretation of psychoanalysis led to discord within the movement. Adler broke away in 1911, followed by Jung in 1914, so that by the following year, three rival groups existed within the psychoanalytic movement. Nevertheless, Freud's views continued to evolve. Impressed by the devastation and tragedy of World War I, Freud came to view aggression, along with sexuality, as a primal instinctual motivation. During the 1920s Freud expanded psychoanalysis from a method of treatment for mentally ill or emotionally disturbed persons to a systematic framework for all human motivation and personality.

In 1923, Freud developed cancer of the jaw and experienced almost constant pain for the remaining 16 years of his life. He underwent 33 operations and had to wear a prosthetic device. Throughout this ordeal he continued to write and see patients, although he shunned public appearances. With the rise of Hitler and the anti-Semitic campaign of the Nazis, Freud's works were singled out, and his books were burned throughout Germany. However, Freud resisted fleeing from Vienna. When Germany and Austria were politically united in 1938, the Gestapo began harassing Freud and his family. President Roosevelt indirectly relayed to the German government that Freud must be protected. Nevertheless, in March of 1938 some Nazi thugs invaded Freud's home. Finally, through the efforts of friends, Freud was granted permission to leave Austria, but only after promising to send for his unsold books in Swiss storage so that they could be destroyed. After he signed a statement saying that he had received good treatment from the police, the German government allowed him to leave for England, where he died shortly after, on September 23, 1939.

An Overview of Freud's System

Freud's views evolved continually throughout his long career. The collective result of his extensive writings is an elaborate system of personality development. Freud described personality in terms of an energy system that seeks an equilibrium of forces. This homeostatic model of human personality was determined by the constant attempt to identify appropriate ways to discharge instinctual energies, which originate in the depths of the unconscious. The structure of personality, according to Freud, consists of a dynamic interchange of activities energized by forces that are present in the person at birth. Freud's homeostatic model was consistent with the prevailing view of nineteenth-century science, which saw the mechanical relations of physical events studied by physics as the epitome of scientific inquiry. Freud's model for psychoanalysis translated physical stimuli to psychic energies or forces and retained an essentially mechanical description of how such forces interact.

Freud posited three specific structures of personality—the id, ego, and superego—which he believed were essentially formed by age 7. These structures may be diagrammatically represented in terms of their accessibility to a person's awareness or extent of consciousness, as in Figure 14–1. The id is the most primitive and least accessible structure of personality. As originally described by Freud, the id is pure libido, or psychic energy of an irrational nature and sexual character, which instinctually determines unconscious processes. The id is not in contact with the environment, but rather relates to the other structures of personality that in turn must mediate between the id's instincts and the external world. Immune from reality and social conventions, the id is guided by the pleasure principle, seeking to gratify instinctual libidinal needs either directly, through a sexual experience, or indirectly, by dreaming or fantasizing. The latter, indirect gratification was called the primary process. The exact object of direct gratification in the pleasure principle is determined by the psychosexual stage of the individual's development, outlined below.

FIGURE 14–1 Diagrammatic representation of the structures of personality according to Freud's formulation. The horizontal line marks the boundary between conscious and unconscious processes, which is penetrated only in dreams, under hypnosis, or during free association.

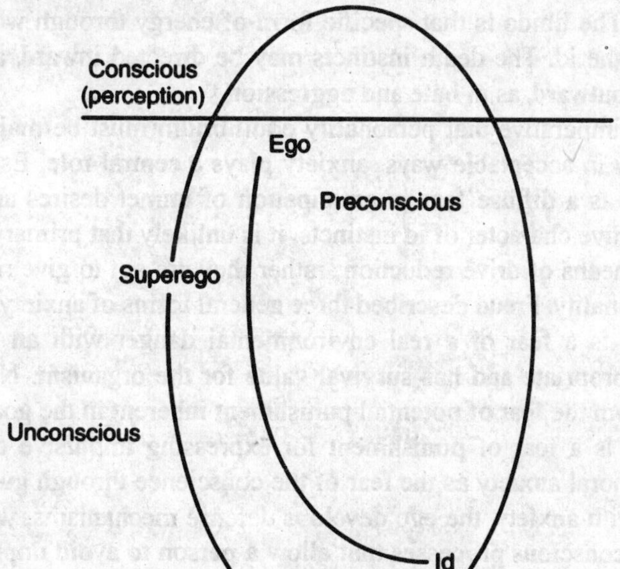

The division or structure of personality that is first differentiated from the id is the ego, often called the "executive" of personality because of its role in channeling id energies into socially acceptable outlets. The development of the ego occurs between the ages of 1 and 2, when the child initially confronts the environment. The ego is governed by the reality principle; it is aware of environmental demands and adjusts behavior so that the instinctual pressures of the id are satisfied in acceptable ways. The attainment of specific objects to reduce libidinal energy in socially appropriate ways was called the secondary process.

The final differentiation of the structures of personality, called the superego, appears by age 5. In contrast to the id and ego, which are internal developments of personality, the superego is an external imposition. That is, the superego is the incorporation of moral standards perceived by the ego from some agent of authority in the environment, usually an assimilation of the parents' views. Both positive and negative aspects of these standards of behavior are represented in the superego. The positive moral code is the ego ideal, a representation of perfect behavior for the individual to emulate. The conscience embodies the negative aspect of the superego, and determines which activities are to be taboo. Conduct that violates the dictates of the conscience produces guilt. The superego and id are in direct conflict, leaving the ego to mediate. Thus the superego imposes a pattern of conduct that results in some degree of self-control through an internalized system of rewards and punishments.

The major motivational construct of Freud's theory of personality was derived from instincts, defined as biological forces that release mental energy. The goal of personality is to reduce the energy drive through some activity acceptable to the constraints of the superego. Freud classed inborn instincts into life (*eros*) and death (*thanatos*) drives. Life instincts involve self-preservation and include hunger, sex, and thirst. The libido is that specific form of energy through which the life instincts arise in the id. The death instincts may be directed inward, as in suicide or masochism, or outward, as in hate and aggression.

With the imperative that personality equilibrium must be maintained by discharging energy in acceptable ways, anxiety plays a central role. Essentially, Freud viewed anxiety as a diffuse fear in anticipation of unmet desires and future evils. Given the primitive character of id instincts, it is unlikely that primary goals are ever an acceptable means of drive reduction; rather they are apt to give rise to continual anxiety in personality. Freud described three general forms of anxiety. Reality, or objective, anxiety is a fear of a real environmental danger with an obvious cause; such fear is appropriate and has survival value for the organism. Neurotic anxiety comes about from the fear of potential punishment inherent in the goal of instinctual gratification. It is a fear of punishment for expressing impulsive desires. Finally, Freud posited moral anxiety as the fear of the conscience through guilt or shame. In order to cope with anxiety, the ego develops defense mechanisms, which are elaborate, largely unconscious processes that allow a person to avoid unpleasantness and anxiety-provoking events. For example, a person may avoid confronting anxiety by self-denial, conversion, or projection, or may repress thoughts that are a source of anxiety to the unconscious. Many defense mechanisms are described in the psychoanalytic literature, which generally agrees that although defense mechanisms are typical ways of handling anxiety, they must be recognized and controlled by the individual for psychological health.

Freud placed great emphasis on the development of the child because he was convinced that neurotic disturbances manifested by his adult patients had origins in childhood experiences. He described psychosexual stages that are characterized by different sources of primary gratification determined by the pleasure principle. Freud wrote that a child is essentially autoerotic. The child derives sexual pleasure from the stimulation of various erogenous zones of the body or by having the mother provide stimulation. Each stage of psychosexual development tends to localize the primary source of gratification to a specific erogenous area. In the oral stage the child seeks primary gratification through sucking, biting, and swallowing. Unsatisfied needs at this stage result in excessive mouth habits, and Freud believed that optimism, sarcasm, and cynicism are all adult behaviors attributable to incidents at this stage. From oral gratification the child progresses to a stage where anal gratification, associated with processes of elimination, is primary. Freud talked about neat, overly clean, and compulsive adults as not having successfully resolved their anal needs. The anal stage is succeeded by the phallic stage of infantile sexuality (ages 3 to 7) where the primary source of gratification is attached to the penis for boys, and for girls, according to Freud, whatever symbolically represents a penis. Following the phallic stage the child enters into an asexual latent period that lasts until the onset of puberty. However, during these psychosexual stages the child is also moving through the Oedipal cycle to eventual adult (that is, what Freud considered appropriate) sexual behavior. At the initial phases of the Oedipal cycle, the young boy has strong sexual desires for his mother. Gradually, this desire is suppressed as the child comes to fear the father and experiences the neurotic anxiety of castration if the father knew of the boy's desires. The boy then identifies with the father as he moves through the latent period and into the pubertal genital stage with a masculine identification. An unresolved Oedipal cycle results in an Oedipus complex—a maladaptive sexual outlook. Thus Freud viewed homosexual behavior as immature sexuality reflecting unresolved Oedipal urges. Later writers attempted a similar description for girls, termed an Electra complex. Corresponding complex psychosexual development in young girls, according to Freud, was complicated by penis envy—the repressed wish to possess masculinity.

This overview of Freud's detailed and complicated theory reflects many of the difficulties with Freud's evolving system. He had few qualms about modifying his views (as long as he did the changing), but often did not change the language, so that the same terminology was frequently used in several ways. Nevertheless, the sheer complexity and uniqueness of his system formed a remarkable achievement. Indeed, he often had to invent new terminology to express his thoughts, and these terms have become an accepted part of our vocabulary.

Freud's Legacy

Relative to the rather single-minded development of psychology as an empirical discipline, whether studied rigorously using an experimental method or less rigorously but still systematically, as in the phenomenological method, Freud's system is most vulnerable. Quite simply, Freud was not a methodologist. His data collection was unsystematic and uncontrolled, consisting primarily of what Freud remembered his patients telling him. He made no independent attempt to confirm the accuracy of his patients' reports. Freud offers only his conclusions; he never discloses how his inferences and conclusions were derived. His variables and constructs are unclear. They are only loosely defined and not quantifiable. His theory emphasizes childhood to the point of asserting that personality is essentially formed by age 7. Yet the only record of Freud's having studied a child concerned a young boy with a phobia, and then Freud worked indirectly through the boy's father. Freud's emphasis on childhood appears to be a deduction from his observations of adults.

Perhaps more seriously, Freud's theory has little predictive value. We examine the role of theory in Chapter 15, but at this point it is important to acknowledge that predictions from a theory permit modification of ongoing events. For example, the theory that cigarette smoking causes cancer allows us confidently to advise younger people not to smoke cigarettes. The person for Freud is over determined. Since a need may be satisfied through any one of many goals, it is difficult to predict adult adjustment by observing the behavior of a child. We just have to wait and see. This criticism is perhaps especially pertinent to Freud, since he advocated both a theory of personality and a treatment for personality disturbances.

As mentioned earlier in this chapter, psychoanalysis has a unique position in the history of psychology. Freud did not develop a theory that generated testable hypotheses or other empirical implications. Yet, on another level, Freud accomplished what few other theorists have: He revolutionized attitudes and created a new set for thinking about personality. It may be said of Freud that his powers of observation allowed him to be right for the wrong reasons. The findings of other more empiricistic theories of personality disturbance have often confirmed many of Freud's observations. If his views do not meet the criteria of empiricistic study, they nevertheless mark a person of genius and insight, whose influence pervades people's thinking about themselves in ways that few others have achieved.

THE DISCIPLES

The psychoanalytic movement was largely the invention of Freud, and his influence far exceeds that of his early followers who subsequently tried to modify psychoanalysis. The major principles of psychoanalysis were redefined and reinterpreted until, by 1930, the movement was fragmented into competing views. Nevertheless, those writers who departed from Freud's speculation retained the basic model of psychoanalysis that conceived of personality in terms of an energy reduction system.

Alfred Adler

Alfred Adler (1870–1937) was born into a wealthy Viennese family but had an unhappy childhood. He was often sick and his accomplishments, compared to those of his oldest brother, failed to fulfill his parents' expectations. He received his doctorate in medicine in 1895, and by 1902 was regularly attending Freud's weekly meetings. He was one of Freud's earliest followers and accompanied him to America in 1909. Gradually, however, Adler began to criticize Freud openly, especially for Freud's emphasis on and literal interpretations of sexuality. By 1911 their split was final, and Adler formed his own circle of followers. Adler was brilliant and a superb speaker who attracted many students with his dynamic and charismatic personality. He lectured widely and in 1934 became a permanent resident in the United States, teaching at Long Island College of Medicine. He continued his extensive lecture trips, and died while on a speaking tour in Europe, in Aberdeen, Scotland, in 1937.

Adler rejected the rigidity of Freud's system. Adler argued, for example, that penis envy in females should not be taken literally, but should rather be viewed as symbolic jealousy of male dominance in society. Indeed, Adler believed that a denial of femininity would be necessarily neurotic. Gradually, Adler developed an alternative to Freud's views, but remained within the psychoanalytic model. His "individual psychology" was not as detailed as the exposition of Freud's theory. Rather, Adler offered a general view of human activity that acknowledged the inferior state of the individual at birth, resulting in the person's continually striving for positive feelings and perfection. By defining a personalistic psychology of the individual, Adler's holistic view of personality emphasized the individual's need for self-unity, perfection, and specifically designed goals. Motivation, in Adler's theory, was not the negative "push" of drive reduction, as Freud had stated, but the positive "pull" of individual striving for self-improvement and superiority. Adler's orientation is similar to the outline of

ALFRED ADLER (1870–1937).
(Courtesy, New York Public Library.)

personality offered by Brentano. The unity of personality is the product of the individual's efforts, so that all psychic phenomena truly originate in the unique creative forces of the individual. The mind itself exhibits definite tendencies to strive for superiority and to attain perfection. Thus, Adler's motivational principle is not reduced to biological instincts, but rather described in psychic, almost spiritual terms of the mind. The striving for superiority, in turn, is a direct reaction to childhood feelings of inferiority, imperfection, and incompleteness. For Adler, a person's present state is guided by his or her future expectations of perfection. These expectations, described as a "finalism" by Adler, are fictional because they are not attainable, but they serve as the collective expression of lifelong goals. Thus, a person's existence is reflected in individual living within a social context and seeking personal harmony in efforts to reach a sense of superiority.

Adler offered case studies of appropriate individual lifestyles that illustrate compensations for inferiority in striving for superiority. However, he was most specific and detailed in his teachings about child rearing. He believed that birth order and the family constellation dramatically affect the development of the individual's lifestyle and creative self. Adler viewed the family as the primary agent of individual socialization, in that subsequent critical patterns of behavior depend on successful upbringing.

Adler introduced an emphasis on the social and creative aspects of human experience into psychoanalysis, and moved away from Freud's rigid emphasis on energy reduction. Adler influenced other important theorists in the psychoanalytic movement, most notably Horney and Fromm. However, the same criticism leveled at Freud—a lack of empiricistic referents and the questionable predictive values of the theory—applies to Adler as well. Moreover, Adler's vagueness, frequent inconsistencies, and lack of a detailed theory of development contribute to an elusiveness that makes him perhaps further removed from an empirical approach than is Freud. Nevertheless, Adler added to psychoanalysis a commonsense approach that made Freud's psychodynamic model more attractive while keeping the model itself intact.

Carl Jung

One of the most fascinating and complicated scholars of this century, Carl Jung (1875–1961) was born to a poor family in a northern Swiss village. He managed to gain entrance to the University of Basel and received a doctorate in medicine in 1900. Jung spent most of the rest of his life in Zurich, teaching, writing, and working with patients. After reading *The Interpretation of Dreams* in 1900, Jung began corresponding with Freud and finally met him in 1907. With Adler, Jung accompanied Freud to America in 1909, where he also lectured and introduced his own work to American audiences. However, Jung began to apply psychoanalytic insights to ancient myths and legends in a search for the key to the nature of the human psyche. Such independent thinking did not meet with Freud's approval, and there is also some speculation that Jung made a critical analysis of Freud's personal life that may have contributed to tensions between them. Freud secured the post of the first president of the International

Psychoanalytic Association for Jung in 1911, but by this time their rift was beyond healing. Finally, in 1914, Jung withdrew from the association and severed all interactions with Freud. Jung continued his own interpretations of psychoanalysis and made several expeditions to study primitive societies in the western United States, Africa, Australia, and Central America. His prolific writings on subjects ranging from anthropology to religion provided novel insights to age-old problems of human existence from a psychoanalytic perspective.

Jung's "analytic psychology" redefined many Freudian concepts; however, Jung retained Freud's terminology, and as a result the same terms often carry different meanings. Jung, like Freud, believed that the central purpose of personality is to achieve a balance between conscious and unconscious forces within the personality. However, Jung described two sources of unconscious forces. One is the personal unconscious, consisting of repressed or forgotten experiences and similar to Freud's preconscious level. The contents of the personal unconscious are accessible to full consciousness. Jung's personal unconscious held complexes, which were groups of feelings with a defined theme that give rise to distorted behavioral responses. For example, a boy who repressed negative emotions about his mother could become an adult with a mother complex, experiencing intense feelings and anxieties when images or stimuli associated with motherhood are encountered. The second source of unconscious forces, unique to Jung's theory, is the collective unconscious, a more powerful source of energy that contains inherited contents shared with other members of an ethnic or racial group. As the personal unconscious has complexes, the collective unconscious has archetypes, defined as primordial images evolved from a primitive tribal ancestry of specific experiences and attitudes passed on over centuries. Jung listed such archetypes as birth, death, unity, power, God, the devil, magic, the old sage, and the earth mother. The notion of a collective unconscious in personality that provides the individual with patterns of behavior, especially at times of life crises, fits well with Jung's preoccupation with myths and symbols. Jung believed that the adequacies of a society's symbols to express archetypal images are an index of the progress of civilization.

Jung focused on the middle years of life, when the pressures of sexual drives supposedly give way to anxiety about the more profound philosophical and religious issues of the meaning of life and death. By reinstating the notion of the spiritual soul, Jung argued that the healthy personality has realized the fullness of human potential to achieve self-unity and complete integration. According to Jung, this realization occurs only after the person has mastered obstacles during the development of personality from infancy to middle age. Failure to grow in this sense results in the disintegration of personality. Accordingly, the person must individualize experiences to achieve a "transcendent function" by which differentiated personality structures are unified to form a fully aware *self*.

Jung redefined libidinal energy as the opposition of introversion–extraversion in personality, bypassing Freud's sexual emphasis. Extraversion forces are directed externally at other people and the environment, and they nurture self-confidence. Introversion leads the person to an inner direction of contemplation, introspection, and stability. The opposing energies must be balanced for the proper psychological functioning of sensation, thinking, feeling, and intuition. An imbalance between extraversion and introversion is partly compensated for in dreams. Indeed, for Jung dreams have important adaptive value in helping the person maintain equilibrium.

As Jung grew older, his writings increasingly came to emphasize mysticism and religious experiences, domains usually ignored by mainstream psychology. Of all the early founders of psychoanalysis, Jung held views in sharpest contrast to those of empiricism. However, he offered a unique treatment of critical human issues that had not been systematically studied by psychologists and still remain in the realm of speculative philosophy. Perhaps Jung was more of a philosopher than a psychologist, and he provoked and confronted issues not readily accommodated in other systems of psychology.

Karen Horney

Born in the German city of Hamburg, Karen Horney (1885–1952) received a medical degree from the University of Berlin in 1913. She was associated with the Berlin Psychoanalytic Institute from 1918 to 1932, and followed the traditional interpretations of Freudian psychoanalysis, having herself been analyzed by Karl Abraham and Hanns Sachs, renowned in Europe for training Freudian psychoanalysts. In 1932, she was named associate director of the Chicago Psychoanalytic Institute and, while there, began to develop a more independent position within the psychoanalytic movement. Two years later she moved to New York City, where she maintained a private practice and taught at the New School for Social Research (New School University). After a few years she was accused of departing radically from orthodox psychoanalysis and was ejected from the New York Psychoanalytic Society. She then founded the American Institute for Psychoanalysis, which she headed until her death.

Horney made significant contributions to the development of a psychology of women. As is clear from the overview of Freud's theory, he reflected his times by giving prime consideration to the goal of attaining an equilibrium of sexual and aggressive energies in human development. Horney, like Adler, rejected such constructs as penis envy as being social standards. Moreover, she offered important insights into the rapidly changing role of women in industrialized society, recognizing that women who had historically been subjected to the repressive burdens of traditional peasant society were undergoing radical changes in the urban work environment.

Horney's views were not acceptable to the entrenched psychoanalytic organization dominated by Freudians. However, despite Horney's expulsion from the psychoanalytic establishment, her revision of Freud's theory remained within the psychoanalytic model. She agreed that human activity is caused by unconscious motivations, and she recognized the primacy of emotional drives. She also saw the value of Freud's description of the defense mechanisms, and shared the Freudian emphasis on transference, free association, and dream analysis in therapy. Although she perceived herself as more of a therapist than a theorist, she asserted important differences with Freud in the structure of personality. She denied the strict distinction and compartmentalization represented by Freud's idea of the id, ego, and superego. If the Oedipus complex exists, according to Horney, it is not a sexual and aggressive interaction between the child and parent, but rather an emotional interplay of anxiety resulting from feelings of insecurity in the child because of rejection, hurt, and overprotection. Horney described libidinal energy in terms of emotional drives rather than the primarily sexual and aggressive energy proposed by Freud. Sexual problems, for Horney, were an effect, not a cause, of emotional distortions.

Horney stressed basic anxiety arising from childhood insecurities that continue throughout life. She argued that humanity has lost the security of medieval society, and neurosis is the natural product of industrialization. Accordingly, psychology is intimately linked with cultural and social values. The total experience that an individual accumulates in life is termed the "character structure," the product of continual development. Horney believed the individual has a great capacity for inner directedness, which may be fully explored through self-analysis, which in turn yields self-knowledge, the prerequisite for psychological growth. The process of proper self-analysis results in the emergence of a strong self-concept, a construct somewhat similar to Freud's idea of the ego ideal. The well-integrated self-concept can effectively combat overreliance on the defense mechanisms that mitigate against self-knowledge by alienating people from themselves. When this alienation does occur, people need the assistance of professional analysis to regain judgment and spontaneity.

Horney described human activity in terms of three modes that are essentially protective and defensive. "Moving toward" is characteristic of infantile behavior and helplessness. For example, if another person is perceived as loving me, then that person will not hurt me. Characteristic activity in adolescence is "moving against"; it is hostile and attempts to control. For example, if I have power, no one will hurt me. Finally, "moving away" is characteristic of isolated adult behavior—if I withdraw, no one can hurt me. These modes of activity are used in pursuing ten neurotic needs described by Horney:

Neurotic Need	***Mode of Activity***
1. Affection and approval	Moving toward
2. A dominant partner in life	Moving toward
3. Seeking a narrowly confined life	Moving away
4. Self-sufficiency and independence	Moving away
5. Perfection	Moving away
6. Power	Moving against
7. Exploiting others	Moving against
8. Prestige	Moving against
9. Ambition	Moving against
10. Personal admiration	Moving against

These neurotic needs can only be overcome by self-analysis. Accordingly, Horney viewed therapy in marked contrast to Freud, who regarded the goal of therapy as the restoration of equilibrium in personality. Rather, Horney believed that the aim of therapy ultimately supports psychological health defined in terms of the continuing process of seeking self-knowledge.

Horney criticized Freud for limiting his observations to children and hysterical women, yet she confined her own perspective to an urban environment. This emphasis leaves her theory without an acceptable concept of normality. According to Horney, personal conflicts do not arise internally but are the product of cultural determinants from industrialization. This weakness, however, is also a strength of her version of psychoanalysis. She recognized the radically changing nature of the social environment and emphasized its great impact on the psychology of the individual. Accordingly, her views are not static but are adjusted to meet the changing demands of society on the roles of both men and women.

SOCIAL PSYCHOANALYSIS

The modifications of Freudian psychoanalysis proposed by Adler and Horney led to a distinct trend in psychoanalysis toward an examination of the social setting of human experience. Two particular theorists, Sullivan and Fromm, are especially representative of this development.

Harry Stack Sullivan

Born in rural New York State, Harry Stack Sullivan (1892–1949) received his medical degree in 1917 from the Chicago College of Medicine and Surgery. Beginning in 1922, he worked in several hospitals conducting research on schizophrenia. In 1933, he became director of a psychiatric foundation, and from 1936 until his death headed its training institute, the Washington School of Psychiatry. He published only one book, *Conceptions of Modern Psychiatry* (1947), but kept extensive notebooks on his work, which were edited and published by his former students after his death. These works form the sources of Sullivan's interpersonal theory of psychiatry.

Sullivan viewed personality, or the self, as an open system interacting with the environment, so that at any given time the individual is defined as the sum of these interacting experiences. Although reminiscent of field theory such as that of Lewin, Sullivan's formulations nevertheless fit into the psychoanalytic movement because he accepted a homeostatic model of anxiety reduction. Tensions emerge from needs and anxieties and require reduction. In his views on development, Sullivan defined several stages that are marked by the nature of social interactions. He suggested various "dynamisms," or propelling social relationships, used as the individual matures to appropriate socialization in adulthood and develops self-esteem.

Sullivan's interpersonal psychology was firmly based on detailed observation, and his views gained wide acceptance because of their specificity as well as their applicability to clinical settings. In many respects Sullivan extended the work of Adler, making a more complete study of the social potential of psychoanalytic theory.

Erich Fromm

Erich Fromm's (1900–1980) idealistic theory is an interesting combination of the psychoanalytic model with existential overtones. He was born in Frankfurt am Main and received his PhD from the University of Heidelberg in 1922, after which he studied at the Psychoanalytic Institute in Berlin. In 1934, he went to America, and taught at several universities in the United States and Mexico.

Fromm consistently stressed the existential view of modern persons as lonely and alienated from themselves and from society (existential influences in psychology are reviewed in Chapter 16). Consistent with Horney's view of the individual searching for security, Fromm viewed the modern world as leaving the individual in a state of loneliness and helplessness. In order to deal with this condition, a person may attempt escape. The methods of escape, analogous to Freud's defense mechanisms, are not satisfactory. Rather, Fromm believed that the essential human freedom of the individual is the key to fulfilling personal needs. Fromm argued that human progress has resulted in five basic needs that go beyond the biological needs of hunger, sex, and thirst.

1. We have a need for relatedness, to establish interpersonal relationships through love and understanding.
2. We have a need for transcendence, to develop the uniquely human capacity of rational and creative thinking.
3. We have a need for rootedness, to belong and become a part of the environment.
4. We have a need for personal identity, to distinguish ourselves from our surroundings.
5. We have a need for a consistent orientation that allows us to understand ourselves and our environment.

Fromm taught that neither capitalism nor communism has succeeded in providing the appropriate social structure for truly human development. He offered his own ideas on a utopia that could facilitate individual growth to meet the five needs. As Fromm continued to develop his views, he moved beyond the traditional role of a psychologist to that of a social philosopher. However, he did try to adjust the psychoanalytic model to respond better to the fluid nature of social change and to recognize the individual dilemma of the modern person attempting to live in a hostile environment.

CONTEMPORARY IMPACT

As mentioned at the beginning of this chapter, psychoanalysis is a unique movement in psychology. It grew out of the same German model of mental activity that produced act psychology and the Gestalt movement. However, psychoanalysis received its immediate expression through the needs of the mentally ill. It was a clinical, not an academic, development. For this reason, psychoanalysis, especially as proposed by writers after Freud, gives the impression of an ad hoc movement that develops as particular problems arise and not a coherent system. Psychoanalysis did not adhere to the commitment to methodology expressed in those systems generated by academic research. Accordingly, there was and is little interaction between psychoanalysis and those systems with comprehensive methods, either empirical or phenomenological. Stated quite simply, psychoanalysis and the other expressions of psychological models do not speak the same language.

The selection of post-Freudian contributors to the psychoanalytic movement presented in this chapter was not intended to be comprehensive, only representative. However, this myriad of psychoanalytic views also reflects the problem of unsystematic methodology. Psychoanalysis has never formulated systematic criteria against which new interpretations may be compared. In a very real sense, there are as many psychoanalytic theories as there are psychoanalysts. This problem plagued the movement in Freud's time and continues to do so. Contemporary psychoanalysis is severely fragmented.

Although not accepted by mainstream psychology, psychoanalysis did assume a dominant role in psychiatry. This is understandable in light of the origins of psychoanalysis as a response to clinical problems. Indeed, psychoanalytic writings enjoyed an almost exclusive position in psychiatry and clinical psychology until the 1960s, when behavior modification began to compete as an alternate model of therapy.

Psychoanalysis also continued to exert a marked influence on art, literature, and philosophy. This influence reflects the major contribution of Freud: his comprehensive analysis of the unconscious. Accordingly, literary and artistic expressions are interpreted in light of the unconscious activities of the artist as well as the unconscious impressions of the perceiver. Psychologists may choose to ignore unconscious motivations or simply to refer to subliminal or subthreshold activities. However, any truly comprehensive theory of psychological activity can no longer be limited to conscious aspects of behavior. Although psychologists may disagree with Freud's interpretation, he did identify some dynamic processes that influence the activity of the individual, processes that psychology cannot ignore.

CHAPTER SUMMARY

The psychoanalytic movement introduced the study of unconscious processes that influence human activity. The movement was fully consistent with the German model of mental activity, going back to the writings of Leibniz and Kant. Although act psychology and the Gestalt movement were also modern expressions of the German model, psychoanalysis emphasized the goal of a homeostatic balance of unconscious energies within personality. Its founder, Sigmund Freud, used his keen powers of observation to devise much-needed therapeutic approaches, and later expanded his formulations to a psychodynamic theory of personality growth dependent on tension reduction. Other theorists modified Freud's model to include cultural influences (Jung) and social needs (Adler and Horney). In addition, scholars have integrated the psychoanalytic model with a field approach (Sullivan) and existential assumptions (Fromm).

As a contemporary movement, psychoanalysis still exerts considerable influence in psychiatry and clinical psychology, although the movement is fragmented owing to a lack of methodological agreement. In addition, Freud's statements on the unconscious have led to new interpretations of artistic expression. However, as a viable model for psychology, psychoanalysis has departed from the empiricistic foundations of psychology and shares little with other systems of psychology that rely on that methodological approach.

BIBLIOGRAPHY

Primary Sources

ADLER, A. (1927). Individual psychology. *Journal of Abnormal and Social Psychology, 22,* 116–122.

ADLER, A. (1956). *The individual psychology of Alfred Adler.* H. L. Ansbacher & R. R. Ansbacher (Eds.). New York: Basic Books.

ADLER, A. (1958). *What life should mean to you.* New York: Capricorn Books.

FREUD, S. (1920). *The psychopathology of everyday life.* New York: Mentor.

FREUD, S. (1938). The history of the psychoanalytic movement. In A. A. Brill (Ed. and Trans.), *The basic writing of Sigmund Freud.* New York: Random House.

FREUD, S. (1955). *The interpretation of dreams.* In J. Strachey (Ed.), *The standard edition of the complete works of Sigmund Freud* (Vols. IV and V). London: Hogarth.

FREUD, S. (1965). *New introductory lectures on psychoanalysis.* New York: W. W. Norton.

FROMM, E. (1941). *Escape from freedom.* New York: Holt, Rinehart & Winston.

FROMM, E. (1947). *Man for himself.* New York: Holt, Rinehart & Winston.

FROMM, E. (1947). *The sane society.* New York: Holt, Rinehart & Winston.

HORNEY, K. (1939). *New ways in psychoanalysis.* New York: W. W. Norton.

JUNG, C. G. (1933). *Modern man in search of a soul.* New York: Harcourt Brace.

JUNG, C. G. (1953). *Psychological reflections* (J. Jacobi, Ed.). New York: Harper & Row.

JUNG, C. G. (1959). *The basic writings of C. G. Jung.* New York: Random House.

SANDLER, J. (Ed.) (1994). *The Harvard lectures of Anna Freud.* Madison, CT: International Universities Press.

SULLIVAN, H. S. (1947). *Conceptions of modern psychiatry.* Washington: W. A. White Foundation.

SULLIVAN, H. S. (1953). *The interpersonal theory of psychiatry.* New York: W. W. Norton.

Studies

ANSBACHER, H. L. (1970). Alfred Adler—A historical perspective. *American Journal of Psychiatry, 127,* 777–782.

ANSBACHER, H. L. (1971). Alfred Adler and G. Stanley Hall: Correspondence and general relationship. *Journal of the History of the Behavioral Sciences, 7,* 337–352.

CAPPS, D. (1970). Hartmann's relationship to Freud: A reappraisal. *Journal of the History of the Behavioral Sciences, 6,* 162–175.

ELLENBERGER, H. F. (1970). *The discovery of the unconscious*. New York: Basic Books.

FORDHAM, F. (1953). *An introduction to Jung's psychology*. London: Penguin.

GAY, P. (1988). *Freud: A life for our time*. New York: Norton.

GRAVITZ, M. A., & GERTON, M. I. (1981). Freud and hypnosis: Report of post-rejection use. *Journal of the History of the Behavioral Sciences, 17,* 68–74.

HALE, N. G. (1971). *Freud and the Americans*. New York: Oxford University Press.

HALL, C. S., & LINDZEY, G. (1970). *Theories of personality* (Rev. ed.). New York: Wiley.

JONES, E. (1955). *The life and work of Sigmund Freud*. New York: Basic Books.

KAINER, R. G. (1984). Art and the canvas of the self: Otto Rank and creative transcendence. *American Imago, 14,* 359–372.

KELMAN, H. (1967). Karen Horney on feminine psychology. *American Journal of Psychoanalysis, 27,* 163–183.

MACMILLAN, M. (1985). Souvenir de la Salpêtrière: M. le Dr. Freud à Paris, 1885. *New Zealand Journal of Psychology, 14,* 41–57.

ORGLER, H. (1963). *Alfred Adler: The man and his works*. New York: Liveright.

RENDON, M. (1984). Karen Horney's biocultural dialectic. *American Journal of Psychoanalysis, 44,* 267–279.

RUBINS, J. L. (1978). *Karen Horney: Gentle rebel of psychoanalysis*. New York: Dial.

SAMUELS, A. (1994). The professionalization of Carl G. Jung's analytical psychology clubs. *Journal of the History of the Behavioral Sciences, 30,* 138–147.

SCHICK, A. (1968–1969). The Vienna of Sigmund Freud. *Psychoanalytic Review, 55,* 529–551.

SIRKIN, M., & Fleming, M. (1982). Freud's "project" and its relationship to psychoanalytic theory. *Journal of the History of the Behavioral Sciences, 18,* 230–241.

STEPANSKY, P. E. (1976). The empiricist as rebel: Jung, Freud, and the burdens of discipleship. *Journal of the History of the Behavioral Sciences, 12,* 216–239.

15

Behaviorism

The system that defines psychology as the study of behavior received firm support in a twentieth-century development that occurred largely in the United States. Observable and quantifiable behavior was assumed to have meaning in itself, rather than simply serving as a manifestation of underlying mental events. This movement was formally initiated by an American psychologist, John Broadus Watson (1878–1958), in a famous paper, "Psychology as the Behaviorist Views It," published in 1913. Watson proposed a radical departure from existing formulations of psychology by asserting that the proper direction for psychology's development is not the study of "inner" consciousness. In fact, he dismissed the entire notion of some nonphysical mental state of consciousness as a pseudoproblem for science. In its place Watson advocated overt, observable behavior as the sole legitimate subject matter for a true science of psychology.

Watson was largely successful in initiating a redirection of the development of psychology. Later in this chapter we examine the various intellectual forces that converged in Watson's time and fostered the acceptance of his views. Although Watson may have been the spokesman for a revolutionary movement defining the scope of psychology, it should be recognized that the subsequent success of the behavioristic movement in psychology is best characterized as evolutionary rather than revolutionary. Behaviorism, especially in the United States, has gradually changed from Watson's initial definition to one that encompasses a wide range of human and infrahuman activity, studied under varieties of empirical methodologies.

The historical trend that led to Watsonian behaviorism may be traced from antiquity to the nineteenth century. The pre-Socratic philosophers, such as the Ionian physicists and Hippocrates (Chapter 2), attempted to explain human activity as mechanical reactions reducible to biological or physical causes. Much later, the French sensationalist tradition, rejecting Descartes' unextended substance in favor of a mechanical system responding to environmental stimuli, served as an important predecessor of twentieth-century behaviorism. Both the sensory reductionism of Condillac and the mechanical physiology of La Mettrie led to the position that mental events are determined completely by sensory input and that the critical level of psychological inquiry concerns sensory processes. Perhaps the British philosophers provided behaviorism with its clearest intellectual foundation. Locke's notion of mental passivity meant that the mind is dependent on the environment for its contents, and the two dominant themes of the British philosophers, empiricism and associationism, contained the major tenets of behaviorism. Behavioristic psychology emerged in the twentieth century as an empirical discipline that studied behavior in terms of adaptation to environmental stimuli. The essential core of behaviorism is that an organism learns behavioral adaptation, and its learning is governed by the principles of association.

A fundamental empirical approach to the examination of associations in behavioristic psychology, although in general agreement with the British philosophers, may be found in the work of a group of predominantly Russian physiologists studying reflexology. Indeed, although important research on the acquisition of reflexes was done prior to Watson's writings, the Russian group had a major impact on behaviorism after Watson's publications and served as a force to expand his original formulation.

IMMEDIATE BACKGROUND OF BEHAVIORISM

Russian Reflexology and Its Dominance in the Soviet Union

We already mentioned in Chapter 10 the advances in the physiology of the brain, most notably in the neurophysiology of Sherrington. In somewhat parallel work during the first years of the twentieth century, a group of Russian physiologists was investigating the physiological basis of behavioral processes. Although Sherrington's work was probably more significant—and indeed it was left for later scientists to examine the full implications of his neurophysiology for behavioristic psychology—the research of the Russian physiologists had a practical direction that was easily adopted into behaviorism as the basic mechanism of learning. However, the Russian researchers were physiologists, not psychologists, and the reduction of psychological processes to physiological mechanisms was inherent in their work. They were not philosophers seeking to articulate a new science of psychology. Rather, they wanted to expand existing knowledge of physiology to include processes that had been labeled psychological. Accordingly, they had little use for a new science of psychology. This tradition continues to this day in Russia and Eastern Europe, where investigations of such processes as learning, sensation, and perception are often included in the study of neurobiology rather than in psychology.

Ivan Mikhailovich Sechenov. Considered the founder of modern Russian physiology, Ivan Mikhailovich Sechenov (1829–1905) received a doctorate in physiology from the University of St. Petersburg. Beginning in 1856, he spent 7 years visiting Western Europe, coming into contact with such eminent physiologists as Hermann von Helmholtz, Johannes Müller, and Dimitri Mendeleyev (1834–1907), the émigré Russian chemist working at Heidelberg. Sechenov held professorships in physiology at universities in St. Petersburg and Odessa; he finished his career in Moscow.

In 1863, Sechenov published *Reflexes of the Brain*, containing his hypothesis that all activities, including the seemingly complex processes of thinking and language, are reducible to reflexes. Moreover, he stressed the excitatory and inhibitory mediational role of the cerebral cortex as the central locus of reflex actions. Sechenov believed that the cause of all intellectual activity, as well as motor activity, involves external stimulation. Thus the entire repertoire of behavior is the result of responses to environmental stimuli, mediated at the cortical level. In a later paper published in 1870, Sechenov dismissed contemporary views of psychology as a collection of surplus concepts that reflected the current state of ignorance of physiology. With further investigation, Sechenov argued, the constructs of psychology would disappear, having been reduced to their proper physiological level of explanation.

Sechenov reduced both psychic and physiological responses to reflexes, so that ideas became associations of reflexes mediated by the central nervous system. Thus, the founder of modern Russian physiology defined *reflexology* as a monistic interpretation of human activity, which equated psychological processes with essential neural processes. Sechenov began an experimental tradition to seek validation of his view of reflexology, which was not very different from that of the successors of Descartes within the French sensationalistic tradition. Interestingly, Sechenov's writings were censored by the imperial government because of their heavy emphasis on materialistic explanations of mental activities. Sechenov did not live to see the advent of Lenin, who established a government based on dialectical materialism, which was more amenable to the reflexology of Sechenov and his successors.

Vladimir Mikhailovich Bekhterev. One of Sechenov's most famous students, Vladimir Mikhailovich Bekhterev (1857–1927), coined the term *reflexology* to describe his work. After studying in St. Petersburg, Bekhterev left Russia to work under Wundt, Du Bois–Reymond, and Jean Martin Charcot, the French neurologist who pioneered the modern use of hypnotism. Bekhterev's interest led him to apply the objective reflexology of Sechenov to psychiatric problems, and in 1907 he founded the St. Petersburg Psychoneurological Institute.

In 1910, Bekhterev published his *Objective Psychology*, which called for discarding mentalistic concepts from the description of psychological events. Although Bekhterev did some innovative experiments on punishment, his major contribution was his extensive writings, which brought greater knowledge and acceptance of reflexology to a wider audience. Moreover, his applications of reflexology to abnormal behavior demonstrated the utility of objective psychology.

Bekhterev was a contemporary and often a rival of Pavlov. Because he was acquainted with Wundt's psychology, he was more sensitive to the issues of concern to psychologists than was Pavlov. Accordingly, his general writings on reflexology earned faster acceptance among psychologists than did the more systematic work of Pavlov.

Rejecting introspection as an acceptable method because it assumes that psychological activity is somehow different from other human activities, Bekhterev stressed the unity of reflexology. Psychological and physiological processes involve the same neural energy, and the observable reflexes, whether inherited or acquired, are governed by lawful relations with internal and external stimulation. The goal of objective psychology is to discover the underlying laws mediating the occurrence of reflexes.

Ivan Petrovich Pavlov. The most comprehensive system of Russian reflexology was offered by Ivan Petrovich Pavlov (1849–1936), whose long productive career was never seriously disrupted despite the upheavals of revolutionary Russia. Pavlov was born in a small town in central Russia, the son of a village orthodox priest. He originally intended to follow his father's vocation, but decided otherwise and went to the university at St. Petersburg in 1870. After several years of tutoring, which provided only a bare subsistence, Pavlov secured a fellowship to the university in 1879 and completed his medical degree in 1883. From 1884 to 1886 he studied in Leipzig and in Breslau (now Wrocław), where he joined a group of scientists investigating pancreatic secretion. In 1890, he became professor of pharmacology at the Military Medical Academy of St. Petersburg, and 5 years later was named professor of physiology. Also in that year, Pavlov helped to start the Imperial Institute of Experimental Medicine and served as its director as well as head of its physiology

IVAN PETROVICH PAVLOV (1849–1936). (Courtesy, New York Public Library.)

department. Along with Marceli Nencki (1849–1901), a pioneering biochemist of Polish origin who had left the University of Berne to lead the biochemistry department of the Institute, Pavlov established a research center of international reputation, moving in the 1930s to new facilities just outside St. Petersburg (renamed Leningrad in 1924 and reverted to St. Petersburg in 1990). Pavlov presided over a vastly expanded institute, and the Pavlovian Institute of Physiology of the Russian Academy of Sciences remains a prestigious center of physiological research on reflexology.

Pavlov was a stern and scholarly person with severe self-discipline. He imposed both his discipline and his rigid expectations on the multitude of students who worked under him during his many productive years. He was a systematic methodologist, for whom data gathering was a serious enterprise. The new laboratory built for Pavlov by Stalin's government was termed the "tower of silence," reflecting both its soundproof construction and the demeanor of the laboratory workers.

Pavlov received the Nobel Prize in 1904 for his work on the neural and glandular bases of digestion. In connection with this research, Pavlov discovered the essential principles of associative conditioning for which he is remembered today. Pavlov had developed a device that he implanted in the cheek of his dog subjects that collected saliva as a measure of digestive processes under investigation. During his careful experimentation, Pavlov noted that the subject reliably salivated in anticipation of receiving food, which was signaled by the approach of an experimenter or the presentation of a food dish. This astute observation launched Pavlov on a program of research that led to the development of conditioning reflexology. He found that he could take a neutral stimulus, such as a metronome beat, a tone, or a light, and after successively pairing it with a primary reward, such as food, a motivated (that is, hungry) dog would respond with salivation to the neutral stimulus presented without the food. He termed the neutral stimulus that acquired the eliciting properties of the primary reward the *conditional stimulus*. In an early translation, *conditional* was rendered as *conditioned*, and so *conditioned stimulus* became the standard terminology. However, the word *conditional* captures Pavlov's meaning better, because he regarded the acquisition of response-eliciting properties as a learned association. To fulfill the criteria of learning, the bond between the conditioned stimulus and the response must be temporary—that is, the bond must be capable of dissipating, resulting in the conditional stimulus losing its response-eliciting properties. Pavlov defined *extinction* as the repeated presentation of the conditioned stimulus in the absence of primary reward, so that the capacity of the conditioned stimulus to elicit the response is diminished. Pavlov specified four experimental events in the acquisition and extinction processes:

1. *Unconditioned Stimulus (US):* an environmental event (such as food) that, by its inherent properties, can elicit an organismic reflex.
2. *Conditioned Stimulus (CS):* an environmental event (such as a tone) that is neutral with respect to the response prior to pairings with the US.
3. *Unconditioned Response (UR):* the natural reflex (such as salivation) elicited autonomically, or involuntarily, by the US.
4. *Conditioned Response (CR):* the acquired reflex (such as salivation) elicited by the CS after association with the US.

Note that the UR and CR are the same response; they differ in their eliciting stimulus and usually in subtle indices of their strength. Pavlov found that different temporal relationships between the CS and the US produced varying rates of acquisition and extinction of CRs. The optimal relationship, making use of the anticipatory response, involves the presentation of the CS just prior to the US and is called *delayed conditioning*:

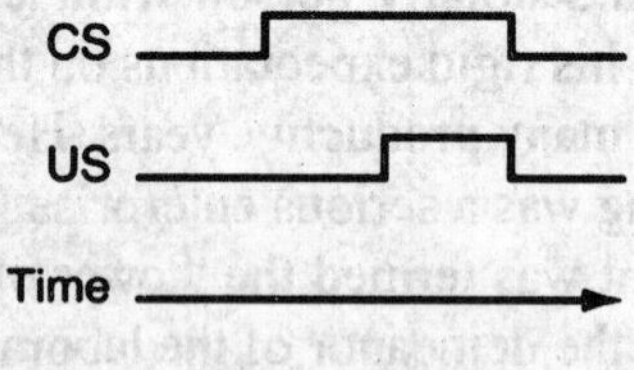

Other temporal relationships also produce conditioning, although at less optimal rates:

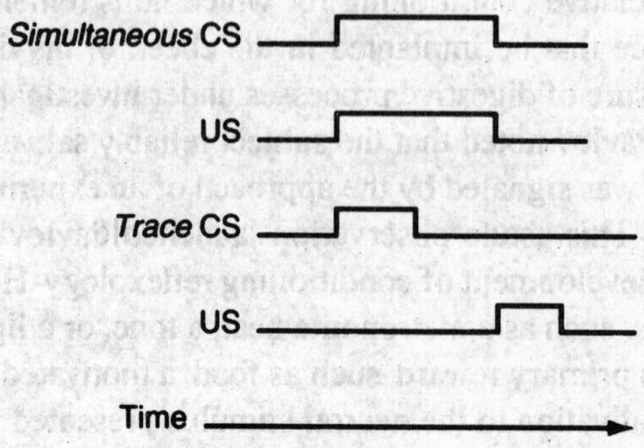

From this basic paradigm, Pavlov drew several principles. First, conditioning procedures represent the quantification and objectification of the acquisition and forgetting of associations. Pavlov had examined under experimental scrutiny the accepted concepts of association theory, discussed by such philosophers as Hume and the Mills, and established what he argued was a complete explanation of the formation of associations on the basis of the materialism of physiological reflexology. In Pavlov's conditioning theory, there was no need for any mental constructs. Rather, the nervous system, and especially the cortex, provided the mechanisms of reflexology. Second, the highly controlled experimental paradigm of conditioning offered the possibility of investigating all of higher nervous activity. Pavlov considered his procedures, involving careful experimenter control of environmental stimuli to produce response changes, to be ideally suited to investigating all types of behavior. Although he later modified his views, Pavlov initially thought that the formation of all associations ultimately involves variants of his basic paradigm. Third, Pavlov was firmly convinced that the temporal relationship, or contiguity, was the basic principle of the acquisition of associations. Again, this view was modified by later theorists, but Pavlov remained adamant that he had discovered the basic form of associations and that all learning reduces to the contiguous relationship between environmental stimuli and the mediational role of the cortex.

Pavlov regularly held laboratory meetings each Wednesday afternoon, presenting coherent overviews of his empirical work. A series of those lectures was published in English in 1927, translated by a former student, G. V. Anrep, under the title *Conditioned Reflexes: An Investigation of the Physiological Activity of the Cerebral Cortex*. In these lectures, representing the first systematic presentation of his views to Western scientists, Pavlov elaborated on his methodological approach to investigating behavioral processes mediated by the cortex. He described the spread of cortical excitation, termed *irradiation*, that results in behavioral generalization among similar environmental stimuli. With his notion of cortical inhibition, Pavlov was able to accommodate discriminative behavior. He discussed such now-common experimental findings as postextinction response elicitation (spontaneous recovery), internal inhibition, and modulation of postasymptotic response levels. In addition, Pavlov described his inducement of "experimental neurosis" in dogs, and devoted five lectures to investigations of cortical pathology.

Pavlov's other works were gradually translated and they, along with the writings of his students, gained him unprecedented eminence in psychology. His reputation remains largely intact to this day. If we rightly acknowledge Pavlov's fundamental position in modern psychology, we pay his memory a paradoxical tribute, for he had little use for psychology. Further, in comparing Pavlov with the founder of American behaviorism, Watson, it is readily obvious that Pavlov's data and interpretations have met the test of time far better. Pavlov, although a physiologist of impeccable competence, was first an experimentalist. He recognized the experimental method as the sole means of finding truth in science. To the extent that behavioristic psychology emulates an experimental approach, Pavlov's objective reflexology remains an honored and unmatched precedent.

American Connectionism: Thorndike

The major American researcher relevant to the precursors of Watsonian behaviorism is Edward Lee Thorndike (1874–1949), although Thorndike is included as a behaviorist somewhat tentatively. Looking at his entire career, which spanned 50 years of productive work in psychology, Thorndike's eclecticism appropriately groups him with the American functionalists. However, his early pioneering work on associations merits study by itself. His experiments on problem-solving behavior resulted in significant findings that were highly regarded at the time of Pavlov and Watson and are still recognized. Thorndike never intended to be a system builder, as Watson did, and his earlier, more theoretical work was later replaced by a shift to more practical problems of human learning and education (see Chapter 12).

Thorndike started graduate work at Harvard under William James and began investigating the intelligence of chickens. However, Cattell offered him a fellowship at Columbia University, and because Thorndike was able to continue his work on animal intelligence there, he accepted Cattell's offer. His doctoral dissertation, *Animal Intelligence: An Experimental Study of the Associative Processes in Animals* (1898), was expanded and published in 1911. In 1899, Columbia University took over the New York College for the Training of Teachers, and Thorndike joined the faculty of the consolidated Columbia Teachers College. He remained there for the rest of his career, pursuing educational issues, especially intelligence testing.

His earlier work on associations concerns us here. Thorndike examined problem-solving strategies in a variety of species, which he tested in puzzle boxes, various chambers designed to reward specific responses. Thorndike was impressed with his subjects' gradual acquisition of successful responses by trial-and-error learning and by accidental success. These observations led him to conclude that there are two basic principles of learning: exercise and effect. The law of exercise stated that associations are strengthened by repetition and dissipated by disuse. Thorndike's original law of effect stated that responses producing reward or satisfaction tend to be repeated, whereas responses producing punishment or annoyance tend to be eliminated. Later, he modified the law of effect to emphasize that reward strengthens associations, whereas punishment results in the subject's moving to another response, rather than weakening the association between the response and the stimulus context. Thus the earlier law of effect postulated that symmetrical reward-and-punishment feedback influences the connective bond; the later, modified version was an asymmetrical statement of the efficacy of reward and relegated punishment effects to a relatively minor role in learning. Thorndike's view of the basis of association differed somewhat from Pavlov's. First, the learning situation was under the control of the subject in Thorndike's procedure; the subject had to emit a response before receiving any reward. Second, the law of effect, or the influences of reinforcement, required recognition of the consequences of a reinforcing event by the subject. Thorndike never satisfactorily explained how reinforcement works. Because the effects presumably feed back to strengthen an associative bond between a response and a stimulus, some mechanism, or principle of realization, is needed for the subject to recognize whether the reinforcement was satisfying or not. This problem, which still plagues reinforcement theory, revolves around the need for the mediation of response-produced effects. Is some postulation of consciousness needed to deal adequately with the judgmental realization in order to act on reinforcement effects? Thorndike suggested that perhaps centers of satisfiers or annoyers exist in the brain. Although this explanation is not supported, Thorndike's principles of repetition and reinforcement, in accounting for learning, remain accepted in current research.

With the hindsight of history, we can recognize that Pavlov and Thorndike were investigating two different paradigms to produce learning, and we review this distinction later in this chapter. However, it is important to note that both Pavlov and Thorndike provided careful empirical documentation of the association process. Although neither directly intended to begin a behavioristic psychology, both contributed to the impulse that resulted in behaviorism. It was Watson who formally proclaimed a systematic context for behaviorism.

WATSONIAN BEHAVIORISM

In 1913, Watson published an article in the *Psychological Review* calling for a behavioristic psychology and changed the course of modern psychology. Watson asserted that a subject's behavior is worthy of study in itself, not for what it might reflect about some underlying state of consciousness. Although he expanded and gave coherence to the arguments favoring the study of behavior rather than consciousness, Watson did not write anything truly original. As mentioned before, the entire French sensationalistic tradition, which reduced mental content to sensory input, established an early version of behaviorism. The reduction of presumed mental events to physical correlates represents a consistent theme in the studies of the French sensationalists, as well as in the later writings of Comte, and finally in Watsonian behaviorism.

The shift from consciousness to behavior as the proper province of psychology received more immediate support from the nineteenth-century evolutionary movement. In particular, Darwin's meticulous observations supporting the principles of evolution by natural selection underscored the importance of the adaptive value of behavior. Supported also by Spencer's hypotheses of "social evolution," behavior as organismic activity came under new scrutiny. The renewed interest in the study of behavior during the latter part of the nineteenth century was immediately translated into an early interest in the value of cross-species comparisons, as we have seen in Chapter 10 from the works of Washburn, Romanes, and Morgan. Watson's emphasis on behavior rather than consciousness was a consistent step toward the development of comparative psychology, resting on the efficacy of homologous and analogous interpretations of behavioral patterns among varying species.

In addition to serving as a catalyst for several converging traditions, Watson's behaviorism provided a strong reaction to the methods of study prevalent in the psychology of consciousness. Watson rebelled against introspection as an appropriate method. Citing the difficulty of finding agreement among introspectionists observing the same processes, he argued that introspection is simply not an objective methodology, and reliance on it would spell disaster for psychology. Accordingly, Watsonian behaviorism "reestablished" the science of psychology. In discarding both the content (consciousness) and the methodology (introspection), Watson was advocating a complete reformulation of psychology.

Watson was born in South Carolina and received his undergraduate degree from Furman University. At the University of Chicago he studied under two important American functionalists, John Dewey and James Rowland Angell. Watson also studied physiology and neurology under H. H. Donaldson and Jacques Loeb and completed his PhD in 1903. Watson's early work on maze learning relied heavily on the methodological practices of physiology, including use of the laboratory rat. In 1908, Watson accepted a position at Johns Hopkins University, where his views on the possibility of an objective psychology took on systematic form as a coherent program. In 1920, Watson divorced his wife and married Rosalie Rayner, his former laboratory assistant. The ensuing scandal forced his resignation from Hopkins, and he never held another academic position. Rather, he applied his expertise to using psychology in advertising with some success, and wrote popular expositions of psychology. Accordingly, Watson's contributions to systematic psychology were completed by the middle 1920s, when he was in his early forties, an age when many scientists are only beginning their most productive period.

Watson's views centered on the premise that the province of psychology is behavior, measured in terms of stimulus and response; accordingly, psychology is concerned with the peripheral elements of stimulus and response impinging on the organism. Every response is determined by a stimulus, so that behavior may be completely analyzed by means of the causal relationship between stimulus and response elements. Watson did not deny the possible existence of central mental states, such as consciousness, but believed that because such alleged central states are nonphysical and cannot be studied scientifically, they are pseudoproblems for psychology.

Watson's view of the nature of the stimulus–response relationship—that is, association—relied primarily on the principle of frequency, or exercise, and secondarily on the principle of recency. Increasingly, he favored the conditioning reflexology of Pavlov and the puzzle box methods of Thorndike. However, Watson never fully appreciated the nature of reinforcement, and was especially skeptical of Thorndike's law of effect, which he criticized as resting on mentalistic inferences, without empirical support. Nevertheless, Watson did have faith in the principles of association as the key to psychological (behavioral) growth, although he acknowledged that his own theory of learning was largely inadequate. Accordingly, all behaviors—locomotive, perceptual, emotive, cognitive, and linguistic—are complexes or sequences of associative stimulus–response bonds.

The parsimony of Watson's proposal for psychology was most appealing. In a debate with William McDougall, later published as the *Battle of Behaviorism* (1929), Watson conceded that McDougall's call for accepting data from varieties of sources to gain a complete view of an individual has an attraction. Yet as soon as a scientist accepts data other than behavioral, the scientific clarity of the investigation begins to deteriorate. Indeed, the use of behavior, defined in terms of stimulus and response elements, gives psychology a unity because of the possibility of consensus yielded through objective observations, and this consensus provided an alternative to the introspective method of structuralism. Accordingly, Watson's system, focusing on behavioral adaptability to environmental stimuli, offered a positive, objective science of psychology, and by 1930 behaviorism was the dominant system of American psychology.

The major criticism of Watsonian behaviorism may be summarized under two points. First, this initial version of behaviorism restricted psychology by limiting behavior solely to the peripheral events of stimulus and response elements. In dismissing mental events, Watson also ignored physical, central mediation of stimulus and response bonds. He seems to have recognized the need for a more thorough elaboration of internal, central mediation by his appreciation of Pavlov. However, by the time Pavlov's views became better known, Watson had been removed from academia and was unable to pursue an integration of his views with those of Pavlov. It was left to his successors to modify the scope of behavioristic psychology by admitting the functions of central mediation, physiological as well as cognitive, to scientific scrutiny.

The second problem with Watsonian behaviorism concerns the issue of reductionism. We can probably say that in 1913 psychology lost its mind. The behavioristic strategy took those functions that had been reserved for the mind since the time of Descartes' speculation and reduced them to behavior. Behavior, in turn, was reducible to environmental stimuli and observable responses. Although Watson did not greatly elaborate on the specifics of the reduced level of stimuli and responses, the logic of Watson's approach is to suggest that behavior really reduces to physics and physiology. As noted, Watson's successors opened the behavioral system and rescued some of the discarded mental functions. Nevertheless, Watson's behaviorism was reductionistic. Carried to an extreme, such reductionism questions whether behavior per se possesses the integrity to warrant a separate and distinct science. On the one hand, if mental functions are added back to psychology, then psychology would once again become a metaphysical, not an empirical, discipline. On the other hand, if psychology is reduced to peripheral stimuli and responses, then it is equated with physics and physiology. So, although Watson's call to behavioristic psychology offered simplicity and clarity, the authenticity of a truly behavioral level of investigation remained questionable.

Other scholars, contemporaries of Watson, contributed to the formulation and acceptance of behaviorism. Like Thorndike, however, they should be considered as transitional figures between American functionalism and behaviorism. They did not intend the systematic upheaval envisioned by Watson, nor were they as committed to the theoretical implications of behaviorism. Nevertheless, their research and attitudes toward emerging behaviorism were critical to its eventual success.

Other Early American Behaviorists

Edwin B. Holt. After completing a doctorate at Harvard in 1901 under the guidance of William James, Edwin B. Holt (1873–1946) taught at both Harvard and Princeton, but devoted most of his career to writing. The titles of his major works,

The Concept of Consciousness (1914), *The Freudian Wish and Its Place in Ethics* (1915), and *Animal Drive and the Learning Process* (1931), all reflect his major contribution to behaviorism, which was to infuse the notion of purpose or motivation in behavior for a more complete system. Holt did not accept the equation of behaviorism and reflexology inherent in the positions of both Pavlov and Watson. Instead of the reduction of behavior to constituent elements, Holt argued that behavior has purpose. Moreover, for Holt, behavior can only be understood from the perspective of the pattern of behavioral acts and sequences of acts. For the psychologist, behavior is more than the sum of stimulus–response bonds. Holt looked to other psychological models emphasizing motivational principles, such as Freudian psychodynamics and theories of instinctual drive, to examine how those views might present a more holistic context for behaviorism. One of his students, Edward Tolman, reviewed later in this chapter, followed Holt's lead and developed a comprehensive cognitive model of behaviorism.

Albert P. Weiss. After immigrating to America as a child, Albert P. Weiss (1879–1931) studied at the University of Missouri under Max Meyer, who had been a student of Stumpf at Berlin. On receiving his doctorate in 1916, Weiss went to the Ohio State University and remained there for the rest of his brief career. His major work, *A Theoretical Basis of Human Behavior* (1925), attempted to deal with many of the complex human activities that were ignored or glossed over by Watson. Weiss concluded that psychology may best be understood as a biosocial interaction; that is, all psychological variables are reducible to either physicochemical or social levels, prompting Boring (1950) to remark that Weiss was a curious mixture of La Mettrie and Comte. However, by modifying reductionistic reflexology through consideration of socially based motivation, Weiss allowed psychology to deal better with complex forms of activity. Accordingly, the integrity of a psychological level of scientific inquiry of behavioral processes was greatly enhanced.

Walter S. Hunter. After receiving his degree from the Chicago functionalist school in 1912, Walter S. Hunter (1889–1954) taught at several universities, then settled at Brown University in 1936. He earned a reputation as a respected researcher rather than a theoretician, and worked mainly on problem-solving behavior in mammals. Some of the behavioral tasks that developed from his experimental work, such as delayed response and double alternation behavior, were assumed to be representative of higher-order problem solving and remain in use today. Interestingly, Hunter, like other behaviorists, disliked the use of mentalistic terms so prevalent in German psychology, and proposed *anthroponomy* as a better term than *behaviorism* to replace *psychology*.

Karl S. Lashley. Karl S. Lashley (1890–1958) was one of the few students to study with Watson during Watson's brief career at Hopkins. After completing his degree in 1915, Lashley taught at several universities before finally affiliating with the Yerkes Laboratory of Primate Biology in 1942. Lashley was a physiological psychologist, introducing the critical role of the physiological correlates of behavior. Moreover, his productive laboratory work served as a model for many psychologists, so that behaviorism became permanently linked to physiological investigation. However, it is important to contrast Lashley's physiological behaviorism with Pavlov's reflexology. Although followers of both scientists might very well have done similar experiments leading to related conclusions, their reasons for doing the experiments were different. Under Pavlov's reflexology, there was no true behavioral level of investigation per se. Rather, presumed psychological events were explained completely through physiological causes. For Lashley, however, the integrity of observable behavior was assumed, then the physiological substrates were investigated. He did not equate the psychological and physiological. Rather, depending on the complexity of the problem under study, he considered the physiological level an explanatory component of psychological events. Thus the integrity of each level was retained.

Operational Positivism

A movement that supported the success of behaviorism was initiated in physics and had wide-ranging influence in all of the sciences. This movement, generally called operationism, was a twentieth-century expression of positivism. In the United States, a Harvard physicist, Percy W. Bridgman, influenced by the work of a group of physicists centered in Copenhagen, published *The Logic of Modern Physics* (1927), in which scientific concepts were defined by the operations employed to observe them. By implication, the concept was equated with the operations, nothing more or less. Any concept that presumably could not be defined operationally was for Bridgman a pseudoproblem—that is, the concept did not have scientific value.

At the same time in Vienna, a group of philosophers was formalizing a broader version of positivism, closely linked to Bridgman's operationism. This group derived its immediate ancestry from the teachings of Ernst Mach and came to be known as the Vienna Circle of Logical Positivists. It sought to supplement Mach's views with contributions from contemporary developments in philosophy and logic. Logical positivism is a comprehensive philosophy of science. This movement essentially stressed the unity of all science because, studied by the methods of empiricism, all science is ultimately physical. Accordingly, all truly scientific issues may be examined through a common language derived from physics and expressed in terms of operationism. Until its members' dispersal in the late 1930s, the Vienna Circle remained a spirited group that aimed to unify science on the basis of the operational character of scientific problems.

The expression of operationism in psychology was an attempt to resolve the conflict between the empirical traditions of psychology and the prevailing metaphysics of the psychology of consciousness. By reinforcing a radical empirical stand within all of science, behaviorism was the sole system within psychology at that time to act as vehicle for logical positivism with its concomitant operationism. The reductionism inherent to the description of psychological events in terms of stimulus and response elements fits nicely into the movement. The movement of the logical positivists and operationists suggested a behavioristic model of psychology and, similarly, behaviorism supported the recognition of the unity of science expressed in an operational approach to sciences. The net result for psychology of this interchange of forces was the further entrenchment of behaviorism.

It is important to try to appreciate the full impact of the development of behaviorism on psychology. Behaviorism fundamentally offered the opportunity for a truly scientific psychology based on an empirical approach, just as in the physical and life sciences. By turning away from the elusive nature of consciousness, behavioral psychology permitted the study of its subject matter through a methodology that had proved successful for nineteenth-century science. Looking back on the historical development of psychology from 1870, Watsonian behaviorism was a bold attempt to reformulate and reestablish the science of psychology.

BROADENING BEHAVIORISM

After Watson's initial formulation of behavioristic psychology and the revisions by the early behaviorists, the movement began an evolution that gradually expanded the scope of behaviorism to include issues relating to the central mediation of behavior. Although the very definition of behaviorism did in fact change, one commonality shared throughout this evolution was the acceptance of an essential empirical methodology in the study of behavior. Perhaps more than Watson's preliminary analysis, it was the positivist character of behaviorism that left a more permanent legacy to psychology.

In the United States, the initial phase of the behavioristic evolution consisted of an intensive effort to build systematic structures of behavior theory. Beginning in the 1930s and lasting for approximately 20 years, eminent psychologists attempted to find a complete theoretical conceptualization for all behavioral processes. In a sense, this phase of theory building reflected an enthusiasm for behaviorism by accepting the possibility that this system of psychology could indeed constitute the definitive model for the new science. This phase of the behavioristic evolution ended with the recognition that an all-encompassing theory of behavior was at least premature and maybe impossible. However, the theory-building phase was critical for the development of behaviorism. In their attempt to formulate a general theory to accommodate the diversity of behavioral processes, these psychologists broadened the scope of the Watsonian version. Moreover, they rendered to the study of behavior a refined scrutiny that entrenched the definition of psychology in terms of behavioral processes.

In the second phase of the behavioristic evolution, the preoccupation with theory building was replaced by data collection. In response to the elusive search for a complete behavioral theory, psychologists began employing data as the guide to progressive research. In this phase, behavioral psychology took on a methodological character that was identical to that of the natural or physical sciences. In many respects, this phase of the behavioristic evolution is still continuing. However, by the 1970s another shift in behaviorism was apparent; it emphasized mini-theory or model building as well as the application of behavioral principles, especially in the development of so-called behavioral technology.

Before examining the details of the behavioristic evolution in America, we should be mindful of what was happening to European psychology during this period. Although the psychological models based on active assumptions of consciousness—such as Gestalt, psychoanalysis, and phenomenology—were European movements, the turmoil created by the volatile events in twentieth-century Europe resulted in the exportation of these views to America. In addition to bringing death and destruction, the world wars completely disrupted intellectual activity. It was no accident that one of the early revisions of Watsonian behaviorism coincided with the emigration of the leaders of the Gestalt movement, who fled the anti-intellectualism and anti-Semitism that followed Hitler's assumption of power in 1933. After the leaders of European proposals for the formulation of psychology fled to the United States, their views modified existing behaviorism to varying extents. However, the European manifestation of the natural science model for psychology—namely, Russian reflexology—remained a developing force in the states of the former Soviet Union. Reflexology exerted a major impact in the early years of American behaviorism, and the more modern expressions of reflexology continue to be an important influence.

Reflexology Expanded

Russian and Eastern European psychology accommodates a full range of theoretical issues and applications broadly based on a variety of perspectives, as in the United States. However, the pioneering work of Russian reflexology, culminating in the work of Pavlov, occupied a preeminent place in Russian psychology. Although seriously disrupted during World War II, Russian science, as rebuilt in the postwar period, supported the continuing development of reflexology.

As mentioned before, Pavlov's reduction of psychological events to physiological materialism was generally consistent with the philosophical underpinnings of the Marxist–Leninist government of the former Soviet Union. After many years of discussion and debate, Marxism–Leninism and Pavlovian reflexology were integrated into a single philosophical basis for psychology. Within this foundation, all mental activity was interpreted as the product of the physiological mechanisms of higher nervous activity centered in the brain. External, overt behavior interacts with the internal, central physiology, so that internal and external processes were considered to be two aspects of the same psychological mechanism. With harmony between science and government, and consistent with the permeation of Marxist–Leninist philosophy throughout Soviet society, the highly centralized Academy of Sciences of the former Soviet Union established research centers of reflexology that applied their research efforts to the full range of psychological issues, such as social psychology, personality, and psychopathology.

One of the more interesting developments in reflexology during the period between the two world wars began with the research of two young medical students at the University of Warsaw, Jerzy Konorski (1903–1973) and Stefan Miller (1902–1941). They became interested in the newly published Russian edition of Pavlov's *Lectures on the Higher Activities of the Brain* (1926), and their ability to read Russian gave them access to primary reports of Pavlov's ongoing research program as well as data from various Russian laboratories. Konorski and Miller argued the novel hypothesis that because Pavlov's account could not completely explain behavioral changes following the reward of certain movements and the punishment of others, there indeed may be two types of conditioning paradigms. They tested their hypothesis in a series of ingenious experiments, which led to the distinction between response-dependent reward, or avoidance of punishment (type II conditioning), and Pavlov's CS–US sequence resulting in behavior change (type I conditioning). The report of their findings to the Warsaw Division of the French Biological Society in 1928 generated considerable interest in Europe, and Pavlov himself extended an invitation to Konorski and Miller to work with him in the Koltushi Laboratory outside Leningrad (now St. Petersburg). Konorski spent 2 years with Pavlov, repeating and confirming the experimental work from Warsaw, later published as *Physiological Bases for the Theory of Acquired Movements* (1933) and in the laboratory journal of the Pavlovian Institute (1936), with an introduction by Pavlov. Because of the political isolation of Russia at that time, the details of Pavlov's research program and Konorski and Miller's revision of the two processes of conditioning went relatively unnoticed in the West. Instead, Skinner's (1935, 1938) similar distinction between Pavlovian and instrumental conditioning, published in English, preempted the earlier work of Konorski and Miller. Although differences in interpretation existed between the view of Skinner and that of Konorski and Miller, the latter researchers may be credited with the important distinction between Pavlovian and instrumental conditioning.

After his work with Pavlov, Konorski returned to Poland and systematically began to approach his major goal in reflexology—the integration of Pavlov's and Sherrington's models of neural processes. His work was halted by the German invasion of Poland on September 1, 1939, and the devastation of the country during the subsequent 6 years. In addition to the destruction of universities and research institutes, the vast majority of Poland's intelligentsia, including Stefan Miller, died during the struggle against Nazism. Konorski spent the war years in the Soviet Union, for the most part working in military hospitals in the Republic of Georgia in the Caucasus.

After the war Konorski was instrumental in the difficult task of rebuilding the Polish scientific establishment. In particular, he founded the Department of Neurophysiology of the Marceli Nencki Institute of Experimental Biology in Warsaw. In addition to carrying out a productive research program in general reflexology, the department served as a center for educating a new generation of Polish neurophysiologists. Moreover, the unique political position of postwar Poland, coupled with the research activity and personal dynamism of Konorski, provided an ideal setting for a dialogue between Western and Eastern scientists. The research direction of Konorski and his coworkers generally concerned brain physiology; in particular, the central mechanisms regulating behavior, especially instrumental conditioned reflexes. In his last systematic contribution, *Integrative Activity of the Brain* (1967), Konorski considered brain activity as a complex cybernetic system controlling activities of the organism as a whole. This work represents a complete synthesis of Sherrington's neurology and Pavlov's reflexology. Accordingly, Konorski viewed higher nervous activity as a dynamic system capable of varieties of adaptive strategies, anticipating the advances in cybernetics and information processing that occurred subsequently in the behavioral sciences. As the most famous of Pavlov's students outside Russia, Konorski followed the reductionistic strategy of reflexology. However, in addition to his rejection of a single conditioning process, his significant contribution consisted of a true interdisciplinary perspective for reflexology, and within this context, Konorski broadened reflexology to accommodate a full range of psychological issues.

Since the end of World War II, Russian reflexology has moved far beyond the study of conditioned associations. Firmly based in materialistic reductionism to the mechanisms of neural activity, the research strategy of reflexology has been employed in all types of psychological research, ranging from the problems of psychiatric disorders to the development of language. A guide for the inclusion of such widespread concern within the materialistic model of reflexology was proposed by a brilliant scientist of the interwar period, L. S. Vygotsky (1896–1934), who influenced many prominent scientists of the postwar period. Essentially, Vygotsky called for the complete application of scientific technology to further the betterment of the individual and society, but at the same time argued for the recognition of the complexity of human nature. Hence, scientific technology should serve the scientist pursuing a holistic understanding of the person. Vygotsky extended Pavlov's reflexology to higher mental functions but insisted that the reductionism of the materialistic methodology must not obscure the complexity of human mental activity.

Perhaps Vygotsky's most famous student was A. L. Luria (1902–1977), whose long career included detailed investigations of such diverse issues as the development of language and thoughts, the neurophysiology of cortical functions, and cross-cultural comparisons of sign systems. Luria's study of language tested Vygotsky's hypothesis that speech forms a critical link in the relationship between external, overt behavior and internal, symbolic thought. While extending Russian psychology beyond the traditional province of reflexology, Luria's view was also consistent with the monistic underpinnings of reflexology in that it stressed the unity of psychological and physiological aspects of experience. Luria identified four distinct and progressive stages in the developmental process of speech functions: activity initiation, activity inhibition, external regulation, and, finally, internal regulation of activity. Internal speech was the foundation of thought processes. In another area, Luria's research on the frontal lobe systems helped to isolate the localization of behavioral patterns, and his studies of recovery of function after brain damage contributed to the understanding of retrieval processes in memory. Luria's productivity and wide-ranging investigations consistently combined clinical and laboratory observations within a coherent theoretical framework of reflexology.

Studies of the neurophysiology of conditioning have continued in Russia. One area of major concern has been the orienting reflex, recognized by Pavlov as a form of external inhibition. In 1958, E. M. Sokolov wrote a classic paper (translated into

English in 1963) that related the orienting reflex to sensory thresholds involved in arousal. Through the use of a variety of physiological and electrophysiological measures, the adaptive significance of the orienting reflex has been extended to such behavioral processes as habituation and attention.

The major focus of contemporary reflexology in Russia has concerned the ongoing theme of the single, materialistic basis of psychological events. Such physicalism, in sharp contrast to idealistic or mentalist constructs, is reflected in the emphasis on measurement of neural mechanisms, especially through electrophysiological recordings. Laboratories, such as the Institute of Higher Nervous Activity and Neurophysiology in Moscow, led for many years by Ezras E. Asratyan (1903–1981), pursued interdisciplinary programs of research. In addition to the active research centers under the direction of the Russian and Siberian Academies of Sciences, research activities are found in nations contiguous to Russia. Especially active are Ukrainian and Armenian Academies of Sciences. Moreover, one of the most creative and productive centers of contemporary reflexology is found in the Georgian Republic. There, the Institute of Physiology of Tbilisi University reached notable achievement under the leadership of Ivan Solomonovich Beritashvili (1885–1974), whose name is sometimes rendered in English as Beritov. After his education in St. Petersburg, Kazan, and Utrecht, Beritashvili returned to post-revolutionary Georgia in 1919 and founded the Institute of Physiology. His various investigative programs included studies of the rhythmic nature of subcortical inhibition, the physiological properties of dendrites, the functions of the reticular activating system, and conditioning and memory.

The movement started by Sechenov, Bekhterev, and Pavlov established the principles of conditioning and evolved into a reliance on the materialism of physiology to explain psychological activity. With the disintegration of the Soviet Union and its replacement by new independent countries, scientific work entered into a period of adjustment. The successor structures of the former Soviet Academy of Sciences are still evolving, so the future direction of reflexology, and psychological research in general, within these countries remains unclear.

The American Behaviorists

As Pavlov's system in the 1930s was undergoing assimilation within Russian Soviet science, behaviorism in the United States was evolving into a theory-building phase. This phase reflected the triumph of behaviorism over the mentalistic psychology of consciousness. Moreover, the identification of psychology as a positive science, similar in approach to the physical sciences, proceeded concurrently. This period was dominated by four important scientists: Edwin R. Guthrie (1886–1959), Clark L. Hull (1884–1952), Edward C. Tolman (1886–1959), and B. F. Skinner (1904–1990). The first three proposed theories of behavior and contributed to this phase of the evolution of behaviorism; Skinner attempted an "antitheory" and marked the end of this phase.

Guthrie's Contiguity Theory. Like Watson, Edwin R. Guthrie advocated a psychology of observable behavior consisting of muscular movements and glandular responses elicited by environmental stimuli. His theory of associations was in the tradition of Pavlov and Thorndike, asserting a single principle to account for learning. Guthrie did not accept Thorndike's reinforcement principle based on the law of effect, but rather viewed Thorndike's secondary notion of associative shifting as the basis of learning.

Guthrie received his doctorate in 1912 from the University of Pennsylvania after earlier training in mathematics and philosophy. He joined the University of Washington in 1914 and remained there for his entire career. Guthrie was not a systematic experimenter and his arguments were mainly based on general observations and anecdotal information. His major experimental work, written in conjunction with G. P. Horton, studied problem-solving behavior of cats and was published as *Cats in a Puzzle Box* (1946). His most influential theoretical work was *The Psychology of Learning* (1935; revised in 1952).

The key to Guthrie's associationistic theory lies in the single principle that contiguity is the foundation of learning. Guthrie viewed behavior in terms of movement rather than responses. By this distinction, he meant that movements are the components of larger response units, or behavioral acts. Accordingly, skilled behaviors may be viewed in terms of a gross response composed of smaller units of movements that are largely muscular. Stimuli were likewise viewed as a complex situation consisting of smaller elements. Guthrie's principle of contiguity stated that when a combination of stimulus elements is accompanied by movement, the movement sequence will recur, given the presence of similar stimulus elements. Guthrie held that learning is a pattern or chain of discrete movements elicited by both environmental and internal stimulus cues.

Because Guthrie's view of associations relied on stimulus and response contiguity, the role of reinforcement received a unique interpretation. Guthrie believed in one-trial learning—in other words, the contiguous relationship between stimulus and response elements immediately produces the associative bond at full strength. The effects of a reinforcing reward or punishment serve to feed back on the stimulus situation, altering that situation and requiring a new bond between the altered stimulus situation and movement. Thus, reinforcement provides a means of changing the stimulus context, requiring movement, and the learning proceeds within the behavior act. Extinction, or forgetting, was interpreted as the result of interference from new associations rather than the decay of stimulus-response bonds caused by the absence of reinforcement. Similarly, practice effects were not seen as affecting stimulus-movement association but rather as improving the coordination of established bonds within the gross behavioral act. In a consistent vein, Guthrie viewed drives not as causal motivational agents but rather as energizers of behavior acts.

Guthrie's arguments and interpretations influenced later behavioristic psychologists. F. D. Sheffield defended Guthrie's views and extended them to include the use of positive reinforcement as a means of refining behavior. Similarly, Virginia Voeks used carefully designed experiments to demonstrate many of the implications of Guthrie's writings. However, the stimulus sampling theory of William Estes (see later Information-Processing and Mathematical Models section) is perhaps the most extensive application of Guthrie's associationism, and statistical models of learning have generally found Guthrie's theory amenable to computer simulation of associative processes.

The major criticism of Guthrie's views may be that they are incomplete and do not deal comprehensively with complex types of learning and memory problems. However, Guthrie's seeming ability to explain, in a parsimonious way, some of the principles of more complicated systems, notably Hull's theory, constitutes his appeal.

Hull's Hypotheticodeductive Theory. Clark L. Hull's systematic theory came closest to a comprehensive treatment of behavioral issues governed by common principles. As a behaviorist, Hull centered his psychological views on habit formation, the accumulations of experiences for effective adaptation. His scientific approach was truly systematic. Recognizing the importance of observation and experimentation, he advocated a hypotheticodeductive structure to guide research. In this strategy, following the approach of Euclidian geometry, a behavior principle or formulation is first deduced from postulates and then rigorously tested. A successful test supports belief in the postulates; failure results in revision of the postulates. Hull's approach was positivist and followed a logical progression, verified through empirical demonstration.

Hull's theory was the product of a lifetime of disciplined work. Born in New York and raised in Michigan, Hull suffered from ill health, had poor eyesight, and was crippled by polio. His education was interrupted at various times because of illness and financial problems. After pursuing mining engineering at Alma College, he transferred to the University of Michigan and completed a degree in psychology. In 1918, he received his PhD from the University of Wisconsin, where he stayed for 10 years as an instructor. During that time he studied the effects of tobacco smoking on performance, reviewed the existing literature on testing, and began research on suggestion and hypnosis, the latter culminating in a widely read book published in 1933. In 1929, he was named to a research position at Yale University and began the serious development of his behavior theory. Until the end of his career Hull and his students dominated behavioristic psychology.

Hull's system is intricate and relies heavily on mathematical predictions. He made detailed modifications as his experimental tests progressed over time. The present survey attempts to highlight his views. Essentially, Hull's theory of learning is centered on the necessity of reinforcement, defined in terms of the reduction of drives arising from motivational states. The behaving organism is viewed in the context of a homeostatic model seeking equilibrium from drive forces. The core of Hull's analysis concerns the notion of *intervening variables*, described as unobservable entities employed by psychologists to account for observable behavior. Thus, from a purely behavioral perspective, Hull extended Watson's conceptualization of behavior in terms of the peripheral (S–R) events to a consideration of central, organismic factors, stimulus–organism–response (S–O–R), intervening variables. This expansion of the behavioral model had been suggested as early as 1918 by Woodworth (see Chapter 12), but it was Hull who systematically articulated organismic variables.

The chief intervening variable for learning in Hull's theory is called habit strength, ($_SH_R$), which depends on two factors for associations. The first is a contiguity principle, meaning that a close temporal relationship must exist between stimulus and reinforcement. The second principle is reinforcement itself, defined in its primary form as drive reduction, but there are also secondary reinforcements, cues that are reliably associated with primary reinforcement and take on reinforcement properties. For example, if a hungry rat is repeatedly given food for correct responses in the presence of a light, the light takes on some of the rewarding characteristics of the food. Hull attempted to integrate Thorndike's law of effect with Pavlovian conditioning, so that the basic procedure in which learning occurs is contiguity of stimulus and response under conditions of reinforcement. Habit strength ($_SH_R$) and drive (D) interact to produce what Hull referred to as reaction potential ($_SE_R$)—the "tendency to produce some reaction under the effect of the stimulus." $_SE_R$ was a theoretical concept for Hull, not synonymous with observable responses; it is the product of $_SH_R$ and D:

$$_SE_R = {_SH_R} \times D$$

Hull's intervening variables represented not only a qualitative conceptualization but also an attempt to define quantitative relationships. For example, on the basis of the above expression, little performance would be observed from a hungry but naive rat: Drive is high, but habit strength is not, resulting in a low tendency to respond. Similarly, accounting for the distinction between learning and performance, a rat with a well-established response to bar-press for food reward would not perform if it is not hungry: Habit strength is high, but drive is minimal, producing little expectation of reaction potential. To complete his framework for the intervening variables mediating performance, Hull included negative, inhibitory factors (I), resulting from fatigue and boredom, as a by-product of performance. He also included the contributions of stimulus magnitude (V)—for example, a faint versus a loud CS; the magnitude of reinforcement (K)—for example, one versus four food pellets per correct response; and the oscillating, momentary threshold of reaction for an individual subject ($_SO_R$). All of these intervening variables were related:

$$_SE_R = {_SH_R} \times D + V + K - I - {_SO_R}$$

It should be noted that this summary equation was itself articulated into more refined components as Hull's theory developed.

The entire detailed structure of Hull's system was applied to the quantification of all possible influences on the acquisition of adaptive behavior. Indeed, empirical tests, largely conducted with laboratory rats, tended to support Hull's conceptualization. This analytic approach assumed that more complex forms of behavior could be derived from these intervening variables. However, the theory as a whole was not successful after all. There were empirical discrepancies, such as the inability of the theory to deal with insightful, rapid acquisition of behavior. Hull's view stressed the importance of practice during training, producing continuous but gradual improvement during acquisition. More importantly, however, the theory fell apart in its attempt to quantify the conceptual relationships among intervening variables. As a model or guide for research, Hull's system was superb; much of our contemporary jargon to describe learning was invented by Hull. However, as the exact, definitive statement of behavior, Hull's views were probably premature, resulting in a fixed, rigid structure not amenable to the variability of human and animal behavior.

Tolman's Cognitive Behaviorism. As a behaviorist, Edward C. Tolman developed a theory that pushed the scheme of Watsonian behaviorism toward further expansion than did either Guthrie or Hull. In his major work, *Purposive Behavior in Animals and Men* (1932), Tolman proposed a consideration of behavior that was molar, as opposed to molecular. He viewed molar behavior as a unified and complete act, which provides the proper unit for psychology. Underlying molecular elements, whether neural, muscular, or glandular processes, were not sufficient to account for the molar act. In this sense, Tolman departed from Watsonian behaviorism by opening psychology to the study of higher, cognitive processes. His approach to molar behavior was not reductionistic. In adhering to the molar level, Tolman argued that reductionism results in the loss of the purely psychological level, and explanations based upon molecular components are not adequate. Thus, for Tolman, molar behavior is more than the sum of the molecular elements.

Like Hull, Tolman originally was interested in engineering, and he received a degree from the Massachusetts Institute of Technology. He switched to psychology and finished his PhD in 1915 at Harvard. After teaching at Northwestern University for 3 years, Tolman joined the University of California at Berkeley, where he contributed considerably to the developing reputation of that institution. Tolman became known as an excellent and warm teacher. In 1950 he led the movement against the California state loyalty oath as an affront to academic freedom, and the oath was eventually discarded. The image of Tolman that comes through is of an open person welcoming new trends and ideas in psychology.

Tolman's view of psychology relies heavily on many of the premises of the Gestalt psychologists. Indeed, he used the term *Gestalt* to describe holistic, insightful learning experiences. Moreover, his conception of behavior as molar and his adoption of mental isomorphism were directly borrowed from Gestalt psychology. He used the latter construct to describe the central product of learning in terms of the acquisition of field maps that exist in the brain as cognitive representations of the learned environment.

Tolman's theoretical orientation was not as systematic in approach as that of Hull. His criticism of the reduction of psychological events to the mechanical elements of stimulus and response caused many researchers of the Hullian orientation to pause and modify their views. Tolman's laws of acquisition essentially focused on practice that builds up sign Gestalts, or expectancies. In maze learning experiments with rats, for example, he described the acquisition of place learning, inferring the acquisition of relationships or cognitive maps in the subject. Similarly, he demonstrated the expectancy of reinforcement in rats trained to one kind of reward and then switched to a more appealing food. Finally, he showed that latent learning occurs in rats, indicating that the quality of reinforcement can exert a differential effect on performance levels. In all of these experiments Tolman used cognitive explanations as intervening variables to show that behavior in organisms is governed by central, mediating processes that go beyond environmental input only.

Tolman was often criticized for his lack of specific explanations of the central mediation of cognitive learning. However, he brought to behaviorism a new perspective that departed from the sterile reductionism of molecular Watsonian behaviorism. Moreover, his repeated demonstration of performance versus learning differences clearly showed that learning is not reducible simply to stimulus-response-reinforcement elements. If he failed to offer a more comprehensive explanation, he nevertheless succeeded in justifying the integrity of molar behavior and stimulated inquiry. Tolman did not leave a systematic school of followers, as did Hull, but he anticipated the entire research theme of cognitive learning prevalent in contemporary psychology.

Skinner's Radical Positivism. In 1950, B. F. Skinner published a paper entitled "Are Theories of Learning Necessary?" His discussion formally signaled the end of the theory-building phase of the behavioristic expansion. Skinner recognized the shortcomings of the attempt at theory building—the inadequacies of the theories and the distortion of behavioral science predicated on questionable a priori assumptions. In the place of theories he advocated a system of behaviorism guided by data. According to Skinner, theory, when psychology's progress permits it, should be confined to loose, descriptive generalizations arrived at through reliance on facts yielded by a positive scientific approach.

Skinner received his doctorate in psychology from Harvard in 1931, after an earlier interest in literature. He taught at the University of Minnesota and Indiana University before returning to Harvard in 1947. In addition to a prolific research record and influence on a generation of neobehaviorists, Skinner popularized his behavioristic principles through novels and commentaries. His novel *Walden II* has sold more than 2 million copies. He was widely known for his opinions on social structures and institutions.

Skinner's positivism consistently advocated a methodological emphasis and a return to the study of behavior defined in terms of peripheral events. He argued against speculating about central mediating agencies of behavior, whether cognitive or physiological. Rather, behavior, for Skinner, was completely subject to environmental determinacy. If the environment is controlled, behavior is controlled. For this reason, Skinner accepted the validity of exhaustive study of a single subject, because variability arises not from individual differences inherent in the organism, but rather from differential environmental contingencies.

The basis for Skinner's research was the study of operant behavior. In contrast to respondent behavior, where responses are elicited by specific stimuli, operant behavior is ongoing without any apparent stimulus. To investigate operant behavior, Skinner devised an environmental chamber where birds could engage in pecking, or rats in bar pressing. In this manner, environmental control is easier to obtain and ongoing, operant rates of responses can be readily recorded. Learning occurs when the operant behavior comes under the control of reinforcement from the environment. At first, the operant responses may be shaped by reinforcement of approximations of the desired operant character. When the refined operant is followed by presentation of the reinforcing event, the probability of the occurrence of the operant is increased. For example, if an operant is defined as bar pressing in a rat, presentation of food following a bar press increases the likelihood of more bar presses. Thus, Skinner's view of reinforcement was defined in terms of the probability of changes in the operant rate. It avoided inferences of satisfiers or annoyers, as in Thorndike's law of effect, or of drive reduction, as in Hull's theory.

Skinner clearly demonstrated the power of reinforcement by showing that characteristic response rates are obtained for particular schedules of reinforcement delivery. Similarly, he translated conditioning processes such as generalization and discrimination to a reinforcement contingency framework. Moreover, he extended the principles of operant control to a consideration of verbal behavior. Skinner used his experimental data to argue that behavior is controlled, and the critical role of the psychologist is to define the parameters of effective control for appropriate social implications.

Skinner's view of behavior has often drawn the harsh criticism of those who are offended by his mechanical conception of human nature. Skinner's view, however, has clear antecedents in the history of psychology—for example, the views of Condillac and some of the pre-Socratic scholars. A more direct antecedent is the German zoologist Jacques Loeb (1859–1924), who taught Watson at the University of Chicago. Loeb suggested a theory of animal tropism and was quite influential in the development of comparative psychology. Whether one considers Skinner's environmental determinacy or Pavlov's physiological reductionism, the net conceptualization of human activity precludes any attributes of personal freedom, self-determinacy, or the dynamics of consciousness. Skinner perhaps earned more of the scorn of critics because he articulated the social controls that are derived from the principles of operant behavior. In the spirit of positivism, Skinner argued that the humanistic characteristics of the species, which presumably set us off from the rest of living evolutionary products, are in fact an illusion, created by us over history to give us a sense of security. In fact, for Skinner, to be truly human means to be in control—to understand and use environmental contingencies to self-benefit. Skinner addressed the place of the person in terms of his own views as well as those of his critics in the following passage from *About Behaviorism*:

> Behavior is the achievement of a person, and we seem to deprive the human organism of something which is his natural due when we point instead to the environmental sources of his behavior. We do not dehumanize him: we dehomunculize him. The essential issue is autonomy. Is man in control of his own destiny or is he not? The point is often made by arguing that a scientific analysis changes man from victor to victim. But man remains what he has always been, and his most conspicuous achievement has been the design and construction of a world which has freed him from constraints and vastly extended his range. (1974, pp. 239–240)

THE ROLE OF THEORY

On leaving the theory-building phase of behaviorism, it is appropriate to pause briefly to examine the uses of theory in science so that the significance of theory building within psychology may be appreciated. Although we are here concerned with theories in behaviorism, we have surveyed many theories so far—among them Wundt's theory of consciousness and Freud's dynamic system. It is interesting to question what theories are supposed to do in the progress of science. Marx defines a theory as

> a provisional explanatory proposition, or set of propositions, concerning some natural phenomena and consisting of symbolic representations of (1) the observed relationships among (measured) events, (2) the mechanisms or structures presumed to underlie such relationships, or (3) inferred relationships and underlying mechanisms intended to account for observed data in the absence of any direct empirical manifestation of the relationships. (1976, p. 237)

Accordingly, theories should provide a framework to test existing hypotheses and generate new hypotheses.

Marx (1963) has provided a useful description of the elements of theory construction that offers three dimensions that distinguish between the arts and sciences; it is presented in Figure 15–1. Marx discussed the three components of theory construction—hypotheses, constructs, and observations—as continua ranging between art and science. The continuum of observations classifies the way a theory reduces the variation in measurement coming from the influence of stimulating conditions.

FIGURE 15–1 The elements of scientific theory construction. (Reprinted by permission of Prentice Hall, Upper Saddle River, NJ, from *Theories in Contemporary Psychology,* 1st and 2nd eds., edited by Melvin H. Marx. Copyright 1963, 1976 by Macmillan Publishing Company.)

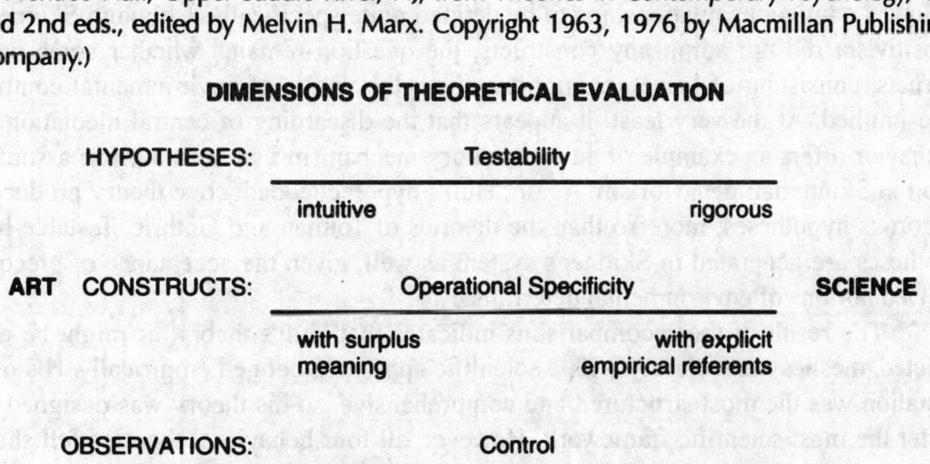

For example, watching spontaneous play activity of children in a schoolyard is a less controlled observational situation than recording the magnitude of the CR upon presentation of the CS. Constructs, the major substance of explanation in theories, vary from general concepts without tight definition to explanatory mechanisms closely defined in terms of readily observable referents. For example, the libidinal energy of Freud has more surplus meaning as an explanation of human motivation than Hull's intervening variable of a drive such as hunger, defined in terms of a specific number of hours of food deprivation. The third continuum of hypotheses is differentiated by the degree of testability. For example, it is more difficult to construct empirical tests of the validity of Freud's stages of psychosexual development than it is to test a hypothesis of frontal lobe mediation of behavioral responses, as derived from Pavlov's theory of conditioning.

It is important to note the bias inherent in Marx's schema of theory evaluation. He has defined science in terms of empirical approaches based on observation. To thus evaluate Freud's largely nonempirical views is perhaps unfair because it is an open question whether a nonempirical approach can be considered scientific rather than necessarily artistic. However, these evaluative dimensions are useful in a comparison of the empirically based orientations of Guthrie, Hull, Tolman, and Skinner.

Relative to Figure 15–1, Hull's and Skinner's theories relied on a more experimental approach than did the theories of Guthrie and, to some extent, Tolman. Hull's experimental observations were dictated by the systematic implications of his hypothesis-testing strategy, whereas Skinner's positivism and reliance on the determinacy of environmental control were readily amenable to experimental investigation. The constructs of the theories of both Hull and Tolman were based on intervening variables. Yet the systems of each may be differentiated along this dimension by comparing Hull's reliance on observable operations, such as the definitions of drive or magnitude of reinforcements, with Tolman's less operationally specific notions, such as cognitive maps or expectancies. Similarly, the parsimonious theory of Guthrie allowed contiguous relationships to be defined quite specifically. Although Skinner's positivism did not admit any constructs, the question remains whether some constructs, consisting of loose assumptions about the nature of environmental control, are implied. At the very least, it appears that the discarding of central mediation in behavior offers an example of an explanatory mechanism that is an a priori assumption in Skinnerian behaviorism. Again, Hull's hypotheticodeductive theory produced rigorous hypotheses, more so than the theories of Tolman and Guthrie. Testable hypotheses are generated in Skinner's system as well, given the acceptance of preconceived notions of environmental determinacy.

The result of these comparisons indicates that Hull's theory, as might be expected, meshes most closely with a scientific approach, defined empirically. His orientation was the most structured and comprehensive, so his theory was designed to offer the most scientific framework. However, all four behavioral theories fell short in terms of the outcomes of the progression of components of theory construction. Specifically, the rigorous testing of hypotheses generated by each orientation fell short of expectations; none of them provided behaviorism with principles of all human and infrahuman activity. Accordingly, behaviorism turned away from this structure of theory building to pursue more modest goals.

POST-THEORY FORMULATIONS

The third phase of the expansion of behaviorism, inaugurated by Skinner's atheoretical positivism, led to a period of mushrooming experimentation. Indeed, with little regard to theoretical guidance, data collection appeared to have become an autonomous project without an overriding rationale. In time, research efforts took on direction, guided by the development of models with a pervasive flavor of applying the principles of behavior. Because of its applied character in this period of the behavioristic expansion, so-called neobehaviorism has also been described as neofunctionalism (see also Chapter 17).

Neobehavioristic Models

As a link to the theory-building phase of the behavioristic evolution, neobehaviorism may be classified in terms of information-processing/mathematical models of learning, neo-Hullian learning models, cognitive models, and operant models. To varying degrees these classifications are derived, respectively, from the views of Guthrie, Hull, Tolman, and Skinner.

Information-Processing and Mathematical Models. The view of intellectual functions in terms of the intricate logic of mathematical and probabilistic relationships reflects a merger between the mathematical underpinning of Hullian theory and the parsimonious orientation of Guthrie's contiguity principle. This movement received a great impetus from the advancing technology of computerized hardware during the postwar period. The potential for quantifiable prediction of learning parameters was initially explored in efforts to simulate learning processes by studies of artificial intelligence conducted in the 1940s and 1950s. For example, the General Problem Solver, developed at Carnegie-Mellon University, used computer programs to simulate human problem-solving behavior in a variety of performance tasks. This work led to the Logic Theorist program, the product of cooperative efforts among the Carnegie group, the Massachusetts Institute of Technology, and RAND Corporation scientists, which provided a model for programs in computer simulation of human intelligence and studies of artificial intelligence.

From the early version of Estes' (1950) stimulus sampling theory that views learning as a statistical process involving the selection of stimulus elements, a wide set of applications of probability functions to predict behavior has been proposed. To define a learning problem, a strategy that started with empirically based assumptions to generate the probability of responses to form predictive learning curves was proposed (see, for example, Estes, 1964). A vast literature developed that portrays human learning in terms of an information-processing system and examines intellectual and motor functions, such as decision making and skilled practice. Moreover, this approach has been extended to the traditional psychological problems of sensory processing (see, for example, Swets, 1961).

As research progressed in the quantitative prediction of learned behavior, more sophisticated programs were developed using noncontinuous assumptions of the acquisition processes. One such program is termed the *Markov model*. In it, acquisition is viewed as a chain process, and each stage of the process may be modified by the effect of a previous trial or stage in the chain. The stimulus elements at a given stage are relatively small in number, but the sampling probabilities associated with each element change from stage to stage.

The information-processing model of learning, based on complex mathematical prediction, has advanced to the point of exhaustive analyses of complex learning processes. Studies of concept formation and linguistic development have yielded detailed views of human learning that transcend earlier theories based on the accumulation of S–R associations. Moreover, the advanced technology developed for such research has helped bridge a gap between simpler learning processes, typically observed in infrahuman subjects, and complex human intellectual activities. This link helps to integrate the neurophysiological substrates examined in infrahuman learning and the sophisticated learning models derived from the information-processing context of human learning.

The Neo-Hullians. Hull's most famous student and later collaborator was Kenneth W. Spence (1907–1967), who spent his most productive years at the University of Iowa. The research of Spence and his many students was characterized by a concern with refining Hull's theory as well as with applying Hull's principles to varieties of behavioral processes, including an analysis of anxiety. Spence's major contribution to the theoretical basis of Hullian behaviorism was his explanation of discrimination learning (Spence, 1937, 1940). Briefly, Spence held that gradients of excitatory potential and inhibitory potential are generated around stimulus values that are reinforced and not reinforced, respectively, during discrimination training. These hypothetical gradients merge algebraically to account for observed performance during assessment of stimulus generalization. Spence and his students also investigated eyelid conditioning and found that certain levels of anxiety facilitate acquisition of that response as well as others, leading them to an examination of the role and the assessment of anxiety (Taylor, 1951, 1953). These studies are important because they represent some of the initial attempts to integrate behavioristic principles and psychopathology, an area of later intensive study.

Another important student of Hull is Neal Miller (1909–2002), whose productive career was characterized by important studies of a variety of psychological issues. His early work (see, for example, Dollard and Miller, 1950) attempted to apply a Hullian analysis to behavioral issues derived from the psychoanalytic literature. Miller's research with Dollard and others on frustration and conflict has become classic, leading to direct support for the contemporary behavior modification trend. In studies of physiological substrates Miller made significant findings concerning the relationship between reinforcement mechanisms and the control of autonomic behavior (Miller, 1969).

A third student of Hull, O. Hobart Mowrer (1907–1982), expressed the distinction between Pavlovian and instrumental conditioning in a 1947 paper. He argued that in avoidance learning the fear of the CS is acquired by Pavlovian principles, and the motor response to that fear is instrumentally acquired through the reinforcing effect of fear reduction. The CS then functions as a sign of impending shock. From this distinction Mowrer asserted a revised two-process theory involving incremental punishment and decremental reward. In incremental reinforcement, stimuli act as signs of fear; in decremental reinforcement, stimuli serve as signs of hope. Applying these principles to psychopathology, Mowrer (1960) helped prepare for the emergence of behavior modification. Subsequent refinements and interpretations of two-process learning by Schoenfeld (1950) and, especially, Rescorla and Solomon (1967) have generated many hypotheses of theoretical and clinical significance that have advanced the use of conditioning principles in behavior modification.

Contemporary research in the neo-Hullian tradition has also extended to questions concerning the physiological basis of learning. Borrowing from the neurophysiological findings of reflexology, these investigations focus on such areas as the ontogeny of learning, consolidation and retrieval processes of memory, and sensory factors of attention (see also Chapter 17). Together with the efforts of Miller and his collaborators working in human psychophysiology, they have rapidly expanded our understanding of learning processes.

Cognitive Models. As mentioned earlier, Tolman did not leave a legacy of followers to pursue his theoretical views. Accordingly, as behaviorism expanded to include a consideration of cognitive learning, there was little of the systematic coherence that was typical of the neo-Hullians. In surveying more contemporary developments in cognitive models, then, diverse research directions are evident.

Egon Brunswik (1903–1955) developed a theory called probabilistic functionalism from his early research on perceptual constancy. Influenced by the Gestalt movement while he lived in Germany, and later by Tolman, who secured a position for him at Berkeley, Brunswik found that subjects tend to perceive consistently, despite distortions in sensory input, through a series of relative adaptive compromises to varying environmental variables, and that these adaptive compromises are self-initiated. Accordingly, Brunswik held that subjects' adaptations in perceptual and behavioral situations are relative and definable in terms of probabilities. His research approach was in contrast to the rigidly controlled behavioristic experiments that typically examined relatively few variables. Brunswik criticized such experimentation as an inadequate and distorted representation of reality. Rather, he advocated a wider sampling of variables as they actually occur in the environment. Although Brunswik's systematic theory was left unfinished because of his early death, he contributed to an acceptance of less analytical and mechanistic approaches to research that allowed cognitive behaviorism to consider the state of organismic–environmental interactions.

A second major development within cognitive models of learning came from the social psychological research of Leon Festinger (1919–1990). His theory of cognitive dissonance essentially asserted that contrasting objectives within a person's value system result in discomfort that must be resolved by adapting behavioral strategies to reduce the dissonance. Of interest is that Festinger's theory is a behavioral view holding an explanatory central mechanism with mentalistic implications. Thus Festinger offered a cognitive model that directly confronts the very basic premise of the peripheral character of Watsonian behaviorism.

Similarly, the more recent study of language processes by Noam Chomsky (b. 1928) falls into the same type of cognitive model as Festinger's theory. Chomsky, who wrote an insightful criticism of Skinner's writings on verbal behavior, argued that the acquisition of syntactical structure of language requires the existence of a mental construct that he calls the language acquisition device. Without this mechanism, Chomsky held that true language cannot emerge. Chomsky's theory is consistent with the cognitive approach of the French psychologist Jean Piaget (1896–1980), who intensively studied the development of knowledge; it also has the support of other American psychologists, such as Jerome Bruner (b. 1915), who has studied the formation of concepts in children (see also Chapter 17).

The development of cognitive models, especially those of Festinger, Chomsky, and Bruner, has been characterized by a radical departure from the initial behavioristic formulation. However, these models accept the significance of examining overt behavior as well as the methodological strategies common to other forms of behaviorism. Nevertheless, the emergence of clear mentalistic constructs has brought behaviorism almost full circle and provides confirmation of the full expansion of the behavioristic model of psychology.

Operant Models. The radical positivist approach of Skinner continues among psychologists who advocate the experimental analysis of behavior into components of environmental and reinforcement contingencies. However, in extending Skinner's initial principles, less rigid interpretations of operant behavior have emerged. In particular, operant research has examined physiological and centrally mediated motivational variables. The applications of operant principles in clinical and teaching settings have also dealt with assumptions of mentalistic constructs underlying observable responses. The *Journal of the Experimental Analysis of Behavior* and the *Journal of Applied Behavior Analysis*, both dominated by the operant approach of Skinner, show increasing diversity and eclecticism in both subject matter and methodology. Moreover, as formulations of behaviorism have evolved and become diffused, the importance of Skinner's methodological innovations has been recognized and used in varieties of laboratory and applied settings.

Applications

Behavioristic principles have wide-ranging applicability in contemporary psychology. Educational policy, military training, and advertising techniques are just a few areas that employ behavioral principles. However, one area of present concern is the application of behavioral findings in clinical and therapeutic settings. The reason for giving this movement more attention than other applications is its relevance to the treatment of mental illness, an issue central to the development of psychology.

During the 1950s and early 1960s there was widespread acceptance of a growing division in psychology between the "experimental" and "clinical" branches of the discipline in the United States. Experimental psychologists were accused of isolating themselves in laboratories, perhaps concerned more with rats than with people; clinical psychologists were seen as relying more closely on the medical attitudes of psychiatry than on any psychological models. One of the factors that helped to ease the division between the theoretical and applied branches of psychology was the advent of a behavioral model of clinical applications, commonly called behavior modification.

The antecedents of behavior modification may be traced to several attempts that related learning principles to psychopathology. In addition to applications proposed as early as Watson (e.g., Watson and Rayner, 1920), we have already mentioned the work of Miller and Dollard and of Mowrer in applying Hullian theory to clinical problems. Moreover, Skinner's notion of environmental control was tried with some success in token economies devised for the closed, controlled setting of mental institutions. However, a book by Joseph Wolpe (1915–1998), entitled *Psychotherapy by Reciprocal Inhibition* (1958), provided the catalytic event that unleashed contemporary behavior modification. Wolpe, a psychiatrist, saw the powerful potential for changing behavior through the application of principles outlined under Pavlovian and instrumental conditioning. Specifically, he took many of the principles of aversive conditioning and extinction and devised such techniques as desensitization and counterconditioning to deal with anxiety-based symptoms.

Behavior modification has generated considerable controversy and criticism, especially criticism aimed at its alleged impersonal and mechanical basis. Consistent with its place in the behavioristic evolution, behavior modification is attacked as the epitome of "mind control" and "brainwashing." However, it has been argued that anxiety reduction techniques are actually adapted to the individual, depending on the hierarchy of anxiety-provoking stimuli, defined by the person. At any rate, it would be premature to evaluate the efficacy of behavior modification per se. For our purposes, we should recognize behavior modification as an important development in the behavioristic evolution. By providing a behavioral framework of clinical application, behavior modification has served the critical function of uniting researchers and clinicians under a common model of psychological activity. Whether behavior modification proves to be a stable force maintaining that unity or simply a temporary bandwagon phenomenon remains to be seen. It has already served an important function in the development of behaviorism.

Concluding this rather involved story of the changes and expansion of behaviorism, we may say that behaviorism has approached the definitive model of psychology more closely than any other conceptualization since the 1870s. Psychology in America was not only recognized in academic and professional settings but was also firmly entrenched in the scientific establishment by the theory-building stage of the behavioral evolution during the 1930s. However, it is important to acknowledge how much has happened during the evolution of behaviorism from the time of Watson. Behaviorism as a model, or a philosophy of psychology, is currently a diffuse system subject to diverse interpretation. Behaviorists may include the radical positivists of the Skinnerian tradition, as well as cognitive psychologists who propose definite mentalistic constructs. Behaviorists may be researchers who investigate behavioral processes exclusively on an infrahuman level, or they may be clinicians preoccupied with therapeutic applications. Neobehaviorism, then, may be a consensus, rather than a system, that recognizes the importance of observable behavior. Beyond that consensus the eclecticism of neobehaviorism appears dominant, precluding further definition.

CHAPTER SUMMARY

The shift in American psychology from the essentially German emphasis on the study of consciousness to a primary focus on behavior was initiated by J. B. Watson in 1913. However, behavioristic psychology had received expression in both the French sensationalistic and British empiricistic traditions. The immediate predecessors of behaviorism were the reflexology of Russian physiology and the associationism of Thorndike. Physiological reflexology received a sound foundation with the works of Sechenov and Bekhterev, but it was Pavlov who refined the reduction of psychological events to behavioral and physiological processes within a comprehensive theory of conditioning. Watson's formulation of psychology was essentially defined in terms of stimulus and response elements. However, in attempting to rid psychology of residual mentalistic constructs, Watson's definition of psychology as solely peripheral events was too confining, and Watson's contemporaries began the process of evolving behaviorism into a more complete system. Such researchers as Holt, Weiss, Hunter, and Lashley restored critical psychological activities to behaviorism. However, it was probably the logical positivist movement, expressing an operational spirit in the unity of science, that insured the initial success of the behaviorist model.

Behavioristic psychology expanded beyond the original formulations of Pavlov and Watson. Russian reflexology continued in the tradition of Pavlov, and one of the more significant developments was found in the work of the Polish scientist Jerzy Konorski, whose goal was to integrate Pavlov's conditioning physiology with Sherrington's neurophysiology. Konorski's early work first drew a sharp distinction between the two paradigms of conditioning, and his career culminated with an insightful discussion of brain physiology supporting a cybernetic system of behavior. Contemporary reflexology in Russia and in nearby countries has greatly expanded to include a wide range of psychological and physiological problems, led by such eminent scientists as Vygotsky, Luria, Asratyan, and Beritashvili.

In the United States, behaviorism has moved through several intellectual stages. In a theory-building phase during the 1930s and 1940s, psychologists such as Guthrie, Tolman, and Hull attempted comprehensive theories of learning. Although receiving their most complete articulation under Hull, comprehensive theories were not adequate, prompting the radical positivism of Skinner. A return to data gathering followed, characterized by the development of models or mini-theories with an applied flavor. Information-processing and mathematical models of learning, neo-Hullian research models, cognitive models, and operant approaches are all examples of recent groupings of behaviorists. A major use of behaviorism has been the behavior modification model of clinical application. Contemporary behaviorism is a dominant force in psychology, but the behaviorism that has evolved is widely based, admitting a wide diversity of assumptions, methodologies, and applications.

BIBLIOGRAPHY

Reflexology

BERITASHVILI, I. S. (1965). *Neural mechanisms of higher vertebrate behavior* (W. T. Libersen, Ed. and Trans.). Boston: Little, Brown.

BERITASHVILI, I. S. (1971). *Vertebrate memory characteristics and origin*. New York: Plenum Press.

BROŽEK, J. (1971). USSR: Current activities in the history of physiology and psychology. *Journal of the History of Biology, 4,* 185–208.

BROŽEK, J. (1972). To test or not to test—Trends in Soviet views. *Journal of the History of the Behavioral Sciences, 8,* 243–248.

BROŽEK, J. (1973). Soviet historiography of psychology: Sources of biographic and bibliographic information. *Journal of the History of the Behavioral Sciences, 9,* 152–161.

BROŽEK, J. (1974). Soviet historiography of psychology. 3. Between philosophy and history. *Journal of the History of the Behavioral Sciences, 10,* 195–201.

BROŽEK, J., & SLOBIN, D. I. (Eds). (1972). *Psychology in the USSR: An historical perspective*. White Plains, NY: International Arts and Sciences Press.

COLEMAN, S. R. (1984). Background and change in B. F. Skinner's metatheory from 1930 to 1938. *Journal of Mind and Behavior, 5,* 471–500.

COLEMAN, S. R. (1985). When historians disagree: B. F. Skinner and E. G. Boring, 1930. *Psychological Record, 35,* 301–314.

KONORSKI, J. (1948). *Conditioned reflexes and neuronal organization*. Cambridge: Harvard University Press.

KONORSKI, J. (1967). *Integrative activity of the brain: An interdisciplinary approach*. Chicago: University of Chicago Press.

KONORSKI, J. (1974). Autobiography. In G. Lindzey (Ed.), *A history of psychology in autobiography* (Vol. 6). Englewood Cliffs, NJ: Prentice Hall.

KONORSKI, J., & MILLER, S. (1933). Podstawy fizjologicznej teorii ruchow nabytych. Ruchowe odruchy warunkowe. *Medyczny Doswiadczalny Spolem, 16,* 95–187.

KONORSKI, J., & MILLER, S. (1937). On two types of conditioned reflex. *Journal of General Psychology, 16,* 264–272.

LURIA, A. R. (1979). *The making of a mind. A person's account of Soviet psychology* (M. Cole & S. Cole, Eds.). Cambridge: Harvard University Press.

MILLER, S., & KONORSKI, J. (1928). Sur une forme particuliere des reflexes conditionels. *Comptes Rendus des Séances de la Société de Biologie, 99,* 1155–1157.

MOWRER, O. H. (1976). "How does the mind work?" Memorial address in honor of Jerzy Konorski. *American Psychologist, 31,* 843–857.

MOWRER, O. H. (1976). Jerzy Konorski Memorial Address. *Acta Neurobiologiae Experimentalis, 36,* 249–276.

PAVLOV, I. P. (1960; orig. 1927). *Conditioned reflexes: An investigation of the physiological activity of the cerebral cortex* (G. V. Anrep, Ed. and Trans.). New York: Dover.

SECHENOV, I. M. (1935). Reflexes of the brain (A. A. Subkov, Trans.). In *I. M. Sechenov, Selected works*. Moscow: State Publishing House for Biological and Medical Literature, 264–322.

SOKOLOV, E. M. (1963). Higher nervous functions: The orienting reflex. *Annual Review of Physiology, 25,* 545–580.

ZIELIŃSKI, K. (1974). Jerzy Konorski (1903–1973). *Acta Neurobiologiae Experimentalis, 34,* 645–653.

Operationalism

BERGMAN, G. (1954). Sense and nonsense in operationalism. *Scientific Monthly, 79,* 210–214.

BRIDGMAN, P. W. (1927). *The logic of modern physics*. New York: Macmillan.

BRIDGMAN, P. W. (1954). Remarks on the present state of operationalism. *Scientific Monthly, 79,* 224–226.

ROGERS, T. (1989). Operationism in psychology: A discussion of contextual antecedents and historical interpretation of its longevity. *Journal of the History of the Behavioral Sciences, 25,* 139–153.

SINGER, E. A. (1911). Mind as an observable object. *Journal of Philosophy, Psychology, and Scientific Methods, 8,* 180–186.

Early Behaviorism

HOLT, E. B. (1915). *The Freudian wish and its place in ethics*. New York: Holt, Rinehart & Winston.

LASHLEY, K. S. (1916). The human salivary reflex and its use in psychology. *Psychological Review, 23,* 446–464.

LASHLEY, K. S. (1923). The behavioristic interpretation of consciousness. *Psychological Review, 30,* 237–272, 329–353.

THORNDIKE, E. L. (1899). The mental life of the monkey. *Psychological Review*, Monograph Supplement, 3, no. 15.

THORNDIKE, E. L. (1936). Edward L. Thorndike. In C. Murchison (Ed.), *A history of psychology in autobiography* (Vol. 3). Worcester, MA: Clark University Press, 263–270.

THORNDIKE, E. L., & HERRICK, C. J. (1915). Watson's behavior. *Journal of Animal Behavior, 5,* 462–470.

WATSON, J. B. (1913). Psychology as the behaviorist views it. *Psychological Review, 20,* 158–177.

WATSON, J. B. (1916). The place of the conditioned reflex in psychology. *Psychological Review, 23,* 89–116.

WATSON, J. B. (1917). An attempted formulation of the scope of behavior psychology. *Psychological Review, 24,* 329–352.

WATSON, J. B. (1919). *Psychology from the standpoint of a behaviorist.* Philadelphia: Lippincott.

WATSON, J. B. (1920). Is thinking merely the action of language mechanisms? *British Journal of Psychology, 11,* 87–104.

WATSON, J. B. (1928). *Psychological care of infant and child.* New York: W. W. Norton & Co.

WATSON, J. B. (1936). Autobiography. In C. Murchison (Ed.), *A history of psychology in autobiography.* (Vol. 3).Worcester, MA: Clark University Press, 271–281.

WATSON, J. B., & MCDOUGALL, W. (1929). *The battle of behaviorism.* New York: Morton.

WATSON, J. B., & RAYNER, R. (1920). Conditioned emotional reactions. *Journal of Experimental Psychology, 3,* 1–7.

WEISS, A. P. (1917). The relation between structural and behavioral psychology. *Psychological Review, 24,* 301–317.

WEISS, A. P. (1925). *A theoretical basis of human behavior.* Columbus, OH: Adams.

YERKES, R. M., & MORGULIS, S. (1909). The method of Pavlov in animal psychology. *Psychological Bulletin, 6,* 257–273.

Broadening Behaviorism

COAN, R. W. (1968). Dimensions of psychological theory. *American Psychologist, 23,* 715–722.

FORSCHER, B. K. (1963). Chaos in the brickyard. *Science, 142,* 339.

GUTHRIE, E. R. (1946). Psychological facts and psychological theory. *Psychological Bulletin, 43,* 1–20.

GUTHRIE, E. R. (1952). *The psychology of learning* (Rev. ed.) New York: Harper & Row.

GUTHRIE, E. R. (1959). Association by contiguity. In S. Koch (Ed.), *Psychology: A study of a science, Vol. 2: General systematic formulations, learning and special processes.* New York: McGraw-Hill, 158–195.

HULL, C. L. (1937). Mind, mechanism, and adaptive behavior. *Psychological Review, 44,* 1–32.

HULL, C. L. (1943). *Principles of behavior: An introduction to behavior theory.* New York: Applcton-Century-Crofts.

KELLER, F. S. (1991). Burrhus Frederic Skinner. *Journal of the History of the Behavioral Sciences, 27,* 3–6.

MARX, M. H. (1963). The general nature of theory construction. In M. H. Marx (Ed.), *Theories in contemporary psychology.* New York: Macmillan, 4–46.

MARX, M. H. (1976). Formal theory. In M. H. Marx and F. E. Goodson (Eds.), *Theories in contemporary psychology,* 2nd ed., New York: Macmillan, 234–260.

SAMUELSON, F. (1985). Organizing for the kingdom of behavior: Academic battles and organizational policies in the twenties. *Journal of the History of the Behavioral Sciences, 21,* 33–47.

SCHNAITTER, R. (1984). Skinner on the "mental" and the "physical." *Behaviorism, 12,* 1–14.

SHEFFIELD, F. D. (1949). Hilgard's critique of Guthrie. *Psychological Review, 56,* 284–291.

SKINNER, B. F. (1935). Two types of conditioned reflex and a pseudo type. *Journal of General Psychology, 12,* 66–77.

SKINNER, B. F. (1937). Two types of conditioned reflex: A reply to Konorski and Miller. *Journal of General Psychology, 16,* 272–279.

SKINNER, B. F. (1966). *The behavior of organisms: An experimental analysis.* Englewood Cliffs, NJ: Prentice Hall.

SKINNER, B. F. (1974). *About behaviorism.* New York: Alfred A. Knopf.

SKINNER, B. F. (1984). Behaviorism at fifty. *Behavioral and Brain Sciences, 7,* 615–667.

SMITH, S., & GUTHRIE, E. R. (1921). *General psychology in terms of behavior.* New York: Appleton.

SPENCE, K. W. (1937). The differential response in animals to stimuli varying within a single dimension. *Psychological Review, 44,* 430–444.

SPENCE, K. W. (1940). Continuous vs. non-continuous interpretations of discrimination learning. *Psychological Review, 47,* 271–288.

TOLMAN, E. C. (1922). A new formula for behaviorism. *Psychological Review, 29,* 44–53.

TOLMAN, E. C. (1948). Cognitive maps in rats and men. *Psychological Review, 55,* 189–208.

TOLMAN, E. C. (1949). *Purposive behavior in animals and men.* New York: Appleton-Century-Crofts, 1932; reprint, University of California Press.

VOEKS, V. W. (1950). Formalization and clarification of a theory of learning. *Journal of Psychology, 30,* 341–363.

VOEKS, V. W. (1954). Acquisition of S-R connections: A test of Hull's and Guthrie's theories. *Journal of Experimental Psychology, 47,* 137–147.

Applications:

ATKINSON, R. C., BOWER, G. H., & GROTHERS, E. J. (1965). *Introduction to mathematical learning theory.* New York: Wiley.

BRUNER, J. S., BRUNSWIK, E., FESTINGER, L., HEIDER, F., MUENZINGER, K. F., OSGOOD, C. E., & RAPAPORT, E. (1957). *Contemporary approaches to cognition: A symposium held at the University of Colorado.* Cambridge, MA: Harvard University Press.

BRUNSWIK, E.(1956). *Perception and the representative design of psychological experiments.* Berkeley: University of California Press.

BUCHANAN, R. D. (1994). The development of the Minnesota Multiphasic Personality Inventory. *Journal of the History of the Behavioral Sciences, 30,* 148–161.

CHOMSKY, N. (1959). Review of verbal behavior by B. F. Skinner. *Language, 35,* 26–58.

CHOMSKY, N. (1972). *Language and mind.* New York: Harcourt Brace Jovanovich.

COFFER, C. N. (1981). The history of the concept of motivation. *Journal of the History of the Behavioral Sciences, 17,* 48–53.

DOLLARD, J., & MILLER, N. E. (1950). *Personality and psychotherapy: An analysis in terms of learning, thinking and culture.* New York: McGraw-Hill.

ESTES, W. K. (1950). Toward a statistical theory of learning. *Psychological Review, 57,* 94–107.

ESTES, W. K. (1959). The statistical approach to learning theory. In S. Koch (Ed.), *Psychology: A study of a science, Vol. 2: General systematic formulations, learning and special processes.* New York: McGraw-Hill, 380–491.

ESTES, W. K. (1964). Probability learning. In A. W. Melton (Ed.), *Categories of human learning.* New York: Academic Press.

FESTINGER, L. (1957). *A theory of cognitive dissonance.* New York: Harper & Row.

GARNER, W. R., & HAKE, W. H. (1951). The amount of information in absolute judgments. *Psychological Review, 58,* 446–459.

MILLER, N. E. (1969). Learning of visceral and glandular responses. *Science, 163,* 434–445.

MOWRER, O. H. (1949). On the dual nature of learning: A reinterpretation of "conditioning" and "problem-solving." *Harvard Educational Review, 17,* 102–148.

MOWRER, O. H. (1960). *Learning theory and behavior.* New York: Wiley.

PIAGET, J. (1952). *The origins of intelligence in children* (M. Cook, Trans.), New York: International Universities Press.

RESCORLA, R. A., & SOLOMON, R. L. (1967). Two-process learning theory: Relationships between Pavlovian conditioning and instrumental learning. *Psychological Review, 74,* 151–182.

SCHOENFELD, W. N. (1950). An experimental approach to anxiety, escape and avoidance behavior. In P. H. Hock & J. Zubin (Eds.), *Anxiety.* New York: Grune and Stratton.

SHEFFIELD, F. D. (1965). Relation between classical conditioning and instrumental learning. In W. F. Prokasy (Ed.), *Classical conditioning: A symposium.* New York: Appleton-Century-Crofts, 302–322.

SWETS, J. A. (1961). Is there a sensory threshold? *Science, 134,* 168–177.

TAYLOR, J. A. (1951). The relationship of anxiety to the conditioned eyelid response. *Journal of Experimental Psychology, 41,* 81–92.

TAYLOR, J. A. (1953). A personality scale of manifest anxiety. *Journal of Abnormal and Social Psychology, 48,* 285–290.

TRIPLETT, R. G. (1982). The relationship of Clark L. Hull's hypnosis research to his later learning theory: The continuity of his life's work. *Journal of the History of the Behavioral Sciences, 18,* 22–31

WEIDMAN, N. (1994). Mental testing and machine intelligence: The Lashley-Hull debate. *Journal of the History of the Behavioral Sciences, 30,* 162–180.

WINKLER, R. C. (1970). Management of chronic psychiatric patients by a token reinforcement system. *Journal of Applied Behavior Analysis, 3,* 47–55.

WOLPE, J. (1958). *Psychotherapy by reciprocal inhibition.* Stanford, CA: Stanford University Press.

WOLPE, J. (1969). *The practice of behavior therapy.* Elmsford, NY: Pergamon.

Studies

BITTERMAN, M. E. (1969). Thorndike and the problem of animal intelligence. *American Psychologist, 24,* 444–453.

BORING, E. G. (1950). *A history of experimental psychology,* 2nd ed. Englewood Cliffs, NJ: Prentice Hall.

BRUCE, D. (1986). Lashley's shift from bacteriology to neurophysiology, 1910–1917, and the influence of Jennings, Watson, and Franz. *Journal of the History of the Behavioral Sciences, 22,* 27–44.

BUCKLEY, K. W. (1982). The selling of a psychologist: John Broadus Watson and the application of behavioral techniques to advertising. *Journal of the History of the Behavioral Sciences, 18,* 207–221.

BURNHAM, J. C. (1968). On the origin of behaviorism. *Journal of the History of the Behavioral Sciences, 4,* 143–151.

BURNHAM, J. C. (1972). Thorndike's puzzle boxes. *Journal of the History of the Behavioral Sciences, 8,* 159–167.

BURNHAM, J. C. (1977). The mind-body problem in the early twentieth century. *Perspectives in Biology and Medicine, 20,* 271–284.

CARMICHAEL, L. (1968). Some historical roots of present-day animal psychology. In B. Wolman (Ed.), *Historical roots of contemporary psychology*. New York: Harper & Row, 47–76.

COHEN, D. (1979). *J. B. Watson: The founder of behaviorism*. Boston: Routledge and Kegan Paul.

COLEMAN, S. R. (1985). The problem of volition and the conditioned reflex: I. Conceptual background, 1900–1940. *Behaviorism, 13,* 99–124.

DANZIGER, K. (1979). The positivist repudiation of Wundt. *Journal of the History of the Behavioral Sciences, 15,* 205–230.

FRANK, P. (1941). *Between physics and philosophy*. Cambridge: Harvard University Press.

HERRNSTEIN, J. R. (1969). Behaviorism. In D. L. Krantz (Ed.), *Schools of psychology: A symposium*. New York: Appleton-Century-Crofts, 51–68.

JONCICH, G. (1968). *The sane positivist: A biography of Edward L. Thorndike*. Middletown, CT: Wesleyan University Press.

LEYS, R. (1984). Meyer, Watson, and the dangers of behaviorism. *Journal of the History of the Behavioral Sciences, 20,* 128–149.

LOWRY, R. (1970). The reflex model in psychology: Origins and evolution. *Journal of the History of the Behavioral Sciences, 6,* 64–69.

MACKENZIE, B. D. (1972). Behaviorism and positivism. *Journal of the History of the Behavioral Sciences, 8,* 222–231.

MCCONNELL, J. V. (1985). Psychology and scientist: LII. John B. Watson: Man and myth. *Psychological Reports, 56,* 683–705.

ROBACK, A. A. (1964). *History of American psychology* (Rev. ed.). New York: Collier.

RUCKMICK, C. A. (1916). The last decade of psychology in review. *Psychological Bulletin, 13,* 109–120.

SAMELSON, F. (1981). Struggle for scientific authority: The reception of Watson's behaviorism, 1913–1920. *Journal of the History of the Behavioral Sciences, 17,* 399–425.

SCHNEIDER, S. M., & MORRIS, E. K. (1987). A history of the term "radical behaviorism": From Watson to Skinner. *Behavior Analyst, 10,* 27–39.

STEININGER, M. (1979). Objectivity and value judgments in the psychologies of E. L. Thorndike and W. McDougall. *Journal of the History of the Behavioral Sciences, 15,* 263–281.

STEVENS, S. S. (1939). Psychology and the science of science. *Psychological Bulletin, 36,* 221–263.

THORNE, F. C. (1976). Reflections on the golden age of Columbia psychology. *Journal of the History of the Behavioral Sciences, 12,* 159–165.

TIBBETTS, P. (1975). The doctrine of "pure experience": The evolution of a concept from Mach to Jones to Tolman. *Journal of the History of the Behavioral Sciences, 11,* 55–66.

TODD, J. T., & MORRIS, E. K. (1986). The early research of John B. Watson: Before the behavioral revolution. *Behavior Analyst, 9,* 71–88.

TURNER, M. B. (1967). *Philosophy and the science of behavior*. New York: Appleton-Century-Crofts.

WASHBURN, M. F. (1917). Some thoughts on the last quarter century in psychology. *Philosophical Review, 27,* 44–55.

WINDHOLZ, G. (1990). Pavlov and the Pavlovians in the laboratory. *Journal of the History of the Behavioral Sciences, 26,* 64–74.

WOODWORTH, R. S. (1959). John Broadus Watson: 1878–1958. *American Journal of Psychology, 72,* 301–310.

YAROSHEVSKI, M. G. (1968). I. M. Sechenov—The founder of objective psychology. In B. J. Wolman (Ed.), *Historical roots of contemporary psychology*. New York: Harper & Row, 77–110.

16

The Third Force Movement

We have seen that the emergence of psychology in Germany during the last quarter of the nineteenth century was presented under the conceptual expressions of a natural science model and a human science model. The reliance of the natural science model on philosophical assumptions of the essential passivity of the mind and the belief in the empirical approach to science were initially reflected in Wundt's structural psychology and later were fully elaborated in America by the behaviorists. In contrast, the human science model had diverse applications, but at the very least it accepted the assumption of mental activity and allowed methods of scientific approach other than empirical strategies only. Gestalt psychology grew out of the German active mind tradition of nonsensory consciousness and recognized the necessity of nonanalytic methods to study psychological processes. In this context, the dynamics of unconscious motivation within the psychoanalytic movement were derived from internal psychic energy and precluded the exclusive reliance on empirical methods of study.

The third force movement in psychology was also derived from the traditions of the human science model of mental activity. The term *third force* is actually a general categorization for several orientations and emphases within psychology. If psychoanalysis is considered the "first force" and behaviorism the "second force" within twentieth-century psychology, then the "third force" may be any movements that are not psychoanalytic or behavioristic. Other labels describe various expressions of the third force movement: *Existential psychology* indicates the applications of existential philosophy to psychological issues. The term *phenomenological psychology* is sometimes used to express characteristic ways of studying psychological events without resorting to reductionism. Finally, *humanistic psychology* describes the approach of a group of psychologists, mainly American personality theorists, who held a view of individuals as seeking the full development of their capacity or potential and who rejected any mechanical or materialistic explanations of psychological processes.

Although the third force movement consists of a diverse collection of psychologists and philosophers, some commonly shared points of view are evident. First, the movement clearly recognizes the importance of personal freedom and responsibility in the lifelong process of decision making to fulfill the human potential. It considers the mind to be an active, dynamic entity through which the individual expresses uniquely human abilities in cognition, willing, and judgment. Second, psychologists of the movement do not accept the reduction of psychological processes to the mechanical laws of physiological events. Rather, they see human beings as separate from other forms of life. Individuals, in the very process of defining their humanity, must go beyond the hedonistic satisfaction of physiological needs in quest of personal values and attitudes of social and philosophical significance. Thus, there is an emphasis on the self in the third force movement, which attempts to foster the development of individually defined, uniquely human personality fulfillment.

The third force movement is not a coherent system of detailed principles that are universally accepted by all followers. Rather, it is an orientation within psychology that has reacted to the inherent reductionism of moving from psychological processes to physiological bases, represented by empirical behavioristic psychology. Like psychoanalysis, the third force movement did not emerge from the academic setting of university research. Rather, its roots may be found in philosophical speculation, literary works, and clinical observations. These sources coalesced after World War II and gave expression to the third force movement in Europe and in America.

EUROPEAN PHILOSOPHICAL BACKGROUND

In the review of European philosophy contained in Chapters 6 through 9, the development of competing models of mental processes was presented in terms of the eventual emergence of distinctive active and passive assumptions about the mind in modern psychology. However, certain philosophical trends continued to develop the notion of mental activity after the initial, formal expressions of nineteenth-century psychology. Collectively, these philosophical trends of existentialism and phenomenology formed the basis of the third force movement.

Existentialism

The core of existential philosophy holds that the individual is free to define life's direction through a continued succession of choices, but that this freedom also gives the individual responsibility for the outcomes of personal decisions, so that freedom is a source of anguish and dread. Before we explore the elaboration and implications of this definition, it is important to recognize that existential themes were common to many of the philosophical views dating back to antiquity. Indeed, it may be argued that all of the superordinate models of dynamic human activity, which stressed a holistic position, were existential. Such philosophers as Socrates, Plato, Aristotle, and Aquinas taught that people are free to decide their individual fate and must also accept the consequences of their choices.

In the nineteenth century modern existentialism began to grow in literature through such notable writers as Fyodor Dostoyevsky (1821–1881) and Friedrich Nietzsche (1844–1900). Dostoyevsky was born and educated in Moscow, but in 1849 was exiled to Siberia for revolutionary activities. Upon his return to European Russia in 1859, he resumed his writing and soon demonstrated his genius as one of the world's greatest novelists. His characters from such novels as *The Idiot* (1869), *The Brothers Karamazov* (1880), and *Crime and Punishment* (1866) confronted and wrestled with the difficult decisions of defining themselves and their feelings about God, social values, and personal ideals. Nietzsche, who wrote on philosophical themes, was born in Saxony and attended the universities at Bonn and Leipzig. At age 24, he was appointed professor of classical philology at the University of Basel. His thinking on the profound issues of life led him to the conclusion that God is dead and the individual is forlorn and alone, unable to rely on God for security. Each individual must confront alone the choices in life and face the consequences of those decisions without recourse to divine reassurance.

Whereas the major themes of existentialism are vividly portrayed in nineteenth-century literature, the formal statement of modern existential principles resulted from a theological controversy surrounding the manner of knowing and experiencing God. As background to this controversy, it should be recalled that the interpretation of dynamic mental activity in the structuring of knowledge, championed by Kant, became the dominant force among the German intelligentsia. This position, known as rationalism, glorified the value of reason in finding ultimate truths. German rationalism of the nineteenth century found an articulate spokesman in Georg Hegel (1770–1831).

As a philosophy, Hegel's rationalism held that intellectual progress proceeds through a sequence in which an idea, or thesis, gives rise to its opposite idea, or antithesis, and the two synthesize into a new unity that in turn becomes a thesis, repeating the cycle. This notion developed into a form of logical argumentation termed "dialectics," which was later adopted by Karl Marx (1818–1883) and Friedrich Engels (1820–1895), who used the dialectical method to formulate their theory of socialism. For present purposes, it is important to note that Hegel's views emphasized the centrality of intellectual progress with the implication of a hierarchy of intellectual activities. Hegel's rationalism found a sympathetic audience among nineteenth-century German theologians who recognized the decline in the strength of dogmatic church authority. Hegel's rationalism offered an alternative to the traditional tenets based on faith; it ordered nature and tried to develop a science of theology based on logical demonstration. Human intellectual activities could be ordered from the primitive level of art to a mediocre level of religion to the highest level of reason and science. The position of religion was relegated to a second-rate belief suitable for second-rate minds. This interpretation was consistent with the prevailing nineteenth-century atmosphere that exalted positivist science above all forms of intellectual activity. Science was viewed as the model that all intellectual endeavors should emulate.

Søren Kierkegaard. A strong reaction to Hegel's rationalism was expressed by a Danish Lutheran clergyman, Søren Kierkegaard (1813–1855). Western civilization had once been Christian but, Kierkegaard insisted, people had lost their faith. He took it upon himself to teach Christianity to Christians and to support the view of the primacy of faith over reason. Kierkegaard perceived the elevation of reason, represented by Hegelian rationalism, as a distortion of human experience. As he confidently put it, "It was intelligence and nothing else that had to be opposed. Presumably that is why I, who had the task, was armed with an immense intelligence" (Kaufman, 1956). Kierkegaard continually questioned the true feelings of Christians and challenged them to demonstrate more than superficial belief.

Kierkegaard was born in Copenhagen, the youngest child of an eventually successful merchant. He was raised in a strict, religious home but seems to have spent his years at the University of Copenhagen rebelling against his father and his religious views. Rejected by the Danish royal guards because of his health, Kierkegaard began seeking a place in life. Around 1835, he went through a religious conversion, which changed his life. In 1837, Kierkegaard met and fell in love with a woman named Regina Olsen. During their engagement he questioned the authenticity of his love for her. He broke off the engagement in 1841 and fled to Berlin, where he immersed himself in philosophical study and finished his first major work, *Either/Or*. He returned to Denmark and spent the remainder of his life attacking the established religious practice in his country and advocating a recommitment to Christianity.

For Kierkegaard, existence is made authentic by a total acceptance of faith. Existence is not to be studied but lived. He described three progressive levels of existence. The first is esthetic and characterizes the childhood stage of living for the moment according to the dictates of temporary pleasure or pain. Although it is an important stage, it is primitive to the extent that the individual is a detached observer of life's events, simply responding to external contingencies according to the needs of the moment. The esthetic gives way to an ethical stage that requires courage on the part of the individual, because the person must make choices regarding the values of life and accept responsibility for those choices. The highest level of existence is the religious. In this stage the individual goes beyond the social morals of the ethical stage to a choice of God, which is an act of faith. In his work *Fear and Trembling* (1843), Kierkegaard recalled the biblical story of Abraham's preparing to sacrifice his son Isaac at God's command. That moment, as Abraham lifted his knife to kill his son, captures Kierkegaard's feelings of faith. Religion is a leap into darkness, accompanied by anguish, fear, and dread. Christianity, for Kierkegaard, must be a total subjective experience directed inward by a fully committed participator with Christ, not by a spectator. Christianity, then, is absurd. Just as it made no sense for the Creator to become a creature in the person of Christ, so a profession of Christianity is unreasonable because the profession of faith runs against the grain of our rational abilities. Christianity requires faith in the unreasonable. Whereas Kierkegaard would disagree with Nietzsche's conclusion that God is dead, he would agree with Nietzsche's feeling of God as dead, because faith requires a person to forsake the security of reason and plunge into the unknown.

Wilhelm Dilthey. Another of the earlier expressions of modern existentialism came from Wilhelm Dilthey (1833–1911), who was mentioned in Chapter 11 as an advocate of alternatives to the natural science model. He brought existential principles to psychological perspective. In 1852, Dilthey began his studies at Heidelberg, initially intending to become a divinity student, but soon turned exclusively to philosophy. After studying Kant's rationalism, Hume's empiricism, and Comte's positivism, Dilthey developed an emphasis on the historical presence of the individual human being. His teaching career took him to several German universities and eventually to Berlin, where he remained until his retirement in 1906.

Dilthey called for a "science of the spirit," as opposed to the natural sciences, to understand the historicity of human beings by discovering what is individual and particular in each person. Historical consciousness is the defining characteristic of each person. In his *Essence of Philosophy* (1907), Dilthey wrote that religion, art, science, and philosophy are all expressions of experiences lived in the world, and these experiences involve not only intellectual functions but individual goals, values, and passions as well. Accordingly, Dilthey's emphasis on lived-through experience asserts the basic individual nature of consciousness that defines existence.

The early expressions of existential philosophy, represented by Kierkegaard and Dilthey, were pursued in the twentieth century by a group of philosophers and writers who moved away from the religious perspective of Kierkegaard to more inclusive statements on the self and the psychology of the individual. Although as a group they gained some prominence between the world wars, it was in the years immediately following World War II that the existentialists exerted influence on Western intellectual life. Their call for a restructuring of human values and respect for individual dignity gained a sympathetic audience among those who had suffered the depersonalization of industrialized warfare.

Modern Expressions of Existentialism

Jean-Paul Sartre. Perhaps the most popular existentialist of the twentieth century, Jean-Paul Sartre (1905–1980) successfully conveyed his existential themes through novels, plays, and philosophical essays. After earning his philosophy degree from the Ecole Normale Supérieure in 1929, he studied in Germany, coming under the influence of the existential and phenomenological teachings of Edmund Husserl and Martin Heidegger. Their influence was reflected in Sartre's first major philosophical work, *Being and Nothingness* (1943). His first successful novel, *Nausea* (1938), was followed by over 15 other novels, plays, and collections of short stories. Sartre was drafted as a private into the French army in 1939. Shortly afterward, he was captured by the Germans in the doomed defense of the Maginot Line, but was released in 1941. He worked in the French Resistance until the end of the war, writing and teaching in the underground. During most of his life he was aligned with leftist and communist politics, and with his long-time companion, the philosopher Simone de Beauvoir (1908–1986), he became a French institution and spokesman for various political and social causes. He refused awards for his writings, including the 1964 Nobel Prize for literature, claiming acceptance would compromise his beliefs.

Basic to Sartre's views is that existence precedes essence. In contrast to the Aristotelian and Scholastic tenet that individual existence is an expression of a general, metaphysical essence or being, Sartre asserted that existence defines the essence of an individual. In this sense, we are what we do. Our existence is not defined by what we might become, but only by what we are in actuality, the collection of our acts. For this reason it is critical that we continually move through choices, for by making decisions we define ourselves and secure personal growth. One is, then, what one wills oneself to be. We are free to choose, but we must take responsibility for our choices. The only compulsion in life is to choose.

The individual lives her or his existence and creates a personal essence. The essence of God, according to Sartre, is a product of humans, who give God existence in their minds. God is reducible to human existence. The qualitative distinction between humans and the rest of nature is our subjectivity. Sartre asserted that human subjectivity is an enormous privilege that provides great dignity, but also condemns us to the freedom of making choices. Accordingly, as individuals, we are filled with anguish. We have total and deep responsibility that rests with us every time we make a decision. For example, if we decide that we shall be truthful, that decision imposes a truthful standard on all people for us. We are forlorn. Sartre concluded that because God does not exist, we are alone and insecure, with everyone potentially possessing the freedom to make up his or her own rules of deportment, with no divinely inspired guidance. We are in despair. According to Sartre, our responsibility is to ourselves and we have only ourselves to rely on. We cannot blame God or "fate" for bad decisions; we can blame only ourselves. Thus, Sartre's psychology is based on the existential premise of the radical freedom of individual existence.

Albert Camus. A novelist–philosopher of the postwar French existential tradition, Albert Camus (1913–1960) took as his main literary theme courage when faced with life's absurdities. Camus was born and raised in poverty in French Algeria. After a serious bout with tuberculosis in 1930, he went on to complete his studies in philosophy at the University of Algiers, although a career in university teaching was precluded because of his medical history. He became involved in theater and journalism in Algiers, and during World War II edited a clandestine newspaper in Lyon, France. Among his many works are the essay *The Myth of Sisyphus* and the novel *The Stranger*, both originally published in 1942. After the war Camus returned to directing theater and writing. He also engaged in a heated debate with Sartre over the application of communist principles to government and society. The circumstances of his death reflected Camus' sense of the absurd. Apparently undecided whether to drive or to take the train to his destination, he died in an automobile accident on January 4, 1960, with a train ticket for that day in his pocket. In his writings Camus continually placed the individual at the mercy of external forces that render the life situation absurd. He attempted to identify individual resources that might allow a person to reorient life to more fulfilling directions by exerting the courage to take control and establish a sense of purpose.

Karl Jaspers. Like Camus, Karl Jaspers (1883–1969) pursued the theme of meaning in existence and the relevance of meaning to psychology. He defined philosophy as the inquiry into freedom, history, and the possibility of meaning in existence. Jaspers studied medicine and law at four German universities prior to joining the staff of a psychiatric hospital in Heidelberg. Specializing in psychology, Jaspers joined the philosophy faculty at the University of Heidelberg in 1913, where he continued to develop his existential basis for psychology. However, because he refused to leave his Jewish wife, the Nazis harassed him increasingly during the 1930s, and by 1938 he had lost his professorship and was forbidden to publish. In 1945, when Heidelberg was liberated by the Americans, Jaspers formed a group to reopen the university and continued with that task until 1949, when he joined the University of Basel in Switzerland.

Jaspers consistently expressed his concern with human existence, which led him to suggest three stages of being. The first stage is *being-there* and places the individual in reference to the external, objective world of reality. *Being-oneself* is the stage that allows the person self-awareness of choices and decisions. *Being-in-itself* is described by Jaspers as the highest stage of existence, characterized by the attainment of the fullness of meaning. This stage is the transcendental world of individual meaning that encompasses and comprehends the totality of meaning; the individual is in effective communication with the social and physical environment, so that existence is fully defined.

Martin Buber. A final representative of modern existential philosophy is Martin Buber (1878–1965), who was born in Vienna and raised by his grandfather, a Hebrew scholar, in the predominantly Polish city of Lvov (now in the Ukraine). Buber received his degree in philosophy from the University of Vienna in 1904, and by this time was involved with the Zionist movement. He spent 5 years in Hasidic communities in Galicia studying the religious, cultural, and mystical traditions of his ancestors. Returning to Germany, he edited *Der Jude* (1916–1924) and with a Catholic and a Protestant coedited *Die Kreatur* (1926–1930). He was professor of comparative religion at the University of Frankfurt from 1923 until his dismissal by the German government in 1933. He went to Palestine in 1938 and taught social philosophy at Hebrew University until his retirement in 1951. He remained active, lecturing in Europe and in America until his death.

Buber's writings are interesting because he did not emphasize consciousness or self-awareness. Rather than "self-dialogue," Buber stressed dialogue between persons and between the person and God, as reflected in his work *I and Thou* (1923). Out of the dual contributors of a dialogue comes a unity, so that individuals define themselves in terms of other persons or in terms of God. Thus, Buber added a critical social dimension of personal growth, which complemented other expressions of self-growth in the existential framework.

Although this brief review of several existential philosophers is certainly not comprehensive, it reflects the diversity of opinion. Existentialists are atheistic as well as religious; pessimistic and optimistic; looking to meaning and relegating life to absurdity. However, they do share common ground in emphasizing the individual's quest for existence and identity. After surveying the phenomenological trend in philosophy, we will review some specific existential interpretations of psychology.

Phenomenology

In Chapter 13, the phenomenological basis of Gestalt psychology was outlined in terms of a general approach in German psychology, which was contrasted with the analytical alternative of other empirical strategies. However, in relation to the third force movement, phenomenology assumes a more critical role, both as a methodology and as an expression of essential assumptions common to many of the positions within the movement. In this context phenomenology developed in a more specific and elaborate way than portrayed as background to Gestalt psychology.

Within the approach of the third force movement, phenomenology concentrates on the study of phenomena as experienced by the individual, with the emphasis on exactly how a phenomenon reveals itself to the experiencing person in all its specificity and concreteness. As a methodology, phenomenology is open to whatever may be significant to the understanding of a phenomenon. The subject experiencing a phenomenon is required to attend to it exactly as it appears in consciousness, without prejudgment, bias, or any predetermined set or orientation. The goals of the method are:

1. The apprehension (literally, the mental grasping) of the structure of the phenomenon as it appears;
2. The investigation of the origins or bases of the phenomenon as experienced; and
3. The emphasis on the possible ways of perceiving all phenomena.

The task of the phenomenologist is to investigate the processes of intuition, reflection, and description. Accordingly, phenomena are not manipulated but, rather, are permitted to reveal themselves.

The substance of phenomenology consists of the data of experience and their meaning for the experiencing individual. Phenomenology rejects the reductionism inherent in the empirical methods of the natural sciences. Rather, phenomenology focuses on the significance and relevance of phenomena in the consciousness and perspective of the whole person.

Edmund Husserl. The founder of modern phenomenology was Edmund Husserl (1859–1938), who was born in Moravia, now a province of the Czech Republic. From 1876 to 1878 he studied at Leipzig, where he heard Wundt's lectures on psychology; in 1881 he transferred to Vienna to study mathematics. While at Vienna he came under the influence of Franz Brentano, whose act psychology became an important part of Husserl's phenomenology. Brentano sent Husserl to Halle in 1886 to study psychology with Stumpf. Accordingly, Husserl's commitment to psychology was by way of the antireductionistic views of Brentano and Stumpf, rather than Wundt's study of the elements composing consciousness. Husserl taught from 1900 to 1916 at Göttingen and then was named to the chair of philosophy at Freiburg, where he remained until his retirement in 1928.

Husserl's goal was to find a philosophy of science, and an associated methodology, which would be as rigorous as the empirical methods but would not require the reduction of subject matter to constituent elements. He distinguished between two general branches of knowledge. One branch includes those disciplines that study the person's experience of the physical world, which involve the person turned outward to the environment. Husserl described these disciplines as the traditional natural sciences. The other branch—philosophy—takes as its subject matter the study of the person's experience of herself or himself, the person turned inward. The major implication of Husserl's distinction between branches of knowledge is that psychology should resolve any differences and study the relationship of the person's inward-directed and outward-directed experiences.

For Husserl, consciousness does not exist as an abstract mental agency or a storehouse of experiences. Rather, consciousness is defined as the individual's being conscious of *something*. That is, consciousness exists as the individual's experiencing of an object. Reflecting Brentano's notion of the intentionality of the person, Husserl stated that every conscious act intends some object. To study consciousness, Husserl introduced the method of phenomenological "reduction," which is not the empirical, elementaristic approach of reducing psychological events to component parts, but a way of grasping the salient images of consciousness by penetrating the "layers" of experience. He described three types of phenomenological reduction:

1. The "bracketing" of being, which specifies relations within an experience between the individual and the object of consciousness while retaining the essential unity of the experience. For example, the experience described by "I see the dog" could be bracketed as follows:

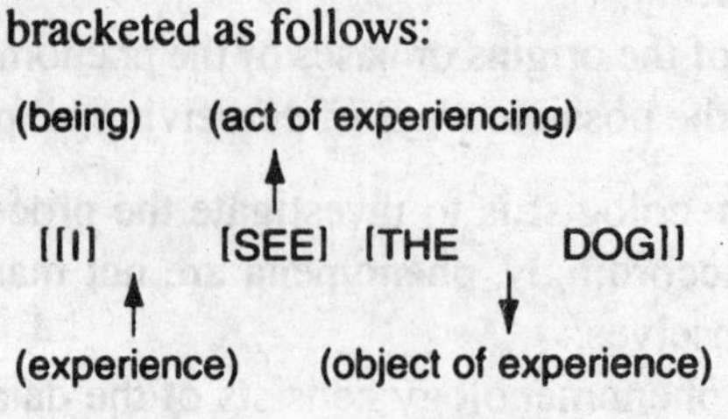

In this procedure, the processes of the experience are articulated and the unity that would be destroyed if any of the processes were isolated and examined separately is underscored.

2. The relationship of the cultural world to an immediate experience. This type of reduction recognizes the assimilation of values and attitudes that people acquire and carry with them, with the result that cultural modes exert a continual contextual set in the appearance of experiences.
3. Transcendental reduction, which leads the person from the phenomenal world of specific experiences to a level of subjectivity that rises above present reality to an integrative level of unifying experience. It is by achieving transcendental subjectivity that we live truly human existences, according to Husserl.

Thus, Husserl attempted to provide an alternative to the elementaristic reductionism of the empirical approaches of the natural science model. He employed a descriptive method that proposed to make psychological inquiry more complete through a consideration of the essential structure of experience and its objects.

Martin Heidegger. One of Husserl's assistants at Freiburg, Martin Heidegger (1889–1976), extended these interpretations of phenomenology. Born in the state of Baden in Germany, Heidegger briefly entered a Jesuit seminary, then spent 2 years training as a diocesan priest in Freiburg. There he was introduced to Brentano's dissertation on the meaning of being in Aristotle, a theme that Heidegger consistently studied throughout his life. In 1909, he began to study philosophy at the University of Freiburg and in 1914 received his degree for a dissertation entitled *The Theory of Judgment in Psychologism*. Shortly after, as a young faculty member, he became Husserl's assistant, marking the start of a productive, although tumultuous, relationship that furthered the development of the phenomenological movement. Heidegger's career from 1933 until his death was clouded by his controversial relationship with the Nazis. Certain pro-Nazi statements by Heidegger were recorded, yet his students during this period have also testified that he was anti-Nazi. After World War II Heidegger retired and made few public appearances, and he denied the more vehement accusations of his alleged collaboration.

His major work, *Being and Time* (1927), was dedicated to Husserl, but this work contains the seeds of their subsequent disagreements. Essentially, Husserl stressed the study of philosophy as an examination of consciousness, whereas Heidegger emphasized philosophy as the study of being. Heidegger wrote that people are estranged from their own being. He distinguished between *being* used as a noun and *being* as a verb. He argued that, throughout history, people have been bound to beings in terms of things or objects, but they have become alienated from being as living. Heidegger used phenomenology as a means of returning to the act of being. Phenomenology (the original Greek word means "to reveal itself") allows phenomena to be understood, if we do not coerce them into preconceived structures. Thus, for Heidegger, the essence of psychology is to study the characteristic modes of a person's being-in-the-world, for if people are estranged from their own being, they go through life alienated and psychologically fragmented, ultimately falling into a psychotic existence.

Heidegger, then, did not refer to an individual or to consciousness, because such terms imply an object. Rather, he categorized human existence by three basic, interacting traits:

1. *Mood or feeling*: People do not have moods; they are moods—we are joy; we are sadness.
2. *Understanding*: Instead of the accumulation of conceptual abstractions, human existence should be examined as the search for understanding our being. Heidegger described this search as standing open before the world so that we can internalize our confirmation of the truth or falsehood of our experience—that is, so we become an authentic self.
3. *Speech:* Rooted in the internal silence of the person, speech as language provides the vehicle for our knowledge of ourselves as beings.

Heidegger suggested that we become truly authentic only after we adjust to the concept of death and internalize the subjective meaning of death. Anxiety is the fear of nonbeing, the very antithesis of being, which is the result of an individual's unwillingness to confront death. By accepting and understanding that we are finite, we begin to penetrate the core of our existence. Thus, the uniqueness of human life lies in our understanding, however dim, of our own being.

The phenomenology of Husserl and Heidegger provided a strategy for studying the individual as an existential person. Together, existentialism and phenomenology gave philosophical substance and methodological direction to the third force movement as a psychological system.

EXISTENTIAL–PHENOMENOLOGICAL PSYCHOLOGY

As an expression of contemporary psychology, existential–phenomenological views are intimately bound up with their respective philosophical underpinnings. Indeed, the boundary separating existential phenomenology as a philosophy from existential phenomenology as a psychology is obscure. Existential–phenomenological psychology is typically an application of the philosophical principles, usually in a therapeutic, clinical setting. These principles may be summarized as follows:

1. The person is viewed as an individual existing as a being-in-the-world. Each person's existence is unique and reflects individual perceptions, attitudes, and values.
2. The individual must be treated as a product of personal development and not as an instance of generalized, human commonalities. Accordingly, psychology must deal with individual experience in consciousness to understand human existence.
3. The person moves through life striving to counteract the depersonalization of existence by society, which has led to subjective alienation, loneliness, and anxiety.
4. Phenomenology as a method permits the examination of the experiencing individual.

We briefly consider two representative psychologists of the European existential–phenomenological movement: Maurice Merleau-Ponty and Ludwig Binswanger. Although both are famous as exponents of existential–phenomenological approaches in psychology, neither can be viewed as a comprehensive system builder. Rather, both Merleau-Ponty and Binswanger reflect the attempt by psychologists to assimilate the basic philosophical tenets of existentialism so as to arrive at successful forms of treatment supporting the individual's search for authenticity.

Maurice Merleau-Ponty

Maurice Merleau-Ponty (1908–1961) received a strong background in philosophy and the empirical sciences, and taught in the most eminent French universities. In 1927, he met Sartre and began a long association that culminated in 1944 with their coediting *Les Temps Moderns*, a journal devoted to philosophical, political, and artistic issues. In 1952, he broke with Sartre over the questionable benefits of Marxist government for France and in the Soviet Union. Also in that year Merleau-Ponty was appointed to the chair of philosophy at the Collège de France, the youngest person ever named to that prestigious position.

In his most famous work, *Phenomenology of Perception* (1944), Merleau-Ponty described psychology as the study of individual and social relationships as they particularly link consciousness and nature. Reflecting the influence of Husserl, Heidegger, and Sartre, Merleau-Ponty maintained that the person is not a consciousness endowed with the characteristics that anatomy, zoology, and empirical psychology traditionally suggest. Rather, the person is the absolute source of existence. The individual does not acquire existence from antecedent physical events. Instead, the person moves toward the environment and sustains physical events by bringing those aspects of the environment into her or his existence. Psychology, then, is the study of individual intentionality. For Merleau-Ponty, every intention is an attention, and we cannot attend to something unless we experience it.

Merleau-Ponty described three major questions confronting modern psychology:

1. Is the human being an active or a reactive organism?
2. Is activity determined internally or externally?
3. Is psychological activity of internal origin, and can subjective experience be reconciled with science?

Merleau-Ponty believed that human processes cannot be accounted for by physics, nor can the empirical, positive method of physics be adequate for psychology. Rather, the primary subject matter of psychology must be experience, which is private and individual, occurring within the person and not subject to public verification and replication. Thus the proper approach of psychology is to learn the secrets of inner perception, which can be accomplished only by the descriptive methods of phenomenology.

Ludwig Binswanger

A second representative scholar of existential–phenomenological psychology is Ludwig Binswanger (1881–1966), who attempted to integrate this movement, especially the works of Husserl and Heidegger, with psychoanalysis. He was born in

Thurgan, Switzerland, and studied at the universities of Lausanne, Heidelberg, and Zurich, receiving his medical degree from Zurich in 1907. Binswanger succeeded his father in 1910 as director of the Swiss Bellevue Sanitarium, which had been founded by his grandfather.

Using Heidegger's notion of the individual's being-in-the-world (signified in German by *Dasein*), Binswanger termed his approach *Daseins-analyse*. Arguing that the reductionism of the natural science methods is inadequate, Binswanger looked to phenomenology to provide a full explanation of mental activity. Binswanger's goal was to have the therapist apprehend the world of the patient as it is experienced by the patient. He restricted his use of analysis to the present experience of the patient represented in consciousness, and he believed the analysis should reveal the structures of phenomena interpreted by each patient's individually defined context of meaning. The structures of phenomenal meaning describe each person's orientation in her or his world with respect to thought processes, fears and anxieties, and social relations.

Binswanger accepted the psychoanalytic emphasis on instinctual manifestations in early development, but maintained their importance only to the extent that they are represented in present consciousness. Accordingly, the past exists only in the present, as contributing to the design of the structure of meaning for each person. Binswanger's psychology and his application of it to psychiatry assumed that phenomenology is the critical tool for discovering the essential self of each person. This apprehension of the phenomenal structures guides the process of helping a patient modify the meanings and interpretations of living.

Both Merleau-Ponty and Binswanger represent the major focus of existential–phenomenological psychology through applications in clinical settings. Such existential themes as forlornness, depersonalization, and absurdity provide the context for accepting individual problems of existence. However, a therapist can expect to understand truly neurotic existence only by meeting individuals at their level of personal meaning.

THE THIRD FORCE MOVEMENT IN AMERICA

Like other systems of psychology imported from Europe, the third force movement in the United States has varied, eclectic expressions. Several psychologists attempted to incorporate some of the tenets and implications of the movement within existing behavioristic or psychoanalytic orientations, whereas a distinct group adhered to a strict existential–phenomenological view. However, in all expressions of the third force movement, the common view poses a contrast to the dominant, reductionistic position of materialistic behaviorism.

American Humanistic Psychology

The expression of the third force movement known as "humanistic psychology" is an eclectic grouping of American psychologists who advocated various interpretations of human personality. The term *humanistic* reflects the focus on defining a human psychology with emphases on individual existence and variability, in sharp distinction to the biological foundation of behaviorism. We shall consider several representatives of the various interpretations within humanistic psychology.

Gordon Allport. The personality theory of Gordon Allport (1897–1967) may be classified under several systems of psychology, but it is presented here as humanistic psychology because Allport, especially later in his career, proposed a framework that was essentially consistent with the existential basis of the third force movement. In his study of personality, Allport distinguished between an idiographic approach, emphasizing the individual and associated variability or uniqueness, and a nomothetic view, stressing groups and minimizing individual differences. Advocating the idiographic approach, Allport continually stressed the uniqueness and complexity of the individual and suggested an underlying unity in personality that ultimately determines consciousness. He emphasized the self, or ego-function, in consciousness, which must be understood as present manifestations of integrated goals with an individual sense of future directions. Reflecting the eclecticism of American humanism, Allport described personality in terms of traits, or predispositions to respond, in a manner similar to Freud's instincts and Horney's needs. As the product of genetic inheritance and acquired learning, Allport's traits are mental structures that account for the consistency of a person's behavior.

Allport's notion of intentions in personality best shows his agreement with the existential–phenomenological position. This construct consists of present and future aspirations and hopes that are individually defined. In Allport's personality theory, intentions account for the continual growth process of becoming. Moreover, intentions provide the ego with unity that results in the development of personal strivings, the sense of individuality, and the attainment of self-knowledge.

Charlotte Bühler. Born in Berlin and educated at several German universities, Charlotte Bühler (1893–1974) was a student of Külpe at the University of Munich when he died suddenly in 1915. A young scholar, Karl Bühler, who had served as an army physician in the German army earlier in World War I, came to Munich to assume supervision of Külpe's graduate students. Charlotte and Karl married in 1916 and, 2 years later, she completed her PhD. Both Bühlers contributed to the growing reputation of psychology in the exciting intellectual atmosphere of Vienna between the wars. From 1924 through 1925, Charlotte Bühler held a fellowship at Columbia University, where she met many of the prominent American psychologists of the period. On her return to Vienna with a 10-year research award, she was well on her way to establishing a reputation as a pioneer in the humanistic approach to lifespan developmental psychology.

The Bühlers' life in Vienna was cruelly interrupted by the Nazi movement in Austria and eventual annexation to Germany. Karl Bühler was briefly imprisoned, and following his release in 1939, the Bühlers went first to Norway and then to the United States. After several years taking short-term clinical positions at several places in the United States, the Bühlers moved to California in 1945, and Charlotte worked as a clinical psychologist at Los Angeles County Hospital and in a part-time academic affiliation with the University of Southern California. From 1953 until 1972, she was in private practice in Los Angeles.

Charlotte Bühler's perspective on development emphasized that healthy growth is psychologically purposive. Central to Bühler's views on personality was her belief in the importance of the harmonious balance of basic tendencies for need satisfaction, self-limiting adaptation, creative expansion, and upholding an internal order. Only the first involves a kind of passivity, and lifelong growth requires active engagement in the latter three tendencies. This conceptualization anticipated Maslow's "hierarchy of needs," described below, and Bühler insisted that this process is continuous throughout life.

Bühler was described by her contemporaries as the living embodiment of her psychological views. Bugental (1975/1976) described her as "a very real and at times a very formidable person who knew her own mind and set about doing things the way she believed they should be done. . . . She was usually on the move, active, doing, involved" (pp. 48–49). Bühler was an active collaborator with Carl Rogers and especially Abraham Maslow in fostering American humanistic psychology. She challenged young scholars and fostered their growth through her activities in the Association for Humanistic Psychology, for which she served as president in 1965–1966.

Abraham Maslow. Another major figure in American humanistic psychology was Abraham Maslow (1908–1970), who is sometimes called the primary mover in the popularization of this approach. He evolved a view of personality that was greatly influenced by European existentialism. Maslow's position was based on a motivational framework consisting of a hierarchy of needs, from primitive biological levels to truly human experience. For example, the physiological needs of thirst and hunger must be met before the needs of safety and security are considered. When those needs are met, individuals proceed to satisfy their need for love and belongingness, then their need for self-esteem, their need for knowledge, and finally their need for beauty. The lifelong process of personal growth through progressive needs satisfaction was termed "self-actualization" by Maslow. The proper result of self-actualization is the harmonized personality, fully utilizing individual talents, intellectual capabilities, and self-awareness.

Rollo May. After obtaining his PhD from Columbia University in 1949, Rollo May (1909–1994) worked in private practice in New York City. *Existence: A New Dimension in Psychology and Psychiatry*, a book May edited in 1958, provided one of the first introductions to the potential for existential principles applied to psychotherapy and personality theory. In the first two chapters of *Existence*, May wrote a detailed argument supporting the notion that existential interpretations of human activity provide a needed direction for psychological inquiry; that is, psychology requires a complete understanding of human experience as experienced in terms of uniquely human issues of willing, choosing, and developing.

Carl Rogers. Perhaps the most popular of the humanistic psychologists, Carl Rogers' (1902–1987) writings on clinical applications are much admired. His "client-centered therapy" holds that the therapist must enter into an intensely personal and subjective relationship with the client, acting not as a scientist or a physician, but as a person interacting with another person. For the client, counseling represents an exploration of strange, unknown, and dangerous feelings, which is possible only if the client is accepted unconditionally by the therapist. Thus the therapist must attempt to sense what the client feels as he or she moves toward self-acceptance. The result of this empathic relationship is that the client becomes increasingly aware of authentic feelings and experiences, and her or his self-concept becomes congruent with the totality of existence.

Rogers' views of personality are basically phenomenological in that he focuses on the experiencing self. The person is seen as existing initially as part of the phenomenal field of experience, and the conceptual structure of the self must become differentiated from the overall field by the acquisition of self-knowledge. The self, then, consists of organized and consistent concepts based on perceptions of the characteristics of the "I" or "me" and perceptions of the relationships of the "I" to others. Once the conceptual structure of the self is known and accepted, the person is truly free from internal tensions and anxieties.

This brief overview of the positions of Allport, Bühler, Maslow, May, and Rogers is intended to show their relationship to the third force movement. Humanistic psychology is primarily a clinical application of a psychology of the individual. Although it accepts the importance of physiological and instinctual influences on personality, humanistic psychology emphasizes individual growth to reach experiences of total realization of the vast potential of personal resources. This goal is achieved by the phenomenological appreciation of self-knowledge.

The Duquesne Group

The most consistent expression of existential–phenomenological psychology in America has come from psychologists centered at Duquesne University in Pittsburgh. Many of the writings of European scholars have been republished through Duquesne University, which initially sponsored publication of current research in the *Review of Existential Psychology and Psychiatry*. Since the early 1970s, the Psychology Department at Duquesne has sponsored the *Journal of Phenomenological Psychology*. As the most active center of existential–phenomenological psychology in America, the Duquesne group represents a rather unique orientation within the prevailing eclecticism of American psychological academia.

The inspirational force of the Duquesne group has been Adrian van Kaam (b. 1920), originally from the Netherlands, who is a member of the order of priests that founded Duquesne. He advocated a revision of psychology based on the principles of existentialism, away from the confining reductionism of natural science models and methods. Having studied with leaders of the American third force movement, such as Rogers and Maslow, van Kaam started an institute at Duquesne designed to explore the development of spirituality. He directed this institute until 1980.

The call for more phenomenological emphases in psychological research is consistent with the definition of psychology as a truly human science. A former member of the Duquesne group is Amedeo Giorgi (b. 1931), who was trained as an experimental psychologist at Fordham University. Giorgi argued for a more open approach to psychology in his 1970 work *Psychology as a Human Science*. He concluded that psychology should have as its subject matter the human person who "must be approached within a frame of reference that is also human, i.e., one that does not do violence to the phenomenon of man as a person" (pp. 224–225). Although a description of research activities of the Duquesne group is beyond our present scope, it is important to recognize that this orientation in American psychology brings the benefit of diverse perspectives on the nature of psychological inquiry.

As we conclude this chapter, it is interesting to note the similarity in the influences of the third force movement and psychoanalysis as systems of psychology. The clearest expressions of each had European origins, and their impact in America has been largely through clinical applications. Both systems lack an empirical base, limiting their appeal to mainstream American psychology. Moreover, both systems are characteristically fragmented in their contemporary expressions. However, the third force movement, unlike psychoanalysis, never had an accepted reference figure, a role provided by Freud in psychoanalysis. Indeed, the philosophical foundation of the third force movement consists of a collection of varied writings, ranging from literary works to comprehensive systems of human existence. Translated into American psychology, the third force movement influenced psychological views, especially in therapeutic applications, but did not become a serious alternative to the dominant behavioristic establishment.

CHAPTER SUMMARY

The third force movement in psychology evolved from the active model of mental processes. Firmly grounded in the principles of existential philosophy, this movement focuses on the individual in quest of identity, values, and authenticity. The nineteenth-century writings of such figures as Kierkegaard, Nietzsche, and Dilthey formed the background for the view of the person as alone and dehumanized. The twentieth-century works of Sartre, Camus, and Jaspers offered further expression to the basic state of anxiety and absurdity in human existence. The methodological writings of Husserl and Heidegger contributed to the development of phenomenology as a means of investigating the holistic character of human experience. The combined existential–phenomenological psychology was an application of a new orientation in clinical settings, represented in Europe by such psychologists as Merleau-Ponty and Binswanger. In America, the humanistic viewpoints of Allport, Bühler, Maslow, May, and Rogers agreed, to varying extents, with the European movement, and a center of existential–phenomenological psychology emerged at Duquesne University. The third force movement is largely a fragmented orientation within contemporary psychology. Although it did not generate a comprehensive alternative to behavioristic formulations, the third force movement has exerted an impact on clinical applications, especially in therapeutic efforts.

BIBLIOGRAPHY

Primary Sources

ALLPORT, G. W. (1947). Scientific models and human morals. *Psychological Review, 54,* 182–192.

ALLPORT, G. W. (1955). *Becoming*. New Haven: Yale University Press.

BINSWANGER, L. (1963). Freud and the Magna Carta of clinical psychiatry. In J. Needleman (Ed.), *Being-in-the-world*. New York: Basic Books.

BINSWANGER, L. (1963). Freud's conception of men in the light of anthropology. In J. Needleman (Ed.), *Being-in-the-world*. New York: Basic Books.

DOSTOYEVSKY, F. (1970). *The idiot* (E. M. Martin, Trans.). London: Everyman's Library.

HEIDEGGER, M. (1949). *Existence and being*. Chicago: Henry Regnery.

HODGES, H. A. (1944). *Wilhelm Dilthey: An introduction*. London: Routledge.

HUSSERL, E. (1962). *Ideas* (W. H. B. Gibson, Trans.). New York: Collier.

KAUFMAN, W. (1955). *The portable Nietzsche*. New York: Viking Press.

KAUFMAN, W. (Ed.). (1956). *Existentialism from Dostoyevsky to Sartre*. Cleveland: Minden Books.

KIERKEGAARD, S. (1954). *Fear and trembling and the sickness unto death* (W. Lowrie, Trans.). Princeton, NJ: Princeton University Press.

KOCKELMANS, J. (Ed.). (1967). *Phenomenology: The philosophy of Edmund Husserl and its interpretations*. Garden City, NY: Doubleday.

MASLOW, A. H. (1962). *Toward a psychology of being*. Princeton, NJ: D. Van Nostrand.

MASLOW, A. (1966). *The psychology of science: A reconnaissance*. New York: Harper & Row.

MERLEAU-PONTY, M. (1962). *Phenomenology of perception* (N. C. Smith, Trans.). New York: Humanities Press.

MERLEAU-PONTY, M. (1963). *The structure of behavior* (A. Fisher, Trans.). Boston: Beacon Press.

ROGERS, C. R. (1951). *Client-centered therapy: Its current practice, implications and theory*. Boston: Houghton Mifflin.

ROGERS, C. R. (1955). Persons or science? A philosophical question. *American Psychologist, 10,* 267–278.

SARTRE, J. P. (1956). *Being and nothingness*. (H. Barnes, Trans.). New York: Philosophical Library.

TILLICH, P. (1952). *The courage to be*. New Haven: Yale University Press.

VAN KAAM, A. (1966). *Existential foundations of psychology*. Pittsburgh: Duquesne University Press.

Studies

BOSS, M. (1962). Anxiety, guilt and psychotherapeutic liberation. *Review of Existential Psychology and Psychiatry, 2,* 173–195.

BRODY, N., & OPPENHEIM, P. (1967). Methodological differences between behaviorism and phenomenology. *Psychological Review, 74,* 330–334.

BUGENTAL, J. F. T. (1963). Humanistic psychology: A new breakthrough. *American Psychologist, 18,* 563–567.

BUGENTAL, J. F. T. (1975/1976). Toward a subjective psychology: Tribute to Charlotte Bühler. *Interpersonal Development, 6,* 48–61.

CARDNO, J. A. (1966). Psychology: Human, humanistic, humane. *Journal of Humanistic Psychology, 6,* 170–177.

CORRENTI, S. (1965). A comparison of behaviorism and psychoanalysis with existentialism. *Journal of Existentialism, 5,* 379–388.

FRANKL, V. E. (1963). *Man's search for meaning.* New York: Washington Square Press.

GAVIN, E. A. (1990). Charlotte M. Bühler (1983–1974). In A. N. O'Connell and N. F. Russo (Eds.), *Women in psychology: A bio-bibliographic sourcebook.* New York: Greenwood Press, 49–56.

GILBERT, A. R. (1951). Recent German theories of stratification of personality. *Journal of Psychology, 31,* 3–19.

GILBERT, A. R. (1970). Whatever happened to the will in American psychology? *Journal of the History of the Behavioral Sciences, 6,* 52–58.

GILBERT, A. R. (1972). Phenomenology of willing in historical view. *Journal of the History of the Behavioral Sciences, 8,* 103–107.

GILBERT, A. R. (1973). Bringing the history of personality theories up to date: German theories of personality stratification. *Journal of the History of the Behavioral Sciences, 9,* 102–114.

GIORGI, A. (1965). Phenomenology and experimental psychology, I. *Review of Existential Psychology and Psychiatry, 5,* 228–238.

GIORGI, A. (1966). Phenomenology and experimental psychology, II. *Review of Existential Psychology and Psychiatry, 6,* 37–50.

GIORGI, A. (1970). *Psychology as a human science: A phenomenologically based approach.* New York: Harper & Row.

KRASNER, L. (1978). The future and the past in the behaviorism-humanism dialogue. *American Psychologist, 33,* 799–804.

KWANT, R. (1963). *The phenomenological philosophy of Merleau-Ponty.* Pittsburgh: Duquesne University Press.

LANGUILLI, N. (1971). *The existentialist tradition.* Garden City, NY: Doubleday.

LUIJPEN, W. (1960). *Existential phenomenology.* Pittsburgh: Duquesne University Press.

MACLEOD, R. B. (1947). The phenomenological approach to social psychology. *Psychological Review, 54,* 193–210.

MAY, R. (Ed.). (1958). *Existence: A new dimension in psychology and psychiatry.* New York: Basic Books.

MCCLELLAND, D. C. (1957). Conscience and the will rediscovered. *Contemporary Psychology, 2,* 177–179.

PERVIN, L. A. (1960). Existentialism, psychology and psychotherapy. *American Psychologist, 15,* 305–309.

SCRIVEN, M. (1965). An essential unpredictability in human behavior. In B. Wolman & E. Nagel (Eds.), *Scientific psychology.* New York: Basic Books, 411–425.

SEVERIN, F. T. (Ed.) (1965). *Humanistic viewpoints in psychology.* New York: McGraw-Hill.

SMITH, D. L. (1983). The history of the graduate program in existential phenomenological psychology at Duquesne University. In A. Giorgi, A. Barton, & C. Maes (Eds.), *Duquesne studies in phenomenological psychology* (Vol. 4). Pittsburgh: Duquesne University Press, 257–331.

SONTAG, F. (1967). Kierkegaard and search for a self. *Journal of Existentialism, 28,* 443–457.

STRASSOR, S. (1963). *Phenomenology and the human sciences.* Pittsburgh: Duquesne University Press.

STRASSOR, S. (1965). Phenomenologies and psychologies. *Review of Existential Psychology and Psychiatry, 5,* 80–105.

STRAUS, E. (1966). *Phenomenological psychology.* New York: Basic Books.

STRUNK, O. (1970). Values move will: The problem of conceptualization. *Journal of the History of the Behavioral Sciences, 6,* 59–63.

17

Contemporary Trends

POSTSYSTEM PSYCHOLOGY

Some of those same forces that led to the intellectual ferment reflected in the existential and humanistic roots of the third force movement described in Chapter 16 also had an impact on the social changes that marked the 1960s and 1970s. Higher education in the United States became more accessible and expanded exponentially. The expanded opportunities for women, in particular, brought both talent and creativity to psychology's research agenda. For obvious reasons, psychology was at the center of these changes within the university arena as well as in the increasingly broadened definition of research. By the turn of the twenty-first century, psychology is interactive with other disciplines as one of many perspectives contributing to the advancement of our understanding of ourselves, and has given rise to entirely new areas of investigation, such as cognitive science and neuroscience.

As psychology moved beyond the middle of the twentieth century, its one hundredth anniversary as a recognized independent discipline, and into the twenty-first century, a transition was evident. Specifically, the psychology that emerged from the period of identifiable, contrasting systems of psychological inquiry evolved toward a greater emphasis on data collection, toward psychology's empirical roots. This transition did not occur all at once, with universal rejection of strict adherence to particular systems. Rather, there was a trend toward investigation of particular issues, which indicated a specific research strategy; the systems themselves were less likely to dictate the issues of importance. Some limited influences from earlier systems of psychology remained and were represented by psychologists who place varying emphases on the underlying philosophical bases of one or more of the systems. Contemporary American psychology may be described as behavioristic only in the broadest sense that accepts observable behavior as a primary, but not exclusive, source of data. Such "watered-down" behaviorism does not prevent psychologists from pursuing research questions beyond the limited scope of traditional behaviorism, in ways that would probably make Watson recoil. Likewise, in the applied sphere, clinical psychologists may value the techniques and research approach of behavior modification, but such views do not preclude their using aspects of the more psychodynamic orientations.

Accordingly, contemporary psychology is characterized as a discipline composed of various areas of study, which might include traditional research psychology issues of learning, perception, development, social activity, and personality. Research efforts in certain of these areas sometimes reflect the earlier dominance of a given research strategy derived from one of the systems. For example, current advances in learning have resulted from studies based in neurophysiology consistent with Pavlovian reflexology and behaviorism in general. The area of developmental psychology has accumulated significant findings from studies based upon the mentalistic assumptions of cognitive approaches and psycholinguistics consistent with the traditions of Gestalt psychology and views emanating from the human science model. However, in most of contemporary psychology, the approach is eclectic, avoiding exclusive commitment to any given systematic framework. Specific issues guide the strategy and direction of research. In this sense, contemporary psychology may be characterized as an empirical, but not a completely experimental, science. The methodological focus of psychology confirms the tradition of sense validation of psychological events—that is, an empirical approach—but further restriction of empiricism in psychology is not universally accepted.

An additional trend in contemporary psychology concerns the tendency toward redefining substantive areas of study. This has occurred by either specializing within psychology or joining part of traditional psychological content with another discipline. New specializations have often evolved because of the demands placed on psychologists for changes in functional roles. Industrial and organizational psychology, community psychology, and sports psychology are examples of specializations defined by new problem areas in which psychologists have found appropriate research issues and applications. Current trends also tend toward inter- and multidisciplinary studies, as opposed to exclusive specializations within psychology. The breakdown of traditional disciplinary barriers as well as the recognition of methodological commonalities has brought two or more disciplinary approaches together to address a given problem. This tendency toward interdisciplinary affinity has been accelerated by advances in the technical sophistication of scientific inquiry, making traditional disciplinary-bound approaches seem outmoded and inadequate. Perhaps the best example of this redefinition of traditional disciplines has occurred in the emergence of the research areas of "cognitive science" and of "neuroscience." Psychology forms an integral part of both interdisciplinary areas. Other specialized research and clinical topics have been similarly defined. For example, Miller (1983) described this intersection of traditional disciplines when he defined the field of behavioral medicine as the "integration of relevant parts of epidemiology, anthropology, sociology, psychology, physiology, pharmacology, nutrition, neuroanatomy, endocrinology, immunology, and the various branches of medicine and public health, as well as related professions such as dentistry, nursing, social work, and health education" (pp. 2–3). In the intervening years, with advances in research against cancer and HIV, this field has expanded almost exponentially (see also Cohen & Herbert, 1996).

Another important trend in this postsystem period has been the increasing "internationalization" of psychology. Immediately after World War II, the United States emerged as a dominant power not only in political and economic terms, but in the intellectual and academic spheres as well. European and Asian universities and research centers were in ruins, so that young scholars from these continents came to American universities to study, bringing back to their countries a particular American perspective, which included American advances in psychology. In the intervening years, these academic and research infrastructures have been restored, and as we enter the twenty-first century, clearly identifiable influences can be described in terms of a triad. In particular, the Americans continue a leadership role in psychology as well as other areas of study, and Asian and European players have assumed coequal places. The Japanese have assumed the lead in Asian science, benefiting from significant government and industry support for research and development, which are viewed as national assets of the greatest importance. Similarly, the economic power of the European Community has fed financial resources for research centers throughout the continent. While scholarly support has fostered the desired goal of unity among the subcultures of European nations, it has also resulted in the emergence of scientific leadership. In this day of instantaneous communication through electronic transference of data, collaboration among scientists approaches a truly international flavor, which has benefited psychology and related disciplines.

In surveying the major areas of contemporary psychology, we attempt to present representative trends in research developments rather than comprehensive reviews of specific fields. The interested student may find more detailed surveys of the contemporary research literature in many general textbooks available in each area of modern psychology. For specialized accounts, consult specific review journals, most notably the *Annual Review of Psychology*, *Psychological Bulletin*, and *Psychological Review*. In a very unscientific classification of the distribution of topics addressed in review articles published in the *Annual Review of Psychology* during the last part of the 1990s, from 1996 through 2000, the relative comparison of areas of psychology reveals interesting trends. A total of 90 articles were classified. The traditional area of learning, motivation, and memory reflected the fertile territory of neuroscience and cognitive science, with 26% of the articles falling within this general area. The second most represented area, with 22% of the total, consisted of review articles on the applications of psychology in schools, health delivery systems, community agencies, and similar institutions. The traditional areas of social psychology and developmental psychology comprised 17% and 12%, respectively. Articles dealing with methodology, statistical inference, and theoretical aspects of assessment made up 8% of the total reviews. Personality theory and applications in psychopathology make up only 6%; the dominant area of nineteenth century psychology—perception—comprised only 5% of the review articles. As mentioned earlier, this classification was arbitrary, but seems to reflect the balance of the traditional areas of psychological inquiry as the distinctions between psychology and related disciplines is blurred in contemporary research and application.

LEARNING, MOTIVATION, AND MEMORY

Theoretical Perspectives on Conditioning

Variations derived from the neobehavioristic formulations continue to guide investigations of learning processes. Although these models do not claim to replace the grandiose designs of the neobehavioristic theories, they serve a cohesive function, binding together lines of empirical research.

The modern statement of the two-process theory of Pavlovian–instrumental interactions in conditioning, elegantly stated by Rescorla and Solomon (1967) and summarized in Chapter 15, gave rise to much research that may collectively be labeled "transfer of control" studies. Essentially, such experiments demonstrate that stimuli with associative value from Pavlovian conditioning can modify instrumentally maintained responses. The data from these experiments showed that more than a simple stimulus-response relationship may be learned during instrumental conditioning, and behavior may ultimately be controlled by two stimuli: the predictive, Pavlovian CS and the instrumental reinforcing stimulus. Experiments reported by such investigators as Hearst and Peterson (1973), Overmier and Bull (1970), and Rescorla and Wagner (1972) confirm the potency of this relationship. The significance of these studies lies with the evidence that organisms do not simply learn to associate responses with stimuli; rather, they learn relationships among environmental stimuli, and these relationships are capable of exerting a powerful influence on various types of responses. Moreover, these studies have provided behavioral evidence in support of the neurophysiology of learning (Rescorla, 1988).

Another postsystem interpretation of incentive motivation was proposed by Bindra (1972, 1974), who discarded the Hullian constructs of habit and drive described in Chapter 15. Instead, Bindra proposed that an organism learns a contingent relationship between two stimuli, such that the presentation of one stimulus evokes the other. These stimuli are in turn represented in a central motive state, according to Bindra. If one stimulus is presented, the second stimulus is anticipated, and if the second stimulus has incentive value for the organism, then specific behaviors will be emitted. In this view of incentive motivation, there is no need for general drive motivation, as in Hull's system, but only for specific motivational states related to the incentive stimuli. The existing inhibitory mechanisms of the central nervous system are sufficient to accommodate the sensory-motor coordination of Bindra's model. He postulated three categories of sensory-motor relationships: regulatory, consummatory, and instrumental mechanisms. Thus, Bindra held that learning is not the acquisition of stimulus-response associations, but rather acquisition of relationships between stimuli as well as between stimuli and the environment.

The third extension of behavioristic formulations of learning processes was the model of Bolles (1967, 1970), who followed in the tradition of "cognitive behaviorism" started by Tolman (Chapter 15). Bolles suggested two types of expectancy in Pavlovian and instrumental learning. The first, similar to that described by Bindra, involves stimuli that predict the occurrence of other biologically important environmental events; the organism learns an expectancy as the result of the relationship between stimuli. The second type of expectancy concerns the learning of the predictive relationship between the organism's own responses and the consequences of those responses. Both types of expectancies occur during Pavlovian and instrumental conditioning; the second type is predominantly involved in instrumental training. To the two types of learned expectancies, Bolles suggested an additional, innate expectancy that imposes constraints, or limits, on a subject's ability to learn. For example, Bolles invoked the concept of species-specific defensive responses (SSDR) to explain avoidance behavior. When frightened, mammals will either freeze or flee. Bolles viewed the SSDR of freezing and flight as innate expectancies because of the subject's expectation that such behavior would successfully remove the source of fear. Thus Bolles integrated both acquired and innate activities within a motivational model based on expectancies.

Biological Predispositions

As reviewed in Chapter 15, one of the weaknesses of early American behaviorism, in its total reliance on environmental determinacy, was its deemphasis of any consideration of the biological inheritance of the organism other than the recognition that an organism is equipped with sensory and motor abilities to acquire experiences. In contrast, Pavlov's reflexology relied on a physiological reductionism to explain psychological events, giving neurophysiological mechanisms a critical role. Neither of these positions, however, placed much emphasis on the role of specific response patterns that might be innate, yet not directly reducible to physiological mechanisms.

The works of such renowned ethologists as Lorenz and Tinbergen, cited earlier in connection with the psychology of William McDougall, established the importance of biological limitations on behavior. Laboratory investigations of such events as the delay of reinforcement and autoshaping have pointed to the importance of the survival value of learning processes. Such factors as the response repertoire of a species and stimulus salience must be included in any comprehensive model of learning. Thus, instinctual patterns, evolutionary background, and social ecology contribute to the acquisition of responses in organisms.

A representative, postsystem model of learning that emphasized biological inheritance focused on the concept of preparedness (Seligman, 1970; Seligman & Hager, 1972). According to Seligman's interpretation of an organism's evolutionary history, members of a given species may be prepared, unprepared, or counter-prepared to associate certain stimulus and response relationships. The ease of acquisition and resistance to forgetting are related to this biologically determined dimension of preparedness. This notion of preparedness for learning has been related to other processes besides simple conditioning, such as language acquisition and phobic neuroses.

An additional area of research in the biological bases of simple learning and motivation deals with the reconsideration of the nature of motivational systems (see Brehm & Self, 1989). In particular, studies have moved away from a focus on neural organization of response patterns to an examination of peripheral activity in sensory control of reinforcement effects. These findings have suggested that organisms acquire representations of reinforcers, which are embedded in the individual motivation systems (White & Milner, 1992). Technical advances allow the study of neurotransmitter systems as well as the manipulation and assay of peptide systems that were largely unknown until relatively recently. Data from behavioral, anatomical, and chemical studies are recognized as critical to a comprehensive view of motivation (see Wise & Rompre, 1989).

Neuroscience and Learning

As described in Chapter 15, the tradition of Russian reflexology has consistently explained learning in terms of physiological reduction. As mentioned in Chapter 15, the investigation of physiological correlates of learning in the United States was initially pursued in the programmatic efforts of Lashley. He influenced a generation of students at several universities before assuming the directorship of the Yerkes Laboratory of Primate Biology in Orange Park, Florida. Lashley was long interested in brain research and developed many of the standard laboratory procedures and experimental designs currently used in physiological psychology.

Since World War II, American research into the neurophysiological basis of learning has mushroomed in a variety of directions. With the demise of the nonphysiological interpretations of traditional behaviorism, many researchers initially turned to the prevailing rationale of Russian reflexology and viewed behavioral psychology as ultimately reducible to the mechanisms of physiology. In addition to intensive study and mapping of functions in the cortex, postwar research employed lesioning and chemical implantation techniques as methods of extirpation to examine subcortical structures and sensory-motor pathways. These methodologies permitted the study of the interactive functions of various neural structures and allowed the recording of many electrophysiological indices of variables measuring acquisition and memory retrieval.

Two major areas of research that began to evolve a more innovative approach to the learning process involved split-brain preparations and state-dependent learning. Both research programs view learning in terms of information processing, as evidenced by the complicated nature of the transmission of neural fibers (see also Gluck & Myers, 1997). The split-brain technique, pioneered largely by Nobel laureate Roger Sperry (b. 1913), attempts to identify neural fiber projections by separating hemispheric input and measuring the effects on learning acquisition and retention. Research groups (e.g., Gazzaniga, 1967) reported impressive evidence of both deficits and excesses of learning, which depended on such factors as the nature of the task, the response requirement, and temporal and sequential factors. The split-brain technique seems to provide an appropriate model for learning and, especially, memory processes. A second major area of research, state-dependent learning, offered an interesting direction in the use of drugs as a tool to facilitate our understanding of acquisition and memory retrieval. Essentially, state-dependent learning studies have indicated that in order for an organism to retrieve a certain set of information, its central nervous system must be in the same physiological state as it was during acquisition. Conversely, disruptions of memory retrieval may be produced by drug-induced differences in organismic states between acquisition and retention testing. Split-brain and state-dependent research both take the approach of isolating functions to study how the intact brain deals with the myriad of information in the organism at any given moment.

Perhaps the most exciting trend in contemporary investigations of learning concerns the study of the neurochemical basis of acquisition, storage, and retrieval of information. Studies of the role of RNA in memory storage led to a consideration of the changes in proteins in the metabolic processes that accompany learning. Investigations of gene expression in several brain structures offer the possibility of direct assessment of neural changes following acquisition, which is described as long-term potentiation (see Martinez, 1989; Matthies & Derrick, 1996). Studies of the role of epinephrine and norepinephrine, hormonal secretions of the adrenal gland, have suggested mediating chemical changes paralleling behavioral changes (e.g., Everitt & Robbins, 1997). Moreover, neural circuitry seems largely responsible as the mechanism whereby a nonspecific chemical change translates into specific reinforcing activity. Although research has not yet progressed to a definitive description of neurochemical storage of information, this direction represents an approach that may lead to the extreme reductionism of psychological processes to neurophysiological and neurochemical elements (see Anokhin & Rose, 1991; Davis & Squire, 1984; Rosenzweig, 1996). A significant area of creative research involves the study of how the brain is able to change its structure and function as a result of experience. Through elegant and sophisticated experiments, investigators have measured changes in neurons through an array of concommitant changes in factors from the learning of new responses to aging to stress (Kolb & Whishaw, 1998).

Cognitive Processes

A final area of current research in learning, motivation and memory, termed *cognitive processes*, concerns broadly defined areas that relate to all those specializations within psychology that deal with explanations of the organization of thinking. "Cognitive science" provides a unifying theme that integrates data collected in disciplinary studies from psychology, anthropology, philosophy, computer science, and neuroscience. In the tradition of Tolman's purposive behavior, and following the neobehavioristic study of human information processing and mathematical models, cognitive psychology includes such global classes of behavior as perception, memory storage and retrieval, categories of social and developmental variables, and social attitudes and traits. In addition, recent studies of animal cognition have suggested such cognitive abilities as time discrimination in rats (Church, 1978), self-awareness in pigeons (Epstein, Lanza, & Skinner, 1981), and cognitive representational processes in invertebrates (Roitblat & von Fersen, 1992). Whereas some reviewers (e.g., Premack, 1983) caution that cognitive processes defined in humans and in animals may not reflect comparable mechanisms, the infrahuman studies show distinctive behaviors and patterns that are not easily accommodated by more traditional conditioning models of learning.

Research on concept formation, decision making, judgments, and attitudes collectively points to entire complexes of activities that fall under the scrutiny of experimental methodologies. Along with advances in neurophysiology and sensory physiology, new techniques are rapidly evolving to measure physical substrates of cognitive processes. For example, Hillyard and Kutas (1983) have reviewed research advances in the use of electrophysiological event markers during stages of various cognitive processes, from selective attention to language processing. Cognitive psychology seems to be emerging as a central explanatory framework for many behavioral activities (e.g., Carpenter, Meyake, & Just, 1995). Although they descended from early theoretical perspectives, contemporary studies of cognitive processes are bound together more by a consensus on issues and methods than by an all-encompassing explanatory and predictive model.

Cognitive science has become a dominant force in contemporary psychology, linking the traditional psychological concerns in learning and memory with other disciplines that investigate thought processes. Moreover, Wellman and Gelman (1992) advanced the view that children as young as 3 or 4 years of age possess a "theory of mind" that explicitly structures their mental world, which challenges more traditional perspectives on cognitive development such as Piaget's theory (discussed later). Quite recent attention has been directed to the accuracy of memory retrieval (Koriat, Goldsmith, & Pansky, 2000; Wenzlaff & Wegner, 2000), which in turn has supported a coherent model of memory accounting for neurological and brain imaging data (Rolls, 2000). Hunt (1989) reviewed prevalent models in cognitive models and argued that the database has advanced to the point that a new paradigm is needed to accommodate new frontiers of research into reasoning, language, and problem solving.

PERCEPTION

Perception is psychology's oldest and most traditional area in terms of the formal expression of psychology as an independent discipline in the nineteenth century. Although the early studies of the psychophysicists and the basic tenets of structural psychology were surveyed for their historical importance, many of the methodological and substantive issues of these movements remain critical in modern psychology. Moreover, the Gestalt movement, which derived its initial formulations from perceptual processes, raised many questions that have current interest in thinking and problem solving.

Recent developments in perception research have distinguished between static and motion perception, with the latter emphasizing perception of an event or events over time. Advances in equipment and measurement techniques, along with the emergence and acceptance of an information-processing model of cognition, have given traditional perceptual research a central place in contemporary psychology. In particular, issues such as sensory detection, filtering, and attention have relevance to those models of learning that view the person in terms of the organization and mediation of sense information (see Haber, 1978). Accordingly, studies of perception have complemented the study of learning processes by focusing on the sensory part of the sensory-motor relationship. Moreover, the relationship between developmental variables and the intersection of perceptual, behavioral, and cognitive maturity defines a promising area of current research (Bertanthal, 1996).

The traditional issues of depth perception and pattern vision have been expanded to include varieties of performance variables. For example, initial work by Pettigrew and his associates (Pettigrew, Nikara, & Bishop, 1968) led to a physiological model of depth perception based on neural differences from cell to cell in the receptor-field locus of area 17 of the visual cortex, when mapped separately from each eye. Although the specifics of the Pettigrew model have been challenged, the neurophysiological techniques developed provide a means to resolve this traditional problem in perception (see DeValois & DeValois, 1980).

The study of perception has been historically linked to specific systems of psychology, and perceptual data have been used to support or refute various formulations, from attention to higher cognitive processes (Kinchla, 1992). Although this linkage has occurred more directly with the structural, Gestalt, and phenomenological systems, all of the twentieth-century systems had interpreted perceptual studies in line with their underlying assumptions. Accordingly, the postsystem period in psychology has freed the study of perception from preconceived assumptions attached to specific systems. In the 1950s textbooks in perception typically began with an introduction to classical psychophysics with overtones of structuralism, or they introduced students to basic Gestalt principles, or reiterated the Lockean assumptions of radical behaviorism. However, by the 1970s perception was defined as an empirical study, devoid of specific assumptions, and this data-based strategy has been apparent in recent research efforts (e.g., Hersh & Watson, 1996). In addition, the integration of various interdisciplinary approaches has afforded research into applied settings that provides insights of a broader perspective (Egeth & Yantis, 1997; Feng & Ratnam, 2000).

DEVELOPMENTAL PSYCHOLOGY

Cognitive Development

Perhaps the dominant model in developmental psychology is derived from cognitive interpretations. Although other approaches are recognized and accepted, it is the cognitive approach, intimately linked to language development, that has shaped the direction of developmental research. The dominant figure in this field has been Jean Piaget (1896–1980). After receiving his doctorate in zoology at age 22, Piaget embarked on a career that centered on the question of how people learn. He considered himself a philosopher concerned with the issue of knowing, as his later work on epistemology and logic amply testify. His methods of study were unorthodox compared to the standards of contemporary empiricism. Indeed, his theory of cognitive development was largely based upon his observations of his own children. Nevertheless, Piaget's influence has been tremendous and his reputation rivals that of Freud in terms of individual contributions to psychology during this century.

Piaget's view of cognitive development posits four distinct periods of intellectual growth that characteristically organize the child's interaction with the environment. Although the rate of intellectual growth may differ from child to child, Piaget maintained that this sequence of development is followed by all children:

1. *Sensorimotor period* (birth to 2 years): This stage is nonverbal and involves the child's initial experiences with environmental relationships, which become internalized in a rudimentary fashion by organization imposed along dimensions of meaning, intentions, causality, and symbolic value.
2. *Preoperational period* (2 to 7 years): During this phase, the child acquires language and deals with time relations, past and future as well as present.
3. *Concrete operations period* (7 to 11 years): At this stage, the child grasps abstract notions represented by complex qualitative and quantitative relations.
4. *Formal operations period* (11 to 15 years): In this final phase of intellectual growth, the child acquires understanding.

Piaget published over 50 books and monographs during his 60 years of active research. In his later years, he concentrated on the issues of the logic underlying the acquisition of knowledge. Although he is best known for his theory of cognitive development, Piaget consistently studied knowledge per se. The potency of the structures of mental development and organization impressed him to argue that education and teaching need not be manipulative but rather should provide the child with opportunities to invent and discover.

With the impetus of Piaget's synthesis of the development of intellectual abilities, the entire range of complex human learning and memory processes received closer attention within developmental psychology. For example, research in concept formation, involving the classification of multiple events or objects into a conceptual category, has been guided by a model of information processing focusing on the critical functions of input, output, and feedback. In this model, conceptual learning is seen as dependent on the learner and on the importance of individual decisions that evaluate stimulus characteristics and environmental contingencies. Moreover, conceptual learning requires a more dynamic interpretation of the complex issues of memory storage and retrieval. That is, the selective attention of the learner to the varying saliencies of stimuli involves flexibility of memory processes.

The emphasis on individual differences in complex learning processes has led contemporary psychology to suggest that intellectual abilities may be formulated within a personally devised *cognitive style,* a term that describes a person's learning strategy or approach to intellectual tasks. Research efforts have attempted to specify characteristic mental strategies. This emphasis on cognitive styles, or the individual imprint on learning strategy, is reminiscent of the trait theories of intelligence proposed in the nineteenth-century British views of Francis Galton and his followers (see Chapter 10). Trait theory was eventually overshadowed in the United States by the advent of Watsonian behaviorism and its successor movements. These views favored a more concrete and restricted definition of intelligence in terms of quantitative stimulus-response associations as reflected by overt performance. It is intriguing to consider that psychology's study of intelligence in the last hundred years has come full circle and research is focused on such dynamic aspects of human intelligence as creativity (e.g., Sternberg & Lubart, 1996).

Contemporary issues in cognitive development include efforts to integrate the traditional frameworks of Piagetian views and rigorous experimental psychology so that more complete models of early development can address motor, cognitive, social, and linguistic variables within the context of the behaving individual. This goal attempts to avoid the pitfalls involved in research that focuses on a stage or phase of development, or is constrained by concern with a single variable, while losing sight of the entire person. For example, recent studies have offered a developmental perspective on social interactions by looking at long-term effects of early experience on later developmental patterns (Cairns & Valsiner, 1984). In summary, two trends evident in contemporary psychology are also seen specifically in cognitive development. Specialization has segmented broad categories of traditional research (e.g., child psychology) to accommodate more refined concerns; at the same time, research is drawing increasingly from interdisciplinary resources (Brainerd, 1996; Flavell, 1996).

Psycholinguistics

Perhaps no other area in psychology benefited more from the genius of Piaget than the study of language. The dominance of the traditional behavioral model of psychology through the first half of the twentieth century had left the study of language in a relatively sterile state. Specifically, behaviorism emphasized the associations between words, and verbal learning was largely viewed as the aggregate of these associations. Indeed, the methods of verbal learning up to the 1950s were essentially unchanged from the approach used by Ebbinghaus late in the nineteenth century (see Chapter 11).

Piaget's theory of cognitive development offered refreshing insights into the development, structure, and use of language. Psycholinguistics may be broadly defined as the study of communication and the characteristics of the persons communicating. The actual shift in research direction to a consideration of the syntax and organization of language was prompted by an article by George Miller that appeared in 1962 in the *American Psychologist*, entitled "Some Psychological Studies of Grammar." This paper introduced the clearly mentalistic constructs of Noam Chomsky, who remains a pioneer in psycholinguistics. Chomsky referred to the developmental necessity for a "language acquisition device" as a mental mechanism to account for the onset of a child's ability to deal with grammar. During the 1960s and 1970s, the study of language development was completely recast. Contemporary psycholinguists portray the child as possessing innate mechanisms for interpreting and organizing the auditory stimulation of speech from the environment. The child's acquisition of grammatical syntax reflects an individual sense of language, permitting the child to mediate the linguistic environment.

Psycholinguistic theory points to the major function of language as the conversion of the varieties of ideas, conceptions, and thoughts into sentence structures. The rules of grammar operate on semantic units to produce a competence in expression. Researchers have extended the range of psycholinguistics by asserting that language is basic to understanding, problem-solving behavior, self-perception, and social relationships. Psycholinguistics, then, focuses on semantics—or the underlying meaning of words, other signs, and sentence structure—and psychologists have designed assessments of semantic value (see, for example, Osgood, Suci, & Tannenbaum, 1957).

Recent trends in psycholinguistic research have emphasized language processes within a comprehensive psychological context. As such, language as an expression among modes of communication assumes a critical role. Investigations of such issues as semantic memory, sentence comprehension, and word processing, including underlying meaning, study contextual variables such as word sensitivity, semantic references, and concept categories (see Foss, 1988; Carpenter, Meyake, & Just, 1995). Moreover, the importance of both biological and cultural determinants of language development and use has generated considerable research in several disciplines. Seen in this light, psycholinguistics and communication in general are best examined through an interdisciplinary approach, merging the approaches of the behavioral sciences, anthropology, sociology, computer science, and philosophy (see also McKoon & Ratcliff, 1998).

Lifespan Development

Traditional developmental psychology has focused on early experience, perhaps because of the compelling and obvious need to educate and socialize children. Such an emphasis has historically been reinforced by theoretical frameworks that view childhood as a period when the critical determinants of adult behavior are acquired. Systems as diverse as behaviorism and psychoanalysis have suggested, for different reasons, the importance of experiences in early development and their profound effects on subsequent maturation. While not ignoring or even de-emphasizing early development, lifespan developmental psychology has recently attempted to describe and explain development as a continual, comprehensive process from conception to death.

One important by-product of this more balanced investigation of lifelong development concerns contemporary interest in the study of aging, a long-neglected topic in psychological research (e.g., Birren & Fisher, 1995).

With the conceptualization of lifespan development, various interpretations have suggested a pluralistic sequence of development that results in critical periods of growth during life. These critical periods tend to generalize their effects across age periods. In other words, developmental functions of specific behavioral changes, such as language acquisition, psychological and biological changes in adolescence, and adult career decisions, may have different patterns of transforming behavior. This rather complicated picture of development has led to the study of lifespan profiles of influences that result from behavioral changes. The profiles, in turn, are designed to show not only individual growth but also how growth patterns interact with biological, environmental, and social determinants. Research in lifespan development includes the investigation of such issues as social development, family constellation, personality, and learning and memory (see Schultz & Heckhausen, 1996).

In this area, as in other contemporary areas of psychology, research efforts are combined because of a common orientation rather than because of a common theoretical heritage. Honzik (1984) categorized contemporary empirical research in terms of age differences, longitudinal changes, precursors of later development, biographical studies, and life satisfaction at a later age. Further, she defined significant developmental areas as health, temperament, intelligence, and issues related to self-concept, self-esteem, and altruism. Within these approaches to the lifespan, there is clearly an emphasis on multidisciplinary study, especially considering current, more comprehensive treatments of the aging part of the lifespan continuum (Baltes, Staudinger, & Lindenberger, 1999).

The study of aging in the context of lifespan development has practical as well as theoretical significance. With the numbers of people reaching old age increasing, society has yet to deal effectively with problems that have immense physical, psychological, and sociological importance for the aged. Psychology has only recently started to generate substantial research on aging, and these initial studies have clearly indicated the often traumatic and largely misunderstood changes that occur. Of theoretical interest is the extent to which the aging process is consistent with earlier, antecedent stages of development. Questions relating to the adjustments in lifestyle that accompany old age suggest the uniqueness of this stage of life, yet the individual's ability to make such adjustments depends on his or her lifelong experiences. Clearly, the investigation of the psychology of aging will provide a major research field for developmental psychology.

SOCIAL PSYCHOLOGY

Social psychology studies the behavioral processes, causal relations, and products of interactions among people and groups. Social activities may be viewed from three perspectives: individual contributions, interpersonal relations, and group behavior. The obvious importance of the social nature of human experience has been recognized since the time of antiquity. The historical antecedents of contemporary social psychology found expression, along with psychology in general, during the nineteenth century. In particular, Comte's positivism viewed the study of social structures and institutions as the most positive of sciences, and he viewed sociology as the culmination of intellectual progress. Darwin wrote on the social character of the evolution of humans, and Herbert Spencer attempted to devise a theory of social evolution. By the beginning of the twentieth century, the influences of Darwin and Spencer had led to the prevailing view that human social activities, in terms of origins as well as articulation, are governed by inherited instincts. The instinctual basis of social behavior was the dominant theme in McDougall's attempt to present a systematic account of social psychology in his textbook written in 1908 (see Chapter 12). However, McDougall's reliance on instinct to explain social processes ran counter to the environmental determinism of early behaviorism, which soon became the dominant American system of psychology. In 1924, Floyd H. Allport published his *Social Psychology*, in which social processes were presented in better conformity with behavioristic principles by avoiding instinctual explanations in favor of what he called "prepotent reflexes," or impulses modified by conditioning. In addition, Allport's book was the first treatment of social psychology that relied entirely on experimental evidence rather than on less controlled observational approaches.

Following Allport's precedent, social psychology has developed a broad base of experimental data. However, just as American behaviorism continually extended its study beyond the narrow confines of Watson's formulations, so, too, did social psychology gradually modify both its content and its methodology. Specifically, social psychology was greatly influenced by the field theory of the Gestalt movement and, to a lesser extent, by phenomenology, so that social psychology evolved into one of the broadest areas of contemporary research. Social psychologists have also studied social influences on individual behavior, investigating such topics as social imitation and learning, attitude and motive development, and social roles. The area of interpersonal relations encompasses the study of social status and communication, and theoretical interpretations have borrowed from other areas of psychology, ranging from stimulus–response learning to cognitive dissonance. The study of groups has concentrated on the development of participation, the formation and maintenance of groups, and the structure and management of organizations.

After World War II, research in social psychology grew tremendously, examining such issues as power, leadership, and social persuasion. For example, Milgram's (1963) classic work on obedience and conformity identified critical variables of social control. Social psychology has evolved into an interdisciplinary study, extending its scope to include cultural, anthropological, and moral questions. The work of cultural anthropologists, such as Margaret Mead's (1949) analysis of social rites in "primitive" societies, was integrated into a more comprehensive social psychology. Similarly, the survey techniques of sociology were employed by social psychologists to examine the development of racial attitudes (see, for example, Pettigrew & Campbell, 1960), and the findings were subsequently used to foster modification of those attitudes during the upheaval and social changes of the 1960s.

Recent studies in social psychology reflect the unsettled theoretical nature of this field. More than in any other area of contemporary psychology, the conceptual basis of social psychology appears to be rapidly evolving, perhaps as the result of the many recent strides at the empirical level with a focus on sophisticated designs (e.g., Kenny, 1996). The disagreements over strategies of study have been summarized within an investigation of social motivation in a review by Brody (1980). Essentially, Brody posed the distinction between phenomenological (defined broadly as nonanalytic) and antiphenomenological approaches in social psychology. A decision about the very basic assumptions of psychological processes seems especially compelling for social psychology, which has as its subject matter the elusive quality of social interactions. It is interesting to note that the dilemma of choosing between phenomenological and nonphenomenological assumptions about social activities, still unresolved, essentially represents the same basic issue with which psychology has struggled since its modern inception in the 1870s.

Nevertheless, on an empirical level, research has generated creative new studies of social interactions. Investigations of group tasks and group problem solving have isolated critical roles played by participants and have analyzed the nature of cooperation and compliance. The use of "games" to examine social influences on resolving conflicts and dilemmas has frequently been reported (see Dawes, 1980) and has provided a means for measuring such social characteristics as leadership, competition, trust, and obedience. Similarly, studies of environmental influences, investigated through the generation of models of social enrichment and deprivation, have provided evidence for a multidimensional reconsideration of individual social background (e.g., Schlenker & Weigold, 1992). The application of these results to the study of not only groups but also formal organizations is an area of growing attention (Harris Bond & Smith, 1996; Snyder & Stukas, 1999). Finally, attribution of perceived causality has become a major research field, with a natural affinity to cognitive, motivational, and personality research. Contemporary social psychology is a very active field, drawing on interdisciplinary resources as it moves toward theoretical coherence.

PERSONALITY

Historically, the study of personality has often served as the vehicle for development of specific systems of twentieth-century psychology. The most obvious example of a personality theory expanding into a system of psychology is psychoanalysis. Further, the close, mutual benefits between a general system and specific implications for personality may be seen in the relationship between Gestalt psychology and field theory, as well as between behavioristic learning theory and behavior modification techniques. Moreover, the phenomenological movement essentially defines psychology itself as the study of the individual personality. Thus personality has offered, and continues to represent, one of those fundamental areas of psychology providing distinctions about human nature among the basic assumptions that compete to guide the approach to psychological inquiry.

Perhaps the most important development in personality research during the contemporary postsystem period of psychology has been the emphasis on the empirical study of personality. The empirical approach, with its requirement of operationally specific variables and tightly controlled observations, reflects the success of neobehaviorism, as this system is most amenable to empirical strategies. The closest approach to behaviorism, of course, concerns the therapeutic application of learning theory in behavior modification (see Chapter 15), wherein personality is literally reduced to learning principles. However, even less extreme approaches conform to the accepted standards of empiricism and are aligned to varying extents with a behavioristic posture.

An empirical approach that has gained wide recognition concerns the orientations collectively called factor theories of personality. One by-product of this perspective on personality is an emphasis on testing and the assessment of personality. Using the statistical procedure of factor analysis to identify common dimensions emerging from many tests, factor theories attempt to identify characteristic traits in personality that not only are of descriptive value but can also be used in the prediction of personality development.

A representative major figure in contemporary factor theories is Hans Jurgen Eysenck (1916–1997). He was born in Germany but spent most of his career at the University of London. Eysenck views personality as composed of a bodily sense and divisions that function as intellectual, affective or emotional, and motivational or striving. In his research, he has identified two fundamental variables of personality: introversion-extraversion and neuroticism. The first is interpreted much as Jung initially proposed, and Eysenck has reported evidence to support the existence of this dimension. Personality highly rated on the neuroticism dimension is characterized by inferior performance in each division of personality, and especially in lowered motivational levels. Eysenck's work is noteworthy because, without accepting a completely behavioristic orientation, he has nonetheless developed an empirically based factor theory that appears to assess quantitatively the dynamic theories of personality.

Factor theories and the study of personality assessment remain major thrusts in contemporary personality research. Various assessments ranging from the measurement of intelligence to the use of projective tests have contributed detailed descriptions of personality. Of interest is the use of tests designed for one purpose and used for another. For example, studies have used the Minnesota Multiphasic Personality Inventory (MMPI) as a general assessment tool for predicting outcomes of varieties of intellectual and performance tasks.

Recent themes in personality research seem influenced by interdisciplinary influences prevalent in other areas of contemporary psychology, particularly cognitive science (see Mischel & Shoda, 1998). Moreover, the framework for personality theory as it influences issues of psychopathology must accommodate the dramatic advances made in neuroscience, from brain imaging studies to neurochemical bases of observable dysfunction. Personality research is presently most active where attempts are made to broaden the empirical basis that was ignored during the period of dominant systems of psychology.

INTERNATIONAL PERSPECTIVES: MODERN ASIAN PSYCHOLOGY

In Chapter 1, alternative intellectual and religious traditions from non-Western cultures were outlined to underscore that, although psychology as an independent discipline emerged from a focused Western perspective on human experience, other cultures had important contributions in dealing with the subject matter of psychology. Although cognizant of the varying traditions, the international character of scientific inquiry is diminishing these differences. That is, in reviewing more recent developments in Asian psychology, we should be mindful that contemporary trends in psychology in Asian countries are quite similar to modern trends in psychology found in Western countries. With the significant governmental investment in science that occurred after World War II in the United States and the former Soviet Union, the two superpowers in politics and economics created spheres of influence in science as well. Both countries opened their universities to students of their respective Asian allies and, by the 1970s, to each other. The models of psychology in both the United States and Russia, although differing in emphases, are founded upon the same rationale as are the European models of psychology. Coupled with the information explosion and increased access to information, an international database of psychology is universally available; access to it depends solely on the availability of the technology needed to retrieve it.

Thus we should not expect contemporary psychology in Asia to be very different from psychology in the United States. Given the survey of the rich traditions of the East in Chapter 1, contemporary Asian psychology as an identifiable discipline is somewhat paradoxical in that it does not rely on indigenous traditions, but rather reflects the Western tradition. Although Asian psychology is not devoid of its native Eastern heritage, that influence is subtle, and the preponderance of Asian psychology follows the same direction as the rest of international psychological inquiry. This review of more recent developments in Asian countries attempts to highlight some of these subtleties.

India

Three centuries of British influence on the educational systems of India, as well as the tradition of sending promising youths from the subcontinent to England for their education, resulted in a systematic imposition of Western thought, including psychology. In 1916 at Calcutta University, N. N. Sangupta became the first professor in the department of psychology. He later went to Lucknow in northern India, where he started a psychology laboratory in 1929. By 1925, Mysore University in southern India also offered a psychology curriculum. The Indian Psychoanalytic Society was founded in 1922, followed 3 years later by the organization of the Indian Psychological Association, which publishes the *Indian Journal of Psychology*. Since the beginning of Indian independence, psychological practitioners have been required to be certified or licensed. India and, especially, China have fewer psychologists in relation to total population than any other country where psychology is recognized or organized.

In recent years there has been renewed interest in the application of yoga in psychology. As mentioned earlier, yoga was a dominant philosophy of ancient India, dating from the Upanishads. Yoga is a system of self-discipline and reflection used to help a person obtain self-knowledge. As taught by the ascetic philosopher Patanjali (ca. 150 B.C.) in his yoga-Sutras, the goal of yoga is to seek independence of the self from all mental context. Achieving this end permits a person to apprehend directly and with certainty the reality and essence of the self. Psychologists have viewed yoga in terms of its implications for behavior therapy, psychophysiological control, and cognitive development.

China

Earlier in the twentieth century, psychology was taught at several universities, and a significant proportion of the professors had been trained in the United States. The first psychology research laboratory was founded at Beijing (Peking) University in 1917, and the first independent psychology department was established at Nanking University. The Chinese Psychological Association was started in 1921 with Zhang Yaoxiang as its first president. Because of their training in the United States, Chinese professors required translations of American textbooks. Also, Chinese psychological journals provided summaries of current American experimental work. At the same time, other behavioral disciplines, especially sociology (Huang, 1995), prospered in China.

When the Japanese invaded five northern provinces of China beginning in 1937, all research activity stopped, not only for the duration of the invasion, but also during the civil war, which lasted until the victory by the Communist forces in 1949. Contact with the West virtually ended then, and early efforts to revitalize psychology were influenced by Soviet advisors who tended to propose models of reflexology (see Chapter 15). The Chinese Academy of Sciences was reopened in 1950, and by 1956 an Institute of Psychological Research was established under the academy's biology section, which perhaps reflected the influence of the Soviet advisors. At any rate, the major role for psychology in the early years of the People's Republic of China was in teacher training, so that teachers' colleges often had the most active departments. In 1955, the Chinese Psychological Association was refounded with Pan Shu, who was Director of the Psychology Institute of the Chinese Academy of Sciences, as its president. During this period the nature of psychology as the study of social relations was expressed in a dialectical approach. This study also attempted to justify a natural science model for those aspects of human experience that are clearly biological.

The Cultural Revolution of the 1960s was a time of great social upheaval, and psychology and psychological research were early targets for attack from the most radical elements. Consciousness as an area of study was condemned as class-determined, and psychological research was denounced as metaphysical and bourgeois. The Chinese Psychological Association was again abolished and its four journals ceased publication. Scientists and professors were often sent to work camps or reeducation centers.

With the normalization of Chinese life in the early 1970s, psychology began to revive. Several national conferences were held in the late 1970s, but psychological study is still largely limited to applications in educational psychology. China's long-range planning to revitalize and upgrade the university system will certainly benefit psychological research and education. Until that is achieved, however, the impact of psychology as a formal disciplinary study is marginal at best.

In summary, psychology in China has been subjected to the turmoil that has characterized all aspects of Chinese society during most of the twentieth century. Little of either the Western models of psychology or the indigenous heritage of Confucian thought has prospered. In fact, it appears that contemporary psychology, in common with other disciplines, is starting a careful process of recovery and rebuilding in China.

Japan

Historically, both Korea and Japan were recipients of Chinese cultural achievements, but in importing Chinese religion, philosophy, and literature, the Japanese impressed a clearly native imprint as they assimilated selected aspects of China. Japanese legend teaches that its sacred islands were created by the gods who gave birth to the first emperor, who was succeeded in unbroken lineage by the present emperor.

Japanese society in the feudal period (lasting from about the year 1000 until the reassertion of imperial authority at the accession of Emperor Meiji in 1868) was rigidly divided into castes. The emperor was largely a figurehead, and real power lay with the shogun, who emerged usually after a power struggle involving fierce fighting and political intrigue. The shogun and his immediate descendants were able to hold on to power for only relatively brief periods before another challenge would begin the struggle again. The contenders for the shogunate involved lords of various ranks whose wealth came from land, peasants, and slaves. Each lord was supported by a class of warriors, called samurai, who numbered more than a million men at various periods in feudal Japan. They followed a rigid code of behavior based upon loyalty, courage, and great sensitivity to personal dignity and honor. The actual work of the country was performed by classes of artisans, peasants, and merchants. There was also a very large class of slaves, almost 5% of the population, who were criminals or people born into, or sold into, slavery. The workers were heavily taxed and were required to donate given periods of labor to the local lord or the state. As in China, the basic social unit was the family, and the first lessons of loyalty and respect were taught in the family context.

The oldest religion in Japan, Shinto, was based upon ancestor worship and had a rather simple creed of respect for tradition with some nationalistic rites and prayers. Shinto did not provide a formal priesthood, elaborate rituals, or a detailed moral code. Aside from requirements for occasional prayers and pilgrimages, little was demanded of the believer. In 522, Buddhism was imported from China and was remarkably successful, seemingly meeting religious needs not accommodated by Shinto. However, the version of Buddhism that succeeded in Japan did not contain the original Buddhist emphasis on agnostic belief and a rigid moral code. Buddhism in Japan became a positive affirmation of belief in gentle gods. With the observance of duty to ritual and the living of a virtuous life of obedience, those who suffered in this life could expect relief in the anticipation of a better lot in the next incarnation. This version of Buddhism fit well into the kind of social control that characterized the hierarchical structure of Japanese society. This, in turn, bolstered the nationalistic aspirations of the society as well.

Confucianism was introduced to Japan in the sixteenth century and provided the first real impetus to and framework for learning. The great Confucian teacher and essayist Hayashi Razan (1583–1657) was widely recognized as a scholar and won converts from Buddhism and the newly introduced Christianity. Although the first Japanese university was founded in Kyoto in the eighth century, a true commitment to higher learning did not emerge until the seventeenth century with the advent of the Tokugawa Shogunate (1603–1867). In 1630 at Yedo, Hayashi Razan started a school for government administration and Confucian philosophy, which later became the University of Tokyo. Thus, Confucianism succeeded in uplifting the intellectual climate and appreciation of learning in Japan. Kaibara Ekken (1630–1714) was perhaps the most famous Confucian philosopher in late feudal Japan. A renowned teacher who emphasized the unity of the person within the environment, he advocated virtuous living to achieve harmony with nature. Japan soon became a center of Confucian study, with rival schools of interpretation.

In terms of specific applications of Confucian thought, Soho Takuan (1573–1645) viewed the individual as a microcosmic reflection of the universe; thus, personal discipline can lead to control of external events. Baigan Ishida (1685–1744) suggested that the mind is physically based and sensitive to environmental inputs. Almost reminiscent of Locke, Ishida taught that the contents of the mind are environmentally dependent, so that personality change accompanies change in the input. Ho Kamada (1753–1821) suggested a total of 14 emotions and proposed a psychological code based on virtuous living to attain personal happiness. Accordingly, Japanese philosophy of the late feudal period was rich in psychological interpretations, many of which were as sophisticated as any views propounded in Europe during the same time period.

The Japanese borrowed liberally from other cultures, especially the Chinese, but restated these imported positions and views to conform to their social and national character. Despite recurring internal strife, limited natural resources, and calamitous earthquakes, the Japanese built a society that eventually accepted the Confucian emphasis on scholarly study. As Japan rapidly moved from a feudal to an industrial society in the late nineteenth century, it built a high-quality educational system based on a deep commitment to learning.

The transition in Japanese society from a feudal to an industrial organization during the latter part of the nineteenth century was truly remarkable. The successful industrialization of Japan was confirmed by the spectacular victories of the Japanese over the Russians during the Russo-Japanese War of 1904–1905. Japanese industrialization retained some of the traditions of Japanese philosophical teachings. Loyalty, affiliation, and family strength were all woven into the structure of Japanese industrial organization. Very clear attitudes toward productivity and education were inculcated into Japanese social values, providing a unique social psychology on a national scale that has served Japanese society well in this century.

Perhaps because of its smaller, more homogeneous population that accepted the utilitarian and functional value of psychological research, the development of modern psychology has fared much better in Japan than in China. Moreover, Japan's close relationship with the United States, imposed on the Japanese by the American victory in World War II, drew the recovery of scholarly activity close to American research directions. Today Japan ranks as a leader in psychological research, and Japanese journals provide a respected and invaluable part of the database of psychology.

The founding of Western models of psychological inquiry in Japan usually counts Yujiro Motora (1858–1912) as Japan's first experimental psychologist. Born in Osaka, he attended Boston University and then received his doctorate under G. Stanley Hall at Johns Hopkins University in 1888. Upon his return to Japan, he became Japan's first professor of psychology at the University of Tokyo and founded a laboratory there. He continued the research that he had started with Hall on dermal sensitivity, and published three books on general and systematic psychology. Motora's successor at the University of Tokyo was Matataro Matsumoto (1865–1943), who had received his PhD at Yale in 1898. He then went to Leipzig to work with Wundt and returned to Japan in 1900. He founded the psychology department and laboratory at Kyoto University and finished his career at the University of Tokyo. Matsumoto referred to his system of psychology as "psychocinematics," or mental works, which was a type of psychophysiological control wherein he studied experimentally the conditions of mental power over bodily motion. Curiously anticipating later developments in Zen psychology, this work led Matsumoto to an association with the Aeronautical Research Institute of the University of Tokyo, where he worked in human engineering psychology.

Kwanichi Tanaka (1882–1962) was instrumental in the introduction of Watsonian behaviorism into Japan. Tanaka publicized and advocated the type of objective methods proposed by Watson. As in the United States, the rigid formulation of the original version of behaviorism was modified and made more flexible to accommodate the data of consciousness and purpose. For example, Koichi Masuda (1883–1947) proposed an early version of animal cognition with his interpretation of animal behavior in terms of consciousness. Similarly, Ryo Kuroda (1890–1947) argued that behavior and consciousness are two aspects of the same experience and are complementary rather than contradictory.

Gestalt psychology was introduced to Japan by Kanae Sakuma (1888–1970), who received his initial education in Japan and then studied in Berlin with Köhler and Lewin. Applying Gestalt principles to the development of language, Sakuma was one of the first psycholinguists. Similarly, after studying in Berlin in 1925 and 1926, Hiroshi Hayami introduced the phenomenological tradition of Husserl as providing more direct access to experience.

Perhaps clinical applications best exemplify the attempt to integrate Eastern and Western traditions in psychotherapy pioneered in Japan. Shoma Morita (1874–1938) argued that the reaction and undue attention to neurotic behavior often exaggerate the problem and result in a vicious cycle. Morita offered a therapeutic alternative that borrowed from a distant version of Buddhism termed Zen Buddhism (Barrett, 1956). Like other Buddhist traditions, Zen is antirationalistic, but it seeks knowledge solely by intuition and interpretation rather than by reliance on traditional Buddhist writings. The goal of Morita's theory was to seek harmony with the universe, not to fight or resist the universe, as taught in Western philosophy. Accordingly, the person simply accepts his or her condition in an attempt to reduce the anxiety associated with undue attention to the condition. Morita employed a four-stage procedure beginning with the patient's yielding to the anxiety and ending with the patient's preparing to return to everyday existence.

This model was developed more fully in the Zen psychology of Koji Sato (1905–1971), a professor of psychology at Kyoto University. Originally influenced by Gestalt psychology, Sato later turned to psychoanalysis and clinical psychology. During the 1950s, he became interested in Morita's work and spent the remainder of his career extending Zen teachings to psychotherapy. Essentially, meditative Zen techniques use physical adaptations in posture and breathing to achieve mental serenity and clarity through the realization of the harmony and integration of the person and the universe. Zen techniques are methods of psychophysiological control that feed back on the mental state to produce an internal sense of well-being. Sato and his teachings in Zen psychology were recognized by Western therapists, such as Karen Horney and Carl Rogers. Zen was a major vehicle in introducing Western psychologists to Eastern thought.

Japanese psychology, in contrast to Chinese, has enjoyed widespread acceptance and prosperity in both Japan and the international community. It is a credit to the industry of Japanese scholars that they were able to recuperate from the devastation and losses of war and the subsequent restructuring of Japan. Interestingly, the relationship with American psychology that predated World War II may have facilitated the rebuilding of Japanese psychology along eclectic lines, a theme common to the traditions of both Japan and the United States.

It appears that contemporary psychology has moved into a phase that characteristically rejects the systems period. Such a development is surely beneficial to the extent that it justifies psychology as a truly open-ended science, without rigid, preconceived assumptions and biases. However, this development must also be qualified, because in rejecting the systems, psychology has substituted a definition of science that places faith in empiricism, especially in the experimental method. Psychology in this sense is aligned with the natural science model of inquiry. Although adherence to this model varies from area to area within the discipline, a general consensus that the natural science model is the optimal approach represents in itself a set of assumptions about the nature of human activity.

CHAPTER SUMMARY

Current trends within psychology are reflected by representative developments in the areas of learning, motivation and memory, perception, development, social psychology, and personality. In addition, from the end of the nineteenth century, when contact with the West became routine, psychology enjoyed relative success in Asia. Indeed, we can point to Japan as a contemporary leader in psychological inquiry in all areas, rivaling the productivity of the United States and Europe. In all areas we see a clear shift away from adherence to the systems of psychology and toward a greater reliance on the data collection of an empirical approach to psychological issues. In general, psychology is guided by a direction best described as eclectic. Although models integrating the diversity of data have emerged, especially in the areas of learning and developmental psychology, the remaining areas of contemporary psychology remain at an empirical level with no universally accepted theoretical views apparent. Such a state is beneficial because empiricism provides psychology with a framework of open-ended study. However, empiricism, especially when articulated as the experimental method, also carries assumptions that ultimately commit psychology to a natural science approach, to the exclusion of other methods of inquiry.

BIBLIOGRAPHY

COHEN, S., & HERBERT, T. B. (1996). Health psychology: Psychological factors and physical disease from the perspective of human psychoneuroimmunology. *Annual Review of Psychology, 47,* 113–142.

MILLER, N. E. (1983). Behavioral medicine: Symbiosis between laboratory and clinic. *Annual Review of Psychology, 34,* 1–31.

Learning, Motivation, and Memory

ANOKHIN, K. V., & ROSE, S. P. R. (1991). Learning-induced increase of immediate early gene messenger RNA in the chick forebrain. *European Journal of Neuroscience, 3,* 162–167.

BINDRA, D. (1972). A unified account of classical conditioning and operant training. In A. H. Black & W. F. Prokasy (Eds.), *Classical conditioning II: Current research and theory.* New York: Appleton-Century-Crofts, 453–481.

BINDRA, D. (1974). A motivational view of learning, performance, and behavior modification. *Psychological Review, 81,* 199–213.

BOLLES, R. C. (1967). *A theory of motivation*. New York: Harper & Row.

BOLLES, R. C. (1970). Species-specific defense reactions and avoidance learning. *Psychological Review, 77,* 32–48.

BREHM, J. W., & SELF, E. A. (1989). The intensity of motivation. *Annual Review of Psychology, 40,* 109–131.

BRUCE, D. (1985). On the origin of the term "neuropsychology." *Neuropsychologia, 23,* 813–814.

CHURCH, R. M. (1978). The internal clock. In S. H. Hulse, H. Fowley, & W. K. Honig (Eds.), *Cognitive processes in animal behavior*. Hillsdale, NJ: Erlbaum.

DAVIS, H. R., & SQUIRE, L. R. (1984). Protein synthesis and memory: A review. *Psychological Bulletin, 96,* 518–559.

DUNN, A. J. (1980). Neurochemistry of learning and memory: An evaluation of recent data. *Annual Review of Psychology, 31,* 343–390.

EPSTEIN, R. (1987). Comparative psychology as the praxist views it. *Journal of Comparative Psychology, 101,* 249–253.

EPSTEIN, R., LANZA, R. P., & SKINNER, B. F. (1981). Self-awareness in the pigeon. *Science, 212,* 695–696.

EVERITT, B. J., & ROBBINS, T. W. (1997). Central cholinergic systems and cognition. *Annual Review of Psychology, 48,* 649–684.

GAZZANIGA, M. S. (1967). The split brain in man. *Scientific American, 217,* 24–29.

GLUCK, M. A., & MYERS, C. E. (1997). Psychobiological models of hippocampal function in learning and memory. *Annual Review of Psychology, 48,* 481–514.

HEARST, E., & PETERSON, G. B. (1973). Transfer of conditioned excitation and inhibition from one operant response to another. *Journal of Experimental Psychology, 99,* 360–368.

HILLYARD, S. A., & KUTAS, M. (1983). Electrophysiology of cognitive processes. *Annual Review of Psychology, 34,* 33–61.

HUNT, E. (1989). Cognitive science: Definition, status, and questions. *Annual Review of Psychology, 40,* 603–629.

KOLB, B., & WHISHAW, I. Q. (1998). Brain plasticity and behavior. *Annual Review of Psychology, 49,* 43–64.

KORIAT, A., GOLDSMITH, M., & PANSKY, A. (2000). Toward a psychology of memory accuracy. *Annual Review of Psychology, 51,* 481–537.

MARTINEZ, J. L., & DERRICK, B. E. (1996). Long-term potentiation and learning. *Annual Review of Psychology, 47,* 173–203.

MATTHIES, H. (1989). Neurobiological aspects of learning and memory. *Annual Review of Psychology, 40,* 381–404.

OVERMIER, J. B., & BULL, J. A. (1970). Influences of appetitive Pavlovian conditioning upon avoidance behavior. In J. H. Reynierse (Ed.), *Current issues in animal learning: A colloquium*. Lincoln: University of Nebraska Press, 117–141.

RESCORLA, R. A. (1988). Behavioral studies of Pavlovian conditioning. *Annual Review of Neuroscience, 11,* 329–352.

RESCORLA, R. A., & SOLOMON, R. L. (1967). Two-process learning theory: Relationships between Pavlovian conditioning and instrumental learning. *Psychological Review, 74,* 151–182.

RESCORLA, R. A., & WAGNER, A. R. (1972). A theory of Pavlovian conditioning: Variations in the effectiveness of reinforcement and nonreinforcement. In A. H. Black & W. F. Prokasy (Eds.), *Classical conditioning II: Current research and theory*. New York: Appleton-Century-Crofts, 64–89.

ROITBLAT, J. L., & VON FERSEN, L. (1992). Comparative cognition: Representations and processes in learning and memory. *Annual Review of Psychology, 43,* 671–710.

ROLLS, E. T. (2000). Memory systems in the brain. *Annual Review of Psychology, 51,* 599–630.

ROSENZWEIG, M. (1996). Aspects of the search for neural mechanisms of memory. *Annual Review of Psychology, 47,* 1–32.

SELIGMAN, M. E. P. (1970). On the generality of the laws of learning. *Psychological Review, 77,* 406–418.

SELIGMAN, M. E. P., & HAGER, J. L. (Eds.) (1972). *Biological boundaries of learning.* Englewood Cliffs, NJ: Prentice Hall.

STERNBERG, R. J., & LUBART, R. I. (1996). Investing in creativity. *American Psychologist, 51,* 677–688.

WENZLAFF, R. M., & WEGNER, D. M. (2000). Thought suppression. *Annual Review of Psychology, 51,* 59–91.

WHITE, N. M., & MILNER, P. M. (1992). The psychobiology of reinforcers. *Annual Review of Psychology, 43,* 443–471.

WISE, R. A., & ROMPRE, P. P. (1989). Brain dopamine and reward. *Annual Review of Psychology, 40,* 191–225.

Perception

BARTENTHAL, B. I. (1996). Origins and early development of perception, action, and representation. *Annual Review of Psychology, 47,* 431–459.

DEVALOIS, R. L., & DEVALOIS, K. K. (1980). Spatial perception. *Annual Review of Psychology, 31,* 309–341.

EGETH, H. E., & YANTIS, S. (1997). Visual attention: Control, representation, and time course. *Annual Review of Psychology, 48,* 269–297.

FENG, A. S., & RATNAM, R. (2000). Neural basis of hearing in real-world situations. *Annual Review of Psychology, 51,* 699–725.

GIBSON, J. J. (1970). *The ecological approach to visual perception*. Boston: Houghton Mifflin.

HABER, R. N. (1978). Visual perception. *Annual Review of Psychology, 29,* 31–59.

HERSH, I. J., & WATSON, C. S. (1996). Auditory psychophysics and perception. *Annual Review of Psychology, 47,* 461–484.

JOHANSSON, G., VON HOFSTEN, C., & JANSSON, G. (1980). Event perception. *Annual Review of Psychology, 31,* 27–63.

KINCHLA, R. A. (1992). Attention. *Annual Review of Psychology, 43,* 742–71.

PETTIGREW, J. D., NIKARA, T., & BISHOP, P. O. (1968). Binocular interaction on single units in cat striate cortex: Simultaneous stimulation by single moving slit with receptive fields in correspondence. *Experimental Brain Research, 6,* 391–410.

Development

BALTES, P. B., REESE, H. W., & LIPSITT, L. P. (1980). Life-span developmental psychology. *Annual Review of Psychology, 31,* 65–110.

BALTES, P. B., STAUDINGER, U. M., & LINDENBERGER, U. (1999). Lifespan psychology: theory and application to intellectual functioning. *Annual Review of Psychology, 50,* 471–507.

BIRREN, J. E. (1983). Psychology of adult development and aging. *Annual Review of Psychology, 34,* 543–575.

BIRREN, J. E., & FISHER, L. M. (1995). Aging and speed of behavior: Possible consequences for psychological functioning. *Annual Review of Psychology, 46,* 329–353.

BRAINERD, C. J. (1996). Piaget: A centennial celebration. *Psychological Science, 7,* 191–195.

BRIM, O. G., & KAGAN, J. (1980). Constancy and change: A view of the issues. In O. G. Brim & J. Kagan (Eds.), *Constancy and change in human development*. Cambridge: Harvard University Press, 1–25.

CAIRNS, L. B., & VALSINER, J. (1984). Child psychology. *Annual Review of Psychology, 35,* 553–577.

CARPENTER, P. A., MEYAKE, A., & JUST, M. A. (1995). Language comprehension: Sentence and discourse processing. *Annual Review of Psychology, 46,* 91–120.

DEESE, J. (1970). *Psycholinguistics*. Boston: Allyn & Bacon.

DENNIS, M. (1985). William Preyer (1841–1897) and his neuropsychology of language of acquisition. *Developmental Neuropsychology, 1,* 287–315.

ELLIS, H. C. (1978). *Fundamentals of human learning, memory and cognition* (2nd ed.) Dubuque, IA: William C. Brown.

FLAVELL, J. H. (1996). Piaget's legacy. *Psychological Science, 7,* 200–203.

FOSS, D. J. (1988). Experimental psycholinguistics. *Annual Review of Psychology, 39,* 301–348.

GLUCKSBERG, S., & DANKS, J. (1975). *Experimental psycholinguistics*. Hillsdale, NJ: Erlbaum.

HONZIK, M. (1984). Life-span development. *Annual Review of Psychology, 35,* 309–331.

JARVIK, L. F. (1975). Thoughts on the psychology of aging. *American Psychologist, 30,* 576–583.

MCKOON, G., & RATCLIFF, R. (1998). Memory-based language processing: Psycholinguistic research in the 1990s. *Annual Review of Psychology, 49,* 25–42.

OSGOOD, C. E., SUCI, C. J., & TANNENBAUM, P. H. (1957). *The measurement of meaning.* Urbana, IL: University of Illinois Press.

PHILLIPS, J. L. (1975). *The origins of intellect: Piaget's theory* (2nd ed.). San Francisco: W. H. Freeman.

PIAGET, J. (1926). *The language and thought of the child*. London: Routledge.

PIAGET, J. (1958). *The growth of logical thinking from childhood to adolescence*. London: Routledge.

PIAGET, J., & INHELDER, B. (1969). T*he psychology of the child*. London: Routledge and Kegan Paul.

PREMACK, D. (1983). Animal cognition. *Annual Review of Psychology, 34,* 351–362.

SCHULZ, R., & HECKHAUSEN, J. (1996). A life span model of successful aging. *American Psychologist, 51,* 702–714.

SNYDER, S. H. (1984). Neurosciences: An integrative discipline. *Science, 225,* 1255–1257.

WELLMAN, H. M., & GELMAN, S. A. (1992). Cognitive development: Foundational theories of core domains. *Annual Review of Psychology, 43,* 337–375.

Social Psychology

ALLPORT, F. H. (1924). *Social psychology*. Boston: Houghton Mifflin.

BRODY, N. (1980). Social motivation. *Annual Review of Psychology, 31,* 143–168.

DAWES, R. M. (1980). Social dilemmas. *Annual Review of Psychology, 31,* 169–193.

HARVEY, S. H., & WEARY, G. (1984). Current issues in attribution theory and research. *Annual Review of Psychology, 35,* 427–459.

HARRIS BOND, M., & SMITH, P. B. (1996). Cross-cultural social and organizational psychology. *Annual Review of Psychology, 47*, 205–235.

KENNY, D. (1996). The design and analyses of social-interactive research. *Annual Review of Psychology, 47*, 59–86.

MEAD, M. (1949). *Male and female: A study of sexes in a changing world*. New York: Morrow.

MILGRAM, S. (1963). Behavioral study of obedience. *Journal of Abnormal and Social Psychology, 67*, 371–378.

PETTIGREW, T. F., & CAMPBELL, E. Q. (1960). Faubus and segregation: An analysis of Arkansas voting. *Public Opinion Quarterly, 24*, 436–447.

SCHLENKER, B. R., & WEIGOLD, M. F. (1992). Interpersonal processes involving impression regulation and management. *Annual Review of Psychology, 43*, 133–168.

SNYDER, M., & STUKAS, A. A. (1999). Interpersonal processes: The interplay of cognitive, motivational, and behavioral activities in social interaction. *Annual Review of Psychology, 50*, 273–303.

Personality

EYSENCK, H. J. (1953). *The structure of human personality*. New York: John Wiley.

MISCHEL, W., & SHODA, Y. (1998). Reconciling processing dynamics and personality dispositions. *Annual Review of Psychology, 49*, 229–258.

International Perspectives

BARRETT, W. (1956). *Zen Buddhism: Selected writings of D. T. Suzuki*. Garden City, NY: Doubleday.

BROWN, L. B. (1981). *Psychology in contemporary China*. Oxford: Pergamon.

CHIN, R., & CHIN, A. L. S. (1975). *Psychological research in Communist China*. New Haven: Yale University Press.

COLEMAN, D. (1981). Buddhist and western psychology: Some commonalities and differences. *Journal of Transpersonal Psychology, 13*, 125–136.

HUANG, SU-J. (1994). Max Weber's *The religion of China*: An interpretation. *Journal of the History of the Behavioral Sciences, 30*, 2–18.

MARX, M. H., & HILLIX, W. A. (1973). *Systems and theories in psychology*, 2nd ed. New York: McGraw-Hill.

MURPHY, G., & MURPHY, L. B. (1968). *Asian psychology*. New York: Basic Books.

NAKAYAMA, S., & SIVIN, N. (Eds.) (1973). *Chinese science: Exploration of an ancient tradition*. Cambridge, MA.: MIT Press.

NEEDHAM, J. (1970). *Clerks and craftsmen in China and the West*. Cambridge, U.K.: Cambridge University Press.

ORLEANS, L. A. (Ed.) (1980). *Science in contemporary China*. Stanford, CA: Stanford University Press.

PETZOLD, M. (1984). The history of Chinese psychology. *History of Psychology Newsletter, 16*, 23–31.

ROSENZWEIG, M. R. (1984). U.S. psychology and world psychology. *American Psychologist, 39* 877–884.

18

Epilogue

The subject of this book has been the vast development of intellectual thought about the nature of humanity, a topic first systematically explored by the Greeks and subsequently elaborated through a gradual but steady focus on psychology as an empirical discipline. For the most part, the development of psychological inquiry since Greek antiquity has been intimately involved with the history of philosophy. Indeed, the central psychological issues debated in the nineteenth century concerned the philosophical basis of psychological study. The assumptions underlying the definition of psychology as well as the proper methodological approach of psychological inquiry are essentially philosophical matters. They address fundamental questions about the nature of people, how they know their environment, how they think, and how they interact with each other. All formulations of psychology ultimately rest on the answers to these questions.

This book opened with a statement recognizing the diversity of contemporary psychology. Psychologists work in many different applied settings, performing in a variety of roles. Even within the halls of academia, contemporary psychology is somewhat difficult to identify. Psychological research and teaching take place in departments of psychobiology, cognitive science, organizational management, and social relations. Psychology appears to be evolving toward greater diversification rather than a cohesive unity. The survey in Chapter 17 of contemporary trends in the traditional research areas of psychology may reflect a consensus of approaches to research issues in specific areas, but it does not indicate agreement about psychology as a unified discipline. Rather, the only major point of agreement appears to be a general consensus that contemporary psychology is an empirical discipline.

At the very least, the systems of psychology developed in the twentieth century offer a reasonable description of how psychology attained its diversity. The systems phase of psychology's development was a necessary part of psychology's evolution. It demonstrated the difficulty of defining psychology as a science and of placing psychology within science. Because the empirical expression of science forms the major commonality among the contemporary areas of psychological inquiry, it is appropriate to update the story of psychology's evolution in Western thought by examining the relationship between psychology and science. However, before addressing that issue, we shall compare the systems of psychology using some basic philosophical assumptions.

THE SYSTEMS OF PSYCHOLOGY: AN INTEGRATION

It is curious that the major twentieth-century systems of American psychology were preceded and succeeded by periods best described as functional. Both the earlier functional period and contemporary neofunctionalism take an eclectic attitude toward specific issues of study. Just as the Greek Sophists abandoned the search for a superordinate framework to guide psychological inquiry and instead sought specific, limited models to contain their utilitarian speculation, both functional periods of twentieth-century psychology have eschewed theory building. During the earlier functional period, the avoidance of systematic overviews for psychology was a reaction to the sterility of the orthodox structural psychology of Wundt and Titchener. The reorientation away from theory and toward eclectic issues and research in contemporary neofunctionalism appears to be a reaction to the futility of the intervening systems phase of twentieth-century psychology. Although the earlier functionalism provided a transition to the systems phase, and to behaviorism in particular, it is premature to ascribe a similar, transitional role to contemporary neofunctionalism. It is certainly tempting to draw a parallel between the two functional periods and speculate that the forces of the *Zeitgeist* are building up to a new phase for psychology, for which the present neofunctionalism is an entrée. However, the historian is on dangerous footing when interpreting the present, so the parallel must remain questionable and tentative.

It is the intervening movement between the two functional periods that concerns us here. Specifically, the evaluative dimensions for considering the mind and its activity, used in Chapter 9 to compare the major traditions in philosophy, are employed below to compare and contrast Gestalt psychology, psychoanalysis, behaviorism, and the third force movement. Before proceeding, two qualifications are in order. The first reiterates what was stated in Chapter 9—namely, that the selection of the evaluative dimensions is arbitrary. Other important dimensions of comparison could be considered and would be informative. The four dimensions offered in Chapter 9 and below represent the kinds of comparisons that may provide some closure on the diversity of the systems of psychology. The second qualification is the reminder that there was diversity within the systems themselves. There was, for example, a Gestalt overview and a general psychoanalytic theory, but within each system, as we have seen, different scholars offered variants of the general scheme. Similarly, behaviorism as a system evolved from a rigid formulation to a more open acceptance of various interpretations, so that Watson's behaviorism was quite different from Tolman's behaviorism. Perhaps most diverse was the third force movement, with its contributors coming from very different backgrounds, including philosophy, science, and literature.

The Mind: Dualistic Activity versus Monistic Passivity

This dimension of evaluation essentially contrasts mind–body dualism with monism. Specifically, a dualistic position holds that the mind is a necessary agent of psychological processes, and the mind functions as an active determinant of psychological outcomes. Mental activity is not synonymous with bodily functions, so that the mind is not reducible to the physical bases and mechanisms of the body. Conversely, materialistic monism holds that all psychological processes are ultimately reducible to bodily or physical processes; therefore, there is no need to speculate about the existence of other agents of psychological activity. The single living entity of the body fully accounts for human experience.

Perhaps the most striking and explicit exponent of dualism among the systems of psychology is psychoanalysis. According to the psychoanalytic position, the major determinant of psychological activity is the largely unconscious forces, or psychic energies, of a characteristic sexual and aggressive nature. For psychoanalysis, the goal of personality is to seek equilibrium and harmony among the forces emanating from the unconscious. The reliance on the mental agency of unconscious personality relegates the physical aspects of the person to a secondary role. Rather, overt behavior and even conscious mental processes have symbolic value beyond their actual representations. The contents of observable behavior and consciousness are manifestations of unconscious forces, so that the bodily functions of the individual assume a reactive stance to the energy forces of unconscious personality. Thus in the psychoanalytic system there is not only an implicit acceptance of a dualistic position but also, within the dualism, an emphasis on the mental, or psychic, aspect over the physical.

Although not as well articulated as the dynamics of the psychoanalytic position, there is also a consensus for an implied dualism among the various writers of the third force movement. In particular, those points of agreement, focusing on the critical nature of decisions, the personal responsibility for individual decisions, the recognition of individual dignity and integrity, and the nurturing of personal freedom for psychological growth, assume the existence of a dynamic mental agency that is neither synonymous with nor reducible to the physical aspect of the body. The extent of the dualism varies within the third force movement with different writers. Certainly, the phenomenological method was developed to study precisely the dynamics of mental acts without the reductionistic limitations of the analytic methods of the physical sciences, which would, if imposed, destroy the mental acts. Rather, the very need for phenomenology rests on the acceptance of a separate, mental kind of activity that is different from physical activity.

The views of the Gestalt movement fall between a completely dualistic position and a monistic assumption. The early formulations of the Gestalt writers, based on their research on perceptual processes, tried to avoid dualistic implications. When possible, they attempted to rely on explanations of phenomena through acquired experiences. Their principle of isomorphism was invoked to provide a physical basis of perceptual phenomena. However, the inadequacies of isomorphism as a physiological explanation and the extension of Gestalt principles to field theory together moved Gestalt psychology closer to an admittedly dualistic position.

Behaviorism, of course, is the major proponent of monism, a single, physically based psychological process. The most extreme positions on monism within behaviorism were the radical view of Watson and the thorough physiological reductionism of Pavlovian reflexology. The readmission of limited dualistic views, through proposals for mental constructs, provided the major factor in the evolution of behaviorism after Watson and Pavlov. Moreover, the diversified views of contemporary behaviorism, ranging from neurophysiological to cognitive types of interpretations, differ in their acceptance of mental activity that is not directly and immediately reducible to underlying physical causes.

The dimension of dualism-monism helps to discriminate among the four major systems. For psychology, the primary implication for acceptance of one or the other assumption lies in the nature of the psychological event that is studied. A dualistic position tends to minimize the significance of observable behavior and conscious thought processes, focusing instead on the inner dynamics of mental activity. Conversely, the monistic position elevates observable, physical behavior as the primary data of psychology. Interestingly, formal psychological study has produced more extreme positions along this dimension in the last hundred years than in all the previous periods. In the philosophical traditions up to the nineteenth century, mental activity, expressed in the German tradition originating with Leibniz, could be contrasted with mental passivity, consistent with the Lockean empirical tradition. The twentieth-century systems produced the behavioral tenet that specifically denied the mind's role in psychology. Accordingly, the dimension of mental activity versus passivity sharply contrasts dualism and monism. The latter position requires a rejection of any need for the concept of the mind, a view that reverts to the position of French sensationalistic philosophy.

Sources of Knowledge: Self-Generative versus Sensory

Another dimension of evaluation concerns the ways in which people acquire knowledge of themselves and their environment. Among the philosophical movements of pre-nineteenth-century psychological thought, this dimension essentially contrasted the empirical position of sensory dependence with the rational thesis of self-generated knowledge—that is, knowledge as the product of dynamic activity. In the twentieth-century systems of psychology, the empirical attitude continued to dominate one side of the dimension, and the notion of internal knowledge was extended.

It was primarily within psychoanalysis that the nature of internally generated knowledge was elaborated beyond the confines of the rationalism derived from Kant and the German tradition. Freud's initial formulation of psychoanalysis showed an appreciation of the writings on unconscious strivings of the will by such nineteenth-century German philosophers as Schopenhauer and von Hartmann. Freud's notion of unconscious motivation, based on psychic energy, added a new interpretation and qualification to the meaning of self-generated knowledge. Specifically, the dynamics of the unconscious are largely unknown to the individual, yet each person's conscious thoughts and other experiences (such as dreams) are structured by unconscious energy forces. Accordingly, the definition of knowledge in Freudian psychoanalysis must be qualified—that is, mental activity is self-generated but largely unknown to the individual and certainly not rational. This qualification of self-generated knowledge was further expanded in Jung's views. According to Jung, we inherit specific conceptual frameworks, stereotypes, and mental structures through the personality construct of archetypes. Again, this "knowledge" is not rational, nor is it understood by the individual, yet such thoughts are present in personality and are self-generated, according to Jung.

Within the third force movement there is a clear reliance on self-generated knowledge. Indeed, one major commonality among the varied contributors to this movement is the insistence on inner, reflective, and deliberate mediation of thought processes as the unique human experience. While acknowledging the person's relationship with the environment and accompanying sensory sources of knowledge, the existential–phenomenological position clearly defines the person–environment relationship as a dynamic interaction. Accordingly, the individual's role in interactions with the environment is not reactive or passive, but rather active and constantly seeking to exert control over the environment so that, in turn, personal actions may be self-governed. Within the positions of the third force movement subjective knowledge is a product of individuals acting on environmental sources of knowledge, and the personalized contribution to the relationship with sensory knowledge is the higher, uniquely human level of knowing.

Perhaps because of its general consistency in appreciating the notion of psychological phenomena, the Gestalt movement shared some of the views on the sources of knowledge advocated within the third force positions. The Gestalt movement relied on the interaction between the perceiving individual and the sensory input of environmental stimuli. Accordingly, the Gestalt principles may be interpreted as a compromise between an empirical basis of sensory knowledge and an active mediation that results in the self-generation of knowledge. The Gestalt movement described the person as predisposed to receive sensory information in characteristic ways. As stated earlier, the major difficulty with this compromise between sensory dependence and inner determination of knowledge lies with the precise manner of accounting for how the interaction is achieved—that is, whether a mental agency or an entirely physical explanation is appropriate. As the Gestalt principles were extended beyond sensory and perceptual issues to the field theory of personality, the basis of knowledge became increasingly obscured by the implicit dependence of the individual on interactions with the environment. More reliance was placed on individual initiative to account for field dynamism.

The empirical position of exclusive dependence on sensory experience as the source of knowledge is the foundation of behaviorism. The basic premise of both the Watsonian formulation and Pavlovian reflexology placed the organism within a radical empirical view of acquiring all knowledge through experience with environmental events. Although this view still finds acceptance among behaviorists who seek to find the mechanisms of learning processes in neurophysiological explanations, the evolution of behaviorism has produced moderation of extreme environmental determinism as well. The initial questioning of the extreme behavioristic position occurred in response to the inadequacies of Thorndike's explanation of reinforcement through his law of effect. Further, reluctance to discard all possibility of subjective mediation led Tolman and later psychologists to offer a cognitive interpretation of behavior, a movement that has a counterpart in recent Russian views, such as Luria's study of language. However, despite the moderation of extreme environmental determinism, behavioristic psychology placed the primary source of knowledge in the acquisition of environmental events.

Sources of knowledge serve as a viable comparative tool because all four major systems differ from one another along this dimension. Moreover, the issue of empiricism as the dominant expression of scientific inquiry comes into clear focus under the scrutiny of this dimension. Specifically, the acceptance of knowledge derived from sources other than the environment makes empiricism untenable.

The Basis of Psychology: Mentalism versus Materialism

A direct implication of the dimension contrasting dualistic and monistic conceptions of the individual is the issue of the physical, materialistic basis versus the psychic, mentalistic basis of psychological processes. The acceptance of either assumption involves perhaps the most basic decision about the definition of psychology. We have seen representatives of both positions among the ancient Greek scholars. The Ionian physicists and the later biologists searched for the basic physical substance responsible for life. Conversely, the teachings of Socrates, Plato, and Aristotle concluded the necessity of some nonphysical, spiritual, life-giving entity that transcends the physical nature of the body and the environment. The Greek notion of the soul was Christianized through the efforts of Saint Augustine and, later, the Scholastic philosophers. This interpretation prevailed through the beginnings of science during the Renaissance. Indeed, Descartes defined psychology as the study of the mind in dualistic contrast to the study of physiology. The philosophical traditions that developed after Descartes essentially differed in their position on mentalism versus physicalism. The German tradition accepted the psychic character of an active entity independent of bodily processes; the French tradition generally advocated the opposite position and relegated all psychological and physiological processes to the materialism of the body. The British empirical tradition attempted to forge a position between these extremes by acknowledging the existence of the mind but ascribing to it the passive role of reacting to environmental input. Although various British empiricist scholars held different views of the exact extent of mental activity–passivity, the empirical view prior to the nineteenth century at least nominally accepted the dualistic assumption.

Mentalism dominated in the systems of psychoanalysis and the third force movement. Both positions were logical descendants of the German philosophical tradition of mental activity, and the psychology of both movements is not reducible to physical processes or mechanisms. The Gestalt movement is also grounded in dualism and mentalism. Given the dynamics of person–environment interactions, a mentalistic assumption underlies Gestalt principles. Although Gestalt psychology is not as dependent on mental activity as psychoanalysis or the third force movement, the empiricistic emphasis of the Gestalt principles alone is not sufficient to account for the dynamics of person–environment interactions.

The most materialistic system is behaviorism. Watson's views were an extension of traditional British empiricism, carrying the logic of empiricism to a final conclusion: If the mind is passive and reactive, and knowledge is derived from the sense information originating in the environment, then there is no need for the construct of mind in psychology. Mentalism interferes with the making of an objective psychology. However, when later behaviorists joined Watson's empiricism to Pavlovian reflexology, the implicit physical assumptions of the latter position transformed empiricism within psychology to materialistic empiricism. That union of empiricism and materialism marked a critical juncture for psychology and a triumph for the natural science approach. By removing mentalism from psychology, despite later efforts by the neobehaviorists to rescue mental constructs, behaviorism took on an attraction as an objective science. Herein lies the justification for infrahuman experimentation; without mentalism, the differences among species of animals is one of complexity, not of quality.

The Acquisition of Knowledge: Internal Mediation versus External Association

A final dimension of evaluation concerns the manner of acquiring knowledge. Intimately related to the other three dimensions, assumptions about the acquisition of knowledge may be contrasted as the dynamics of mental organization or the mechanics of associations. The systems of psychoanalysis, the third force movement, and Gestalt psychology depend on various expressions of inner activity, ranging from the innate forces and concepts of psychoanalysis, to the internal characteristics of growth and responsible decisions of the third force movement, to the inner organization of mental activity of Gestalt psychology. In all three systems the acquisition of knowledge is influenced and structured according to forces or patterns arising from within the individual. In contrast, behaviorism relies on mechanical principles of association based on the contingencies of environmental events to explain the acquisition of knowledge.

To summarize briefly, it is interesting that these representative dimensions of critical evaluation in psychology, which successfully discriminated among the pre-nineteenth-century philosophical views, also served as an effective discriminative tool for the twentieth-century systems of psychology. This observation confirms that the critical issues of psychology were not resolved during the systems phase. Indeed, it may be suggested that the systems drew sharper contrasts, because the formal application of diverse assumptions to explicit systems of psychology produced even greater fragmentation within psychology in general. As both the specific descriptions of the systems and the eclectic flavor of contemporary neofunctionalism indicate, the major point of agreement is the acceptance of some version of empiricism—which takes us to the relationship between science and psychology.

THE PROBLEM OF SCIENCE

There was a striking parallel between psychology's development as an independent discipline and the development of empirical science itself, dating back to the origins of modern psychology during the Renaissance. By the nineteenth century, empiricism had demonstrated its benefits in the physical sciences by successfully generating new knowledge with utilitarian applications. The carefully controlled methods of empirical investigation justified a faith in scientific study to improve society and improve the quality of life. Accordingly, the methods of biology, chemistry, and physics offered the optimal model for psychology to emulate. Copleston (1956) has argued that the rise of empirical science is one of the major intellectual accomplishments of the post-Renaissance world, and this period is remarkable because of the enormous advances in empirical discoveries. Moreover, the empirical sciences nurtured the development of the applied sciences, or technologies, with beneficial consequences for our industrialized civilization.

In contrast, the development of organized inquiry outside of empirical science has not prospered as much. Speculative study largely deteriorated to personal accounts. Without empirical verification, it is difficult to offer convincing arguments and win acceptance. For example, within the twentieth-century systems of psychology, the psychoanalytic movement underwent a marked fragmentation brought about by the contributions of divergent scholars who did not conform to any common form of rigorous empiricism. Similarly, despite the development of the phenomenological method, the characteristic diversity among writers of the third force movement has made it difficult to find specific points of universal agreement. This situation offers further support of empiricism, to the point where empiricism and science are generally equated. As a result, empiricism has become the dominant perspective of contemporary psychology, gaining almost universal acceptance. There seems to be widespread agreement that scientific advances are optimally produced and conveyed under the procedures of empirical verification; other forms of inquiry do not appear to offer the compelling attraction of empiricism.

When the twentieth-century systems were contrasted along the dimension of mentalism versus materialism, one of the important developments in modern behaviorism was seen to be the link between empiricism and materialism, brought about through the union of Watson's formulations and Pavlov's reflexology. Materialistic empiricism was, in turn, reinforced by the logical positivism of the Vienna Circle (Chapter 15), which provided a philosophy of science to bolster an objective, behavioral psychology. By relying on the semantics of logical positivism, behaviorism was able to define its subject matter operationally and presumably discard, once and for all, the metaphysics of mentalism.

In evaluating the union of materialistic empiricism, it is important to consider the possibility of nonmaterialistic empiricism. Examining the works of John Locke, the founder of modern empiricism in psychology, we should recall that Locke did not discard mentalistic activities. Acknowledging the dependency on sensory input, he nevertheless recognized two ways of knowing: associations and reflection. The latter process consists of the mental activity of compounding ideas, which represents a mental function. Subsequent refinements within the British empirical tradition, such as the proposal for mental induction by John Stuart Mill, served to separate empiricism and materialism.

This survey of the twentieth-century systems of psychology dramatically underscores one of the major implications of materialistic empiricism: Behaviorism is clearly set apart from the other formulations of psychology. Expressed succinctly, behaviorism, with its reliance on materialistic empiricism, has evolved a definition and methodology of psychology that contrast sharply with other systems. Although the phenomenologists recognized the need to devise a nonmaterialistic empirical method, the difficulty in applying their procedures resulted in an obscurity and elusiveness when compared to the readily quantifiable methods of objective empiricism. Accordingly, the evaluation of the systems leads to the dichotomy of either accepting or rejecting the underlying materialism of behaviorism.

A second implication of the materialism of empirical behaviorism may be seen in contemporary neofunctional trends and concerns the logical conclusions of consistent investigations; namely, that a purely psychological level of analysis, having an integrity in itself, becomes lost in the application of materialistic empiricism to its ultimate end. The distinction between psychological processes and underlying physical explanations is blurred, resulting in the equating of psychology with physiology or other underlying levels, such as cellular or neurochemical biology. By studying psychology as defined through materialistic empiricism, the ultimate conclusion is the rather startling implication that psychology may not be necessary. Consequently, we can observe contemporary trends toward identifying psychology in terms of an interdisciplinary subject matter, reflecting this inherent reduction, such as in psychobiology or in neuropsychology. Whereas these areas of study may indicate an appropriate scientific approach to given issues by circumventing the artificiality of disciplinary barriers, such labels also identify the vulnerability of psychology equated with materialistic empiricism.

In considering the question of psychology's place within empirical science, it is informative to re-examine the contrasting models of Wundt and Brentano, offered over a century ago. Essentially, Wundt, and later Titchener, proposed a model of psychology that had a similarity to materialistic empiricism. Although they recognized the necessity of a mental construct, they argued that the contents of the mind could be reduced to the elements of sensations. However, this analytic model of psychology ultimately led to a reduction of the sensations to their corresponding stimuli. The integrity of psychology itself was lost in Titchener's analysis, so that psychology was logically reduced to physics. The sterility of Titchener's model for psychology's development resulted in the complete failure of his structural psychology. In contrast, Brentano proposed an open-ended model of empirical psychology. His less articulated views recognized a distinctive psychological area of inquiry. Certain psychological events are phenomenal and are destroyed if reduced. Unfortunately, Brentano's act psychology has never been developed to its fullest extent. Certainly, his thoughts influenced both the Gestalt and the phenomenological movements, but the full impact of his nonmaterialistic empiricism has not been explored systematically.

Of interest in the recent neofunctional trends is the growth of certain areas of psychology that seem to accept an implied nonmaterialistic empiricism underlying their research approach. In particular, the study of psycholinguistics and cognitive views of learning, as well as certain trends in social psychological research, suggests a reaction to materialistic empiricism. Although these developments grew out of certain neobehavioristic implications, it is probably no longer correct to label as behavioristic such areas as psycholinguistics and cognitive psychology, as they rely on modest mentalistic overtones in their empirical approach. An expansion of these areas to a systematic and comprehensive framework for psychology in general is premature at this time. However, these developments offer an indication that empiricism may be broadened to include appropriate mentalistic as well as materialistic assumptions. At the very least, it seems that Sperry's (1995) observation that psychology may be in the midst of a Kuhnian paradigm shift is certainly compelling.

CONCLUSIONS

The history of psychology offers a fascinating reflection of the evolution of intellectual thought in general. Because its traditional subject matter is human activity, psychology's past reflects the larger view of the course of Western civilization. For this reason, psychology's development cannot be separated from the evolution of all knowledge. Moreover, as students of psychology, we must accept and tolerate the dissonance, contradictions, and inconsistencies in the history of psychology, for such factors have been present in the often turbulent history of Western civilization. From the turmoil of disagreement and controversy comes greater clarity of the issues, and knowledge is advanced.

In Chapter 2, Comte's description of the historical progress of civilization was cited, and it was noted that he regarded ancient Greek thought as the transition between a reliance on theological explanations and a view that looked either internally to the person or externally to the environment for causal explanations. The philosophical studies of the ancient Greeks identified basic issues of psychology, focusing on the necessary assumptions about the nature of human activity. These critical issues, which the Greek scholars did not resolve, still baffle psychologists. Can human beings explain psychological activity in terms of physical matter only, or is some proposal of mental life necessary? The progress of Greek thought led to the development of the concept of soul, as represented in the comprehensive philosophies of Plato and Aristotle. Despite the almost 2,500 years since the flowering of Greek thought, very little of truly original quality has been added. Changes, modifications, and reinterpretations have been offered, but essentially science as we know it today is a study based on an Aristotelian framework of knowledge. Indeed, Scholastic philosophy, reaching a pinnacle under the genius of Saint Thomas Aquinas, marked the resurrection of civilization after centuries of social deterioration and provided a reintroduction to Aristotelian philosophy, interpreted in the light of Christianity.

The modern age of empirical science began with Descartes, whose interpretation of knowledge was based on the Aristotelian system. The philosophical justification of empiricism, as well as the first statements of how knowledge is acquired by the association of ideas, was introduced by Hobbes and Locke. From their writings, the possibility of an empirical science of psychology was explored. However, the empirical strategy of psychology was not universally accepted, and rival conceptualizations were proposed. One tradition, centered in France, rejected the need for a psychology, presenting the modern monistic argument by reducing mental activity to the elements of sensory physiology. Alternatively, the tradition that began with Leibniz was inspired by the Greek concept of the activity of the soul, and he proposed a psychology determined by the activity of the mind. This tradition culminated in the rationalism of Kant, and mental activity was the prevailing theme in German philosophy well into the twentieth century.

Nineteenth-century psychology inherited the competing models of psychology. It was the empiricist framework that propelled psychology into a separation from philosophy, physics, and physiology, so that by the 1870s psychology started to gain recognition as an independent discipline. However, even within the empiricist orientation there were disagreements about the scope and methods of psychology. As we have seen, Wundt and Brentano disagreed over the type of empiricism, and their argument essentially involved the issue of mental versus materialistic empiricism.

From this survey of the origins of psychology, we see that the history of psychology can hardly be described as a smooth, even flow of progressive developments. The new science of psychology inherited some violent disagreements over the most fundamental assumptions of the discipline. On the basis of psychology's past, the subsequent turmoil of psychology's first century was easily predicted. The systems phase of twentieth-century psychology attempted to deal with basically different conceptualizations of the nature of psychological activity. The systems phase, however, did not produce the definitive model of psychology. Behaviorism, the dominant system in the United States, evolved so drastically that it is barely recognized now as a consistent system. Rather, behaviorism dissolved into an eclectic attitude emphasizing empiricism. As the systems phase of twentieth-century psychology was preceded by a functional psychology, so too has it been succeeded by a neofunctionalism. Accordingly, we must conclude that contemporary psychology is deficient as a theoretical discipline. The major agreement of contemporary neofunctionalism lies with the consensus that psychology is an empirical science, which is itself an atheoretical statement. Thus we must reserve judgment about psychology's theoretical future and await the forces of the *Zeitgeist*.

On an applied level, we can safely conclude that psychology has been successful over the last century. The empirical developments in psychology have added to our knowledge of very different areas, from psychopathology to advertising to interethnic understanding. In this sense, the functional character of psychology has produced results. Moreover, using the criteria of the utilitarian and the eclectic, we may feel confident about the future of psychology.

We started this inquiry into psychology's past by noting the diversity and apparent confusion of opinions in contemporary psychology. It was not our purpose to resolve the disparity. It was our aim, rather, to clarify the confusion by using the knowledge of history to discover where the present myriad of views on psychology originated. To survey the present state of psychology quite simply, psychology is an active and exciting discipline, despite its failures, regressions, and false starts. This statement readily admits that psychology is not an easy discipline to study. The student must cope with some very basic decisions before moving on to systematic investigation of the issues. However, such dissonance is perhaps appropriate, for unlike other disciplines, psychology has as its subject matter the most complex of questions to answer—namely, why are we what we are, and why do we do what we do?

CHAPTER SUMMARY

The scope of this book is summarized in terms of the basic issues that have historically confronted psychology. The four systems of psychology—psychoanalysis, Gestalt psychology, the third force movement, and behaviorism—have been compared along four critical dimensions: mental dualism versus monism, self-generative versus sensory sources of knowledge, mentalism versus materialism, and internal mediation versus external association in the acquisition of knowledge. The evaluation points to the relationship between psychology and science, and in particular, the problems resulting from a reliance on materialistic empiricism. Psychology as a theoretical discipline has suffered from the disagreements and controversies of the first century of its existence. Yet in its first hundred years, psychology has been successful as an applied science.

BIBLIOGRAPHY

General Sources

COPLESTON, F. (1956). *Contemporary philosophy: Studies of logical positivism and existentialism*. Westminster, MD: Newman Press.

GIORGI, A. (1970). *Psychology as a human science*. New York: Harper & Row.

KOCH, S. (Ed.) (1959). *Psychology: A study of a science, Vol. II: General systematic formulations*. New York: McGraw-Hill.

SPERRY, R. (1995). The future of psychology. *American Psychologist, 50*, 505–506.

TURNER, M. B. (1967). *Philosophy and the science of behavior*. New York: Appleton-Century-Crofts.

Glossary

The following terms and concepts are defined in the context of their use in this book. In selecting items for this glossary, one goal was to include terms and concepts derived primarily from disciplines other than psychology. The discipline of psychology evolved from philosophy, theology, and the natural sciences, whose terminology may not be familiar to the student of psychology. This glossary therefore concentrates on definitions of terms drawn from other disciplines that are appropriate to the history of psychology. Italicized words within each definition refer to terms and concepts defined separately in the glossary. Numbers in parentheses at the end of each definition refer to chapters in which the terms are elaborated.

Act psychology. In general, those versions of psychology that emphasize the unity and cohesion of individual interactions with the environment and recognize the multifaceted levels of human experience. In particular, the act psychology proposed by Brentano, Stumpf, and the Würzburg school contrasted the integrative unity of human experience with the elementarism of Wundt's *structural psychology*. Consistent with the German philosophical tradition of *mental activity*, Brentano especially argued that the psychological act is intentional, and those methodologies that are rigidly *analytical* result in the destruction of the act under study. A truly descriptive approach to psychology must be *empirical* yet at the same time deal with the phenomenal character of psychological acts (11, 13, 16).

Analysis. The general strategy of study, or methodology, that seeks to explain psychological events in terms of contributing parts or underlying elements. The experimental method, as applied in the *natural sciences* and in some models of psychology, is an analytic method. Such methods are usually *reductionistic*; they may also be described as elementaristic, *atomistic*, and *molecular* in character. Analytic methods are in contrast to views of psychological events in terms of unity and *phenomenal* quality, which require a holistic or a *molar* strategy of study (9, 11, 18).

Analytic psychology. The label given to Jung's modification of Freud's *psychoanalysis* (14).

Associationism. The view that interprets higher mental processes as resulting from combinations of *sensory* and/or mental elements. A reliance on associationistic processes was an outcome of *empirical* models of psychology, because associations provided a *mechanism* of environmental adaptation and learning. Among the explanations for the formation of associations, the principles of *contiguity*, *contingency*, and *similarity* were most prominent in the British philosophical tradition of *mental passivity*. The quantification of association principles was realized in Pavlov's theory of conditioning, adopted by the behaviorists (7, 15).

Atomism. The philosophical view that complex events can be *reduced* to component elements. Applied to psychology, atomistic views support the *analysis* of experience into constituent parts—for example, an idea reduced to stimulus-response associations. Moreover, an atomistic framework admits that little is lost in the reductionistic analysis, so that the simpler components completely explain the complex psychological event (2, 6, 7, 11, 15).

Behaviorism. The system of psychology that admits as its subject matter overt, *observable*, and measurable behavior. In its most rigid form, offered initially by J. B. Watson and later by B. F. Skinner, behaviorism denies the admission of traditional issues of mental events. Contemporary behaviorism has evolved to a broadly based *eclectic* system of psychology, which emphasizes to varying extents the study of behavioral processes that may be mediated by nonobservable *mechanisms* or agencies (12, 15, 17, 18).

Buddhism. A religion or philosophy that originated in India with Buddha (sixth century B.C.) and spread throughout Asia. Buddhism generally prescribes a life of reflection and self-denial that will enable the individual to reach a level of bliss and release from the bonds of earthly desires (1, 17).

Confucianism. A system of ethics originating with the Chinese philosopher Confucius (ca. 551–478 B.C.), which emphasized social values, individual loyalty, and family ties (1, 17).

Consciousness. The personal awareness of subjective experience at any point in time. In general usage, consciousness has meant the individual's total awareness of past experiences and future aspirations, as well as ongoing self-knowledge, and this usage implies active self-reflection. Specific definitions of consciousness have ranged from global meanings, such as William James's interpretation of consciousness as continuous and transcendent of time, to Freud's limited view of consciousness as confined to distorted reflections of the more active and encompassing *unconscious* (6–9, 11, 12, 14, 16).

Constructs. In models and theories, explanatory devices that tend to be only distantly related to specific *empirical* referents. For example, the notion of *consciousness* is a construct because it is used to explain several psychological processes, but it is not directly *observed*, nor is it defined in terms of specific observable and measurable events. Constructs are often contrasted with *intervening variables*, with the former having less *observational* reference than the latter (16, 18).

Contiguity. A general principle of *associations* stating that two or more events occurring together in time tend to be associated and retained together. The British philosophers Hume, Hartley, and James Mill gave contiguity a primary role in accounting for the acquisition of associations; other philosophers, most notably J. S. Mill and Brown, argued that principles in addition to contiguity (e.g., *contingency* and *similarity*) were needed to explain associations. More recently, Pavlov's conditioning theory has advocated the temporal relationship between the unconditioned stimulus and the conditioned stimulus as the primary determinant of successful acquisition, and Guthrie's learning theory emphasized contiguity as the central principle of association (7, 15).

Contingency. The principle of association referring to the extent of the dependent relationship between two or more events that accounts for the association of the events. Although contingency was recognized by the British *empirical* philosophers, this principle has received attention in more recent revisions of Pavlovian conditioning theory that stress the contingent, predictive value of the conditioned stimulus in relation to the unconditioned stimulus, such as in acquisition training through the delayed conditioning procedure (7, 15).

Cosmology. The branch of *metaphysics* that studies the total universe and the general properties of nature in terms of ultimate principles. Those early Greek scholars who attempted to find the basic substance of life in the physical environment may be described as cosmologists. Modern astronomers who search for ultimate theoretical explanations of the origins and workings of the universe continue the tradition of cosmological study (2, 4, 10).

Deduction. The *logical* process or sequence of reasoning involving a progression from a known principle or premise to an unknown; from the general to the particular. Deductive reasoning is a complement to *inductive logic*. Systematically presented by Aristotle, deductive reasoning assumed a dominance in the methodology of *Scholasticism*. Its value was later questioned as *empiricism*, based primarily on *induction*, assumed importance in the rise of *Renaissance* science (2, 4, 5, 10).

Determinism. In psychology, the doctrine or philosophical assumption that a given psychological event or process is completely governed by specified factors, usually beyond individual control. For example, the Roman Stoics resigned themselves to the belief that all of life's events are determined by fate, independent of individual desires or intentions. Similarly, contemporary operant conditioning theory accepts the possibility that complete control of environmental events will lead to perfect control of behavior, reflecting a deterministic position. Complete determinism is in opposition to views emphasizing personal freedom, individual variability, and free *will*. The issue of determinism occupies an important role in the historical development of psychology and in the expressions of contemporary systems (2, 3, 9, 15, 18).

Dialectics. In general, any extended, detailed *logical* argument in reasoning. In particular, the nineteenth-century philosopher Hegel devised a dialectic method of logic, proposing that an event or an idea (thesis) gives rise to its opposite (antithesis), leading to a reconciliation of the opposition (synthesis). Historical progress is explained by repetitions of the cycle, and Hegel's dialectic method was adapted by Marx and Engels to explain changes in society and in nature (3, 16).

Dualism. Any of the philosophical assumptions underlying psychology that accept the position that human beings possess two basic aspects, mental and physical. Since postulated by the ancient Greeks, dualistic positions have consistently dominated Western preconceptions of human psychology. The modern development of psychology was spurred by the dualism of Descartes, who proposed that physiology is the study of the physical and *material* body; psychology, the study of the nonphysical immaterial mind. Dualism is in contrast to *monism*, which postulates only a single physical or mental aspect of human existence. As a recurring theme in psychological inquiry, dualistic positions are also represented in such twentieth-century systems as *psychoanalysis* and the third force movement (all chapters).

Early Middle Ages. The period lasting roughly from the removal of the imperial capital from Rome (476) until the eleventh century, during which western Europe was torn by war, disease, and ignorance. The ancient great cities of the Roman Empire were largely abandoned, and European life became predominantly rural. Intellectual life first stagnated, then regressed to the point of almost complete eradication. The social fabric became feudal and the sole international or intertribal institution of any authority was the Church, headed by an increasingly powerful papacy (4).

Eclecticism. An attitude in psychology that supports the selection of positions and interpretations of psychological issues from various diverse systems. The resulting system attempts to blend the valued interpretations picked up from other theories into a harmonious, consistent perspective on psychology. The prevailing theme giving unity within an eclectic selection typically involves some limited goal of explaining psychological processes. For example, the ancient Greek Sophists were eclectic in their attempt to eschew the search for an all-encompassing principle of life and settled instead for a view that bonded together natural, physical events only. Similarly, the twentieth-century American *functionalists* were eclectic to the extent that they abandoned the search for a theoretically coherent psychology of *consciousness* in favor of gathering *empirical* data of an applied value (2, 12).

Empiricism. A philosophy of *science* that accepts experience as the sole source of knowledge. As a result, *sensory* processes constitute the critical link between the environment and subjective knowledge, and *observation* through the senses becomes the standard criterion for the validity of empirical science. Scientific empiricism and psychology developed together in post-*Renaissance* Western thought. Empirical psychology received its initial comprehensive expression within the British philosophical tradition, and subsequently, in the latter part of the nineteenth century in Germany, two empirical models competed for the definitive framework of psychology. Twentieth-century psychological systems have differed primarily in their commitment to empiricism (1, 7, 9, 11, 18).

Epistemology. The branch of *metaphysics* that studies the origins, characteristics, modes, and limits of knowledge. Aristotle's human psychology was basically epistemological because his views on mind and body were mostly descriptions of how we acquire knowledge. Piaget's developmental psychology is a modern example of epistemology because Piaget was ultimately concerned with knowing and how we know (2, 4, 11, 16, 17, 18).

Essence. A philosophical issue derived from the Greek scholars, especially Aristotle, who viewed all living and nonliving objects as manifestations of some defining property or characteristic common to all similar objects. As applied to humans, each person has an individual *existence*, but all people share in the common essence of the *soul*. The adoption of the Greek concept of essence into Christian thinking by the *Scholastic* philosophers of the late *medieval* period resulted in the very definition of psychology as synonymous with the Christian goal of perfection of the essence shared by each person—namely, eternal salvation of the soul. Thus, for the Scholastics, essence precedes existence. In the nineteenth and twentieth centuries this Scholastic notion of essence was seriously challenged by the philosophy of existentialism, reflected in the third force movement within psychology (2, 4, 16).

Evolution. A process of orderly development and growth. Within psychology, evolution implies the contemporary systems that tend toward *monistic* assumptions, such as *behaviorism*, and accept that human beings are phylogenetically continuous with the rest of living nature. In general, Darwin's theory of natural selection forged an inseparable link between evolutionary data and psychology (10).

Existence. In the Aristotelian and *Scholastic* sense, the individual expression of each object *observed* in nature. For people, existence is the individual state of being that gives expression to the universal essence shared by all people. Existential philosophy challenged the Scholastic conception by proposing that individual existence, the act of being, defines a person's *essence*, so that existence precedes essence (2, 4, 16).

Faculty psychology. The view of the mind or of mental processes that holds that human psychology is derived from a number of mental abilities or specific mental agencies, such as memory, reason, and the *will*. Although a justification for faculty psychology may be found in ancient Greek thought, the more modern expression evolved within the German philosophical tradition of *mental activity*. The emphasis on mental predispositions, exemplified in Kant's categories, gave rise to a distinct proposal that the mind has functions that underlie similar mental activities. Contemporary expressions consistent with faculty psychology include trait theories of intelligence and personality (2, 8, 10, 15, 17).

Functionalism. An attitude that has dominated American psychology during most of the twentieth century. Functionalism emphasizes and values the utilitarian and applied aspects of psychological activities, as opposed to describing psychological *structures* and contents. Philosophically supported by American *pragmatism*, varying degrees of functionalism may be ascribed to all systematic expressions of twentieth-century psychology in America, because all of the schools, including those imported from Europe, were subjected to criteria of applicability and usefulness (12, 17).

Geocentric, or **Ptolemaic, system.** A system of the universe holding that the earth is the fixed center around which all heavenly bodies revolve. This position was systematized by the astronomer Claudius Ptolemy (ca. second century) of Alexandria in Egypt and was essentially accepted by the early Christian philosophers, such as Saint Augustine, and the later *Scholastics*. The earth-centered system was not seriously challenged until Copernicus proposed the *heliocentric* alternative (4).

Gestalt. A German word literally meaning shape or form. As used in psychology, Gestalt implies any of the unified patterns or *structures* that make up experience and have properties that cannot be *reduced* to the component parts or elements of the whole, so that the unity of the whole is more than the sum of the parts. Gestalt psychology, which can be readily traced to the German philosophical tradition of *mental activity*, advocated an emphasis on psychological events as *phenomena* that cannot be reduced without loss of the integrity and unity of the event (8, 13, 15, 18).

Great man theory. Interpretation of historical progress suggesting that significant events of historical and social importance occur because of the efforts of outstanding persons. This position is in contrast to the *Zeitgeist*, or "spirit of the times," interpretation (1, 18).

Hedonism. In psychology, the view that individual activity is governed by the seeking of pleasure and the avoidance of pain. For example, the Epicurean philosophy of ancient Rome advocated ethical values based on hedonistic principles; in contemporary psychology, the role of *reinforcement*, especially as expressed in Thorndike's law of effect, provides an example of hedonistic implications (3, 6, 16).

Heliocentric, or **Copernican, system.** A system of planetary organization that holds that the sun is the center of the universe, and all heavenly bodies revolve around the sun. Moreover, the turning of the earth on its axis is responsible for the apparent rising and setting of the sun. Championed by the Polish astronomer Copernicus, the heliocentric position initially argued from the parsimony of logic against the prevailing *geocentric*, or *Ptolemaic*, view. Copernicus's heliocentric system was successful in challenging the Ptolemaic system that was supported by the authority of the Church (4).

Hinduism. The varied indigenous systems of Indian philosophy that teach that the person should seek to move from the level of specific individuality to the level of unity and harmony with the universal (1).

Humanism. Those trends in Western thought that tended to view individual people in terms of their dignity, ideals, and interests. Humanistic themes throughout history have elevated the value of human intellectual powers and downgraded *deterministic* interpretations of life, whether arising from theistic or environmental control. Humanistic attitudes may be found in the development of the concept of the rational *soul* in Greek scholarship, in the art and literature of the *Renaissance*, and in the *humanistic psychology* of the contemporary third force movement (2, 4, 11, 16).

Humanistic psychology. A label given to a group of mostly American psychologists within the third force movement who have devised eclectic criteria for individual growth, applied primarily in clinical settings. They hold that each individual is capable of a truly psychological level of well-being, characterized by self-fulfillment and a complete integration of subjective goals, desires, and expectancies within an honest *perception* of reality (16).

Human science model. In general, a series of assumptions about the definition and the methods of psychology, advocating recognition of the complexity of human motivation and dynamic activity and suggesting that human psychology is qualitatively different from other forms of life. Moreover, the human science model is open to several strategies of methodology, all within an open-ended definition of *empirical science*.

The human science model is in contrast to the *natural science model*, which defines a psychology of observable activity in a relatively restricted sense and fails to differentiate the types of psychological events emanating from various species, instead seeking to model psychology on the methodology of the natural or physical sciences (11, 18).

Hypotheticodeductive approach. A methodological approach in *science* that may be *logically* or *empirically* based and that involves a series of initial principles that are tentatively accepted and subsequently tested for all implications before their final inclusion in a general theorem. The Greek mathematician Pythagoras provided an early example of the logical rigor of this method; in contemporary psychology the empirical use of this approach is seen in the systematic *behaviorism* of Hull (2, 15).

Idiographic. Descriptions of psychological events and processes from the perspective of the singular individual, in contrast to *nomothetic*, which emphasizes group or normative description. An idiographic approach is a characteristic approach of *humanistic psychologists* (9, 16, 17, 18).

Individual psychology. The term used to describe Adler's modifications of Freudian *psychoanalysis* (14).

Induction. The *logical* process or sequence of reasoning involving the progression of inference of known principles to some general statement covering all instances applicable to the principle; reasoning from the particular to the general. Inductive logic is the complement of a *deductive* progression in logic. Induction forms the logical basis of *empirical* methodologies in science, because conclusions about particular *observations* are tested for generalizations to cover all possible cases of similar observations (2, 4, 5, 10).

Innate ideas. A philosophical assumption that the mind is born with content. The innate content of the mind may be specific, such as Descartes' suggestion that each person is born with the knowledge of God or Jung's notion of the archetypes of the collective *unconscious*. The view that the mind is predisposed to certain modes of knowing, such as the position of Kant and, more recently, *Gestalt* psychology, is consistent with belief in innate ideas and, more generally, the doctrine of *nativism* (8, 9, 11, 13, 14, 18).

Interactionism. The philosophical interpretation of mental and bodily processes suggesting that the mind and body, although separate entities, influence each other. This view contrasts with *parallelism*, which asserts no interaction, but rather processes that grow in the same direction yet are independent. A contemporary example of interactionism is *psychoanalysis*, which holds the psychic energy of the *unconscious* mind exerts dominant direction over all human processes, including bodily responses (6, 7, 8, 9, 11, 13, 18).

Intervening variable. A relatively specific process or event that is assumed to link directly *observable* events. For example, a Pavlovian-conditioned *association* may be used as an intervening variable to account for the relationship produced between a conditioned stimulus (CS) and a conditioned response (CR) following appropriate acquisition. Both the CS and the CR are observable events, and the conditioned associative bond is directly inferred from these events. Intervening variables are usually contrasted to *constructs*; both devices are used as explanatory *mechanisms*, but intervening variables have more direct *operational*, *empirical* referents than constructs (15).

Introspection. In general, an individual's reflection or contemplation of subjective experience. Recorded introspective reports abound in literature, with perhaps Saint Augustine's *Confessions* being the most famous example. In a specific sense, introspection was the experimental method used by Wundt for the study of psychology. Within his *structural psychology*, introspection became a highly controlled method by which a trained psychologist could presumably study the contents of immediate experience (3, 5, 11).

Logic. The study and method of correct reasoning. Logic as systematized by Aristotle contains the criteria for the validity of arguments derived through orderly, sequential processes of *deduction* and *induction*. Although the *empirical* methods are logical, not all logical methods are empirical, because logical arguments may be exclusively based in rational abstractions, as indeed was the case prior to the post-*Renaissance* emergence of empirical *science* (2, 4, 5, 9, 18).

Materialism. The philosophical position that matter is the only reality and that all objects and events, including psychological processes of thinking, *willing*, and feeling, can be explained in terms of matter. Materialistic assumptions are often equated with physicalism, because the latter holds the similar view that psychological events are based on the physical mechanics of bodily processes. Materialism is opposed to those views of psychology that stress mentalistic *constructs* (2, 4, 11, 15, 18).

Mechanism. A philosophical position that assumed the systematic operation of physical and mental processes to produce all psychological experiences; also referred to as a mechanical view of psychology. Usually, a mechanical interpretation accepts that bodily mechanics in particular underlie all psychological events, precluding the need for mental *constructs*. A belief in mechanism is in contrast to *vitalism*, which insists on the need for some living agency distinct from physical mechanics. A mechanical view of learning is inherent in contemporary conditioning theory, which interprets *associations* as accounted for and explained by sensory-motor relationships in the nervous system (2, 9, 18).

Medical model. An approach to psychological problems or deviancy that adapts the strategy of physicians and proposes that *behavioral* manifestations are symptomatic of some underlying cause. For example, *psychoanalysis* interprets overt activities as important only for what they reflect of unobservable, largely *unconscious* psychodynamics. A medical model approach to psychology may be contrasted to a behavioral model, which values *observable* activity as the critical approach to treatment (14, 15).

Medieval period. The epoch in European history existing between the *Early Middle Ages* and the *Renaissance* and lasting approximately from 1000 to 1500. This period was characterized by social and political dominance of the papacy, the emergence of the nation-states of Europe, and the gradual revival of scholarship. *Scholasticism* was the primary intellectual achievement of medieval Europe (3, 4).

Mental activity. The core of assumptions underlying the psychology that dominated the German philosophical tradition. This view basically holds that some agency, such as the mind, is separate and distinct from the body and is responsible for the higher psychological processes of thinking, *willing*, and perceiving. The mind is active and dynamic, and, to varying extents, the mind generates knowledge and is not entirely dependent on environmental input for its contents. Contemporary psychological views that emphasize mental activity include *psychoanalysis*, the third force movement, and, to a limited extent, *Gestalt* psychology (8, 9, 11, 13, 14, 16, 18).

Mental passivity. The basis for the *empirical* psychology of the British philosophical tradition, in which the mind is viewed largely as a reactive receptive agency, and the contents of the mind are dependent on environmental input. Mental passivity justified the study of psychological processes in terms of the *observable* events of the environment as the substance of *sensory* processes. Moreover, the acquisition of knowledge about the environment is valued for its role in assisting the adaptation of the organism. Mental passivity is the converse of mental activity, which posits the mind as acting by its own abilities on input from the environment. Mental passivity is implicit in *behavioristic* psychology in general (6, 7, 9, 11, 12, 15, 17, 18).

Metaphysics. The branch of philosophy dealing with the ultimate explanations and first principles of being or reality (*ontology*), the nature and structures of the universe (*cosmology*), and the study of knowing (*epistemology*). Metaphysics and *empirical* psychology have often been set in opposition in the history of science, because the development of *empiricism*, based on *observation*, was viewed as essentially replacing metaphysical explanations of psychological events with demonstrated facts (2, 16).

Molar behavior. The view of total behavior defined in relatively large units, emphasizing behavior as unified and purposive. Molar behavior is contrasted with *molecular behavior*, which tends to lead to *atomism* and *reductionism*. Tolman exemplified a molar approach in *behaviorism* when he incorporated some of the *phenomenological* emphases of *Gestalt* psychology into behaviorism (15).

Molecular behavior. Behavior defined through the collection of small segmented units, such as muscular or glandular activities. The *logic* of *atomism* in molecular interpretations of behavior leads to the reduction of units of behavior to the neurophysiological or neurochemical levels of *analysis*. Molecular behavior is in contrast to *molar behavior* (15).

Monadology. The philosophy of Leibniz, who postulated the existence of "monads," defined as life-giving entities that are ultimate units of life. Monads essentially provide the agency of activity in the world. The human *soul* is the most active monad. As the founder of the German philosophical tradition of psychology, Leibniz gave an active imprint to the direction of this tradition, leading to the emergence of a psychology based on the assumption of dynamic, self-generating *mental activity* (8, 9).

Monism. Any of the philosophical assumptions underlying psychology that admit to only a single substance or principle of psychology, usually in the *materialism* of the body. Monistic systems of psychology reject the mental activity of *dualistic* positions. The growth of monism has paralleled the development of *empirical* psychology itself, so that materialistic monism is the predominant perspective in contemporary psychology, expressed in the teachings of neobehaviorism and *reflexology* (2, 6, 7, 9, 15, 17, 18).

Nativism. Specifically refers to the doctrine of *innate ideas* and is used in psychology to indicate any inherited ability, predisposition, or attitude. Nativism is in contrast to *empiricism*, and is represented in such diverse contemporary constructs as Jung's collective *unconscious*, *perceptual* predispositions of *Gestalt* psychology, and Chomsky's language acquisition device (all chapters).

Naturalism. The philosophical position that the natural world as we know and experience it is the only certain reality. Standards of ethics are derived from the lawful relationships in nature. If God or other supernatural forces exist, such influences are not responsible for the direction of human activities. Naturalism is a recurrent theme in Western psychology, from the early Greek explanations of life by specific natural events to Spinoza's equation of the forces of God with the forces of nature (2, 5).

Natural science model. In general, a set of assumptions about the definition and the study of psychology that supports the notion of psychological events as *materialistically* or physically based *observable* processes. These assumptions translate to a methodological *empiricism* that is essentially the same as that used to study such sciences as biology, chemistry, and physics. The natural science model stands in contrast to the *human science model*, which sees psychological activity in human beings as qualitatively different from other forms of life (9, 11, 18).

Nomothetic. Used to describe psychological events and processes in terms of general laws that tend to apply to group or normative standards. A nomothetic emphasis contrasts with an *idiographic* approach, which stresses the individual rather than the group (9, 16, 18).

Observation. The deliberate act of examining particulars of an event through *sensory* awareness on the part of the observer. Observation may involve direct sensory experience of an event under scrutiny or may employ instrumentation that serves as a medium between an event and the observer's sensory processes. Observational methods, which form the core of *empirical science*, may range from rather informal descriptions to formal controlled procedures, such as the experimental method. In psychology, observation is the critical dimension in the definition of psychology's subject matter. For example, a process not directly observed, such as the mediation of id energy by the ego in *psychoanalytic* theory, may be readily distinguished from directly observed events, such as the magnitude of a conditioned response in conditioning *reflexology* (1, 2, 4, 7, 9, 11, 18).

Ontology. The branch of *metaphysics* that studies the ultimate nature and relations of being. Ontology seeks to discover the abstractions that define a thing to be what it is. Those philosophical speculations that proposed the existence of an immaterial *soul* as the ultimate life-giving element are examples of ontological study (2, 3, 4).

Operationism. The scientific view that requires the definition of a scientific event in terms of the identifiable and repeatable procedures that produce the event, and nothing more. As a formal doctrine, operationism grew out of the logical *positivism* movement of the early twentieth century, which attempted to rid scientific issues of surplus meaning and pseudoproblems. Adapted to psychology, operationism advocated the definition of psychological events in terms of the procedures required to produce *observation* of such events. For example, hunger may be operationally defined as the motivational state that results from 24 hours without food (2, 5, 11, 15, 18).

Paradigm. In *science*, a model or pattern that accommodates all forms of diversity and variability related to a given issue. For example, if the association of ideas is viewed as synonymous with conditioning, the model of conditioning serves as a framework to explain all forms of *associations* (1, 11, 18).

Parallelism. A general interpretation of mental and bodily processes asserting that mind and body are independent entities functioning separately but in parallel. As an explanation of *dualism*, parallelism is in contrast to *interactionism*, which stresses the mind acting on the body and vice versa. A contemporary example of parallelism is the *Gestalt* psychologists' assertion of psychophysical parallelism through their principle of isomorphism, which suggests a dynamic correspondence between the excitation in the physical brain field and a perceived field of experience (6, 7, 8, 9, 11, 13, 18).

Perception. Any experience that is primarily dependent on *sensory* input but has content and organization usually derived from previous experience or predisposition. Perception is typically interpreted as a cognitive process, in contrast to *sensation*, which is usually defined as sensory experience only. The exact distinction between sensory processes and perceptual processes is not clear and varies with diverse models of psychology; as a result, the two terms are often used interchangeably or to imply different levels of complexity along the same dimension. The study of perception is one of the fundamental areas of psychological inquiry, and has provided critical issues for all systems and theories throughout psychology's history (9, 11, 13, 17, 18).

Phenomenology. A methodological approach to psychology and other disciplines that focuses on the unity and integrity of events and experiences. Phenomenology may be informal to the extent that it is an attitude of study that allows free expression to the various ways that events appear to the observer, enabling the observer to apprehend the integrity of an event as going beyond its component elements. The *observational* approach of Purkinje's physiological research is an example of informal phenomenology. An example of systematic, formal phenomenology is Husserl's method of specific procedures of observation (9, 10, 11, 13, 16, 18).

Phenomenon. Literally, that which appears. In psychology, phenomena are usually described as events that are experienced as unified and unanalyzed. The view of psychological experience as phenomenal contrasts with assumptions that hold experience to be capable of *analysis* into components. As an example of the study of phenomena in psychology, the *Gestalt* psychologists accepted the principle that *perception* of environmental objects is an experience that is not reducible to *sensory* elements, but rather consists of a whole process of person–environment interactions that produce the integrity and unity of experience (10, 11, 13, 16).

Phrenology. An attempt in the eighteenth and nineteenth centuries to relate specific mental abilities or traits to the size and skull contours of brain areas where particular capacities were assumed to reside. Essentially, phrenology tried to establish a physical basis for the prevailing philosophical view of *faculty psychology*. However, the movement was ridden with quackery, and subsequent neurophysiological research showed that phrenology was completely untenable (10).

Platonism. The collective philosophical views based on the writings of Plato. For psychology, Platonic belief stressed the unreliability of *sensory* knowledge and the inherent evil of bodily passions; only the wisdom of the human *soul* can provide the key to truth, knowledge, and understanding. Revived interest in Plato's views during the ascendancy of the Roman Empire, termed Neoplatonism, led to the assimilation of Platonic thought into Christian teachings, primarily through the writings of Saint Augustine (2, 3).

Positivism. The philosophical systems that consider knowledge as derived exclusively from *sense* experience, rejecting all *theological* and *metaphysical* sources of knowledge. Popularized in the nineteenth century by Comte, positivism asserts that scientific study must be centered on *observation* and avoid speculation. Positivism and *empiricism* are essentially compatible because both rely on the experiencing observer for the data of science. In the twentieth century, a movement called logical positivism placed a more radical emphasis on observation through a proposal for the unity of all sciences to be achieved by operationally defining true scientific issues (2, 7, 9, 10, 11, 15, 18).

Pragmatism. The American philosophical view, initially formulated by William James and Charles Peirce, that asserts the interpretation of the truth and meaning of concepts in terms of the practicality and utility of results. Pragmatism forms the intellectual justification for *functional* approaches to psychology (7, 12).

Psychoanalysis. A system of psychology, initially developed by Freud and later modified by others, based on the notion of *unconscious* motivation. Psychoanalysis also refers to the therapeutic activities, mainly free association and dream analysis, that are a product of the Freudian system and deal with the symbols and behavioral manifestations of underlying unconscious conflict (14).

Psychophysics. The study of relationships between the physical properties of environmental stimuli and the intensity of *sensory* experience. Historically, psychophysics emerged as a movement that immediately preceded the formal recognition of psychology as an independent discipline in the 1870s. Psychophysicists developed innovative methodologies that have successfully survived the last hundred years of psychology (10).

Rational psychology. Those psychological systems and theories that assume the existence of a mind or *soul* having such characteristics as immortality and spiritual aspirations.

In this sense, the *Scholastic* psychology of Aquinas and Descartes are examples of rational psychology because both views accepted the essential features of a human soul created and nurtured by God. Rational psychology is also used to describe those *dualistic* interpretations that stress *mental activity* as providing knowledge, independent of *sensory* input. In this sense, rational psychology is similar to *faculty psychology*, the psychological views arising out of the German philosophical tradition, as exemplified in the writings of Kant and Wolff (8).

Reductionism. Any scientific method that explains complex processes or *observations* in terms of simpler, underlying elements. Such *analyses* accept that the complex level may be completely understood at the simpler level. Reductionism is opposed to all of the *phenomenological* approaches. The *reflexology* interpretation of *associations* as the product of the relationships of conditioning components is an example of reductionism. Reductionism is an inherent part of many applications of *empiricism*, especially those that stress *operationism* (9, 11, 15, 18).

Reflexology. The view that psychological processes can be explained by the biologically based *associations* of *sensory*-motor relationships. Modern reflexology received a firm basis in the neurophysiology of Sherrington and later was developed systematically by Pavlov and his successors (10, 15, 17).

Reformation. The sixteenth-century movement that initially attempted to correct abuses within the Roman Catholic Church, but later disputed doctrinal matters as well, resulting in the establishment of Protestant sects. The Catholic Church attempted to recoup its losses by a Counter-Reformation that reorganized Church practices and clerical discipline (4).

Reinforcement. The principle of learning asserting that positive or rewarding events facilitate acquisition of and negative or punishment events inhibit acquisition of an *associative* bond. In Pavlovian conditioning, the unconditioned stimulus serves as the reinforcing agent; in instrumental conditioning, the consequences of a response serve as the reinforcement. Behaviorists disagree as to the specific impact of reinforcement on learning processes, but all generally accept the necessity of some agent of reinforcement for learning to occur (15).

Renaissance. Literally "rebirth"; used to describe the revival of art, literature, and scholarly pursuits during the fourteenth to sixteenth centuries in Europe. Starting in Italy, the Renaissance spread throughout Europe and resulted in the elevation of interest in humanity per se and a fascination with human intellectual capacities, largely at the expense of concern for the spiritual aspects of life (4).

Rhetoric. The study of effective use of words in oral and written communication. Considered a basic subject in the academies of ancient Greece, rhetoric assumed a prominent place in the curricula of the *medieval* universities (3).

Scholasticism. The system of Christian philosophy based on the essential teachings of Aristotle. Originating in the late medieval universities and culminating in the writings of Saint Thomas Aquinas, Scholasticism succeeded in elevating human reason, in addition to faith, as a source of knowledge by Church authorities. Scholastic philosophy produced a *rational psychology* that viewed reason and *will* as the ultimate sources of psychological processes (4).

Science. In its most general sense, the acquisition of knowledge, or knowing. However, the term has evolved to mean the systematic study of the natural world, primarily through methods based on *observation*. With the rise of *empiricism*, logical observation assumed a critical characteristic of science, and for psychology, empiricism has dictated many of the assumptions underlying the definition of the science of psychology. Those methods of controlled observation produced a *natural science model* of psychological inquiry; a *human* *science model* was also proposed, involving an open-ended, less rigid form of empirical study (1, 9, 11, 18).

Sensation. The basic unit of experience made up of input from the senses. Sensation is properly distinguished from *perception* in that the former is an unanalyzed element of the latter, although the terms are often used interchangeably. Sensory processes have offered challenging issues that discriminate among the varied systems of psychology (9, 11, 13, 17, 18).

Similarity. The principle of *association* referring to the extent to which a specific event recalls another event that it resembles along some dimension. Long recognized as a fundamental principle by associationists of the British *empiricist* tradition, similarity retains its importance in contemporary conditioning theory. For example, Osgood's transfer surface is a quantitative model that predicts the strength of associations based on the dimension of similarity of stimulus and response elements of verbal associations. Moreover, the basic learning processes of stimulus generalization and Pavlov's views on irradiation result from responses made to the similarity dimension of a stimulus (7, 15).

Skepticism. The philosophical belief that all knowledge must be continually questioned and that the process of intellectual inquiry begins with doubt. A skeptical attitude challenges the validity of knowledge that is given on the basis of some authority. Historically, *empiricism* was supported by a skeptical attitude on the part of scientists who declared that existing knowledge is questionable and tenuous until demonstrated through *sensory observation* (2, 5).

Soul. Regarded as the life-giving entity of living beings, and of people, the soul is viewed as the immortal and spiritual aspect of a person, having no physical or material existence, but accounting for the psychological processes of thinking and *willing*. The soul is perhaps the oldest subject matter of psychology, with the very name of the discipline derived from the Greek *psyche*. From the initial search for some essential common basis to all of life, Greek scholars eventually postulated the concept of soul, which was systematized in Western thought by Aristotle. This concept was eventually Christianized and emerged intact through *Scholasticism* to post-*Renaissance* Europe, at which time Descartes suggested the mind–body *dualism*, with psychology as the study of the former and physiology the study of the latter aspects of human experience. The concept of the soul in psychology was not seriously challenged until the nineteenth and twentieth centuries with the rise of *materialism* in *empiricistic science* (2, 3, 4).

Structural psychology. (Also known as **content psychology.**) The system of psychology that defined its subject matter as the study of immediate experience in the *consciousness* of the normal adult mind. Moreover, psychology was to use as its exclusive methodology the highly controlled procedure of introspection by a trained observer. Largely invented by Wundt and championed in the United States by Titchener, structural psychology was limited in both its definition of psychology and its method of inquiry. From its very inception, structural psychology's limitations were recognized by critics who argued that the study of immediate experience produced observations of *sensations* based on the physical properties of stimuli. Under such constraints, psychology was vulnerable to *logical reduction* to physics. Structural psychology was largely a force against which alternative models reacted. By 1930, structural psychology had ceased to be a viable conceptualization of psychology (11, 12, 13).

Theology. The study of God and the relationship of God to the universe. In contrast to knowledge of God based on faith or revelation, theology uses the rationality of human intellectual abilities to inquire about God. Formally presented by Aquinas in the *Scholastic* movement, systematic theology has logically argued for the necessity of a first principle (an impersonal God, at the very least) that may be described as the "uncaused cause" or the "prime mover." Under the dominance of Scholasticism, Western psychology and theology were virtually synonymous. Another later, *rationalistic* systematization of theology was developed by the nineteenth-century followers of Hegel, against whom Kierkegaard reacted (2, 3, 4, 16).

Topology. The branch of mathematics that studies those dimensional and abstract properties of geometric figures that remain constant even when under distortion. Lewin used the analogy from topology to describe the relationships of person–environment interactions (13).

Unconscious. A general term used to describe those levels of psychological activity that are not within a person's awareness or *conscious* accessibility. Emphasized in nineteenth-century German philosophy, the unconscious received interpretations ranging from its being responsible for unconscious strivings to suggestions that it was capable of subthreshold psychophysical *sensory* detection. However, it was Freud who built an entire system of personality on unconscious motivation through his model of energy exchange. The unconscious tends toward primarily *psychoanalytic* interpretation in contemporary psychology (8, 10, 14).

Vitalism. The philosophical position that views life and psychological processes as caused and maintained by a living force or agency, separate and distinct from the physical *mechanisms* of the body. Most *dualistic* assumptions about psychology are vitalistic, ranging from traditional theories of the dynamic *soul* contained in Aristotelian and *Scholastic* philosophies to contemporary expressions in the third force movement, emphasizing individual definition of *existence* and the importance of the *will* (2, 4, 9, 11, 16, 18).

Will. In general, the motivational capacity to pursue goals and aspirations voluntarily. Within the Christianized concept of the *soul*, the will and the intellect were viewed as the primary mental functions. In this context, good impulses were pursued and evil desires were inhibited to achieve human perfection. Certain suggestions of the will as irrational and *unconsciously* determined provided the philosophical bases of Freud's concept of id energy as the principle of motivation. More recent formulations have questioned the existence of the will and the notion of freedom in personal motivation (e.g., Skinner), and these views are consistent with *materialism* (2, 3, 4, 8, 11, 14, 15).

Zeitgeist. Literally, "spirit of the times." An interpretation of historical trends suggesting that the intellectual and social forces of a given period create a momentum for progress and produce individuals to express the changes within the times. This interpretation of historical developments is in contrast to the *great man* interpretation (1, 18).

Name Index

Subject Index

培文书系心理学英文影印系列 同期推出

心理学：大脑，人，世界(第 2 版)

Psychology: the brain, the person, the world

Stephen M. Kosslyn　Robin S. Rosenberg

哈佛大学心理学教材

■内容简介

这是一本非常成功的心理学基础教材，它从大脑、个人以及现实世界的概念入手，帮助学生积极地将心理学的概念应用到生活中去。作者通过自己的研究、临床实验和经验积累探究出了一套心理学研究的概念和方法。从大脑（生物组成要素角度）、个人（信仰、愿望和情绪的角度）、现实世界（社会、文化和环境角度）的概念出发，通过与学生的互动，将概念融会贯通，达到应用概念解决现实问题的目的。

ISBN 7-301-07416-6
定价：88.00 元
(16 开　880 页)
原著书号：020537609
(Allyn & Bacon 2004)
原著定价：$91.60

心理学史：观点与背景(第 3 版)

History of Psychology: Ideas and Context

Wayne Viney　D. Brett King

■内容简介

本书全面描述了心理学从远古到 20 世纪末各个阶段的发展历程，对有助于心理学发展的智识内容进行了广泛的考察。开阔了读者的视野，启迪了人们的批判和鉴赏能力。在依据传统的方式探求心理学的发展历史的同时，对其中技术发展的关键角色进行了强调，认为心理学作为一门正式学科建立于一系列人本主义改革运动的背景中。

ISBN 7-301-07367-4
定价：53.00 元
(16 开　526 页)
原著书号：0205335829
(Allyn & Bacon　2003)
原著定价：$86.40

教育心理学：理论与实践(第 7 版)

Educational Psychology: Theory and Practice

Robert E. Slavin(Johns Hopkins University)

■内容简介

本书由著名的教育心理学家 Robert Slavin 撰写，将深奥的心理学理论与实际生活相结合，语言通俗易懂，使教师能够在课堂上使用这些实际案例。作者用一系列的问题去引导教师，使他们能够通过随堂案例模拟实践活动。是一本独具特色的教育心理学教材。

ISBN 7-301-07369-0
定价：65.00 元
(16 开　654 页)
原著书号：0205351433
(Allyn & Bacon 2003)
原著定价：$90.00

培文书系人文科学英文影印系列 同期推出

语言心理学

The Psychology of Language

Timothy B. Jay (Massachusetts College of Liberal Arts)

■内容简介

本书适合语言心理学的大学高年级课程，概括了最前沿的研究成果，展示了人们是怎样理解、产生、获得语言的，描述了语言演变的过程及其在现实生活中所起的作用。本书着眼于在社会发展中语言演变的情感层面，不但论述了“语言是什么”，而且论述了“语言能够做什么”。本书具如下特色：将语言心理学的研究广泛地应用到每天的生活中去；强调情感；侧重于语言在社会情景和交流中的应用；每章开头部分设置了开篇思考题；通过章末总结来呼应开篇的思考题；章节练习和特色栏目贯穿始终。

ISBN 7-301-07370-4
定价：65.00 元
（16 开 636 页）
原著书号：0130266094
（Prentice Hall 2003）
原著定价：$93.00

社会心理学(第 11 版)

Social Psychology

Shelley E. Taylor　Letitia Anne Peplau　David O. Sears

■内容简介

本书适合本科生的社会心理学课程，涵盖了社会心理学的基本内容。本书将该学术领域内的最前沿的研究成果和最经典的理论进行了完美的结合，对社会心理学的方法和理论在日常生活和处理当前社会问题时的应用情况进行了展示。

为适应教学与社会发展的需要，本书在旧版的基础上对一些知识进行了修改和更新，如：新增加了社会心理学和法律一章（第 16 章）；在每章的开头新增加了“新闻透视”栏目；增加了自我评估方法；对社会感知一章的内容进行了修改；新增加了暴力和攻击以及恐怖管理理论方面的内容；拓展了有关自我调整、健康、疾病以及两性方面的内容；强化了包括种族和种族划分内容的社会文化方面的知识；本书还鼓励学生多从人类经历差异性的角度考虑问题；“Cultural highlights”栏目使学生能够对不同文化的异同点进行比较；“Research close-up” 栏目给学生提供了当前热点问题的研究概要。

ISBN 7-301-07366-6
定价：62.00 元
（16 开 620 页）
原著书号：013099006X
（Prentice Hall 2003）
原著定价：$102.67

市场营销中心联系名录

主任:张涛

副主任:潘建　刘宗彦

地区	联系人	电话
北京	经理:潘建(兼) 经理:刘梓盈 助理:宋诗安 助理:黄英	62752018 62767313 62752018 62767313
陕西、河南、天津	经理:饶勇 助理:陈志国	62757439 62752935
广东、浙江、湖南	经理:刘宗彦(兼) 助理:宗秀菊	62759712 62757317
江苏、东三省、内蒙古	经理:王林冲 助理:郗雨	62767314 62757438
上海 安徽 四川 重庆	经理:谢尚楹 助理:李瑞芳	62757299 62752954
山东、福建、海南、江西	经理:张志国 助理:许秀文	62752013 62757295
河北、山西、新疆、甘肃、青海、宁夏	经理:梁滨 助理:陈志国	62757298 62752935
湖北、云南、贵州、广西	经理:张继承 助理:许秀文	62757295 62750694

北京大学出版社培文教育文化公司

地址:北京市海淀区中关村北大街 118 号 1 号楼 1209 室

邮编:100871　　联系人:孙明卉

网址:http://cbs.pku.edu.cn　　电子信箱:pw@pup.pku.edu.cn

电话:010-58874097　　传真:010-58874098

北京大学出版社经管类图书经销商名录

省份	客户名称	联系人	电话	邮编	单位地址
甘肃	甘肃纸中城邦书业有限公司	文群	0931-8831085	730000	兰州市城关区东岗西路462号
	甘肃新华西北书城	何严萍、李涛	0931-4960050/4950049	730030	兰州市庆阳路
	兰州新华图书大厦	蔚凯	0931-2333360	730050	兰州市西津西路89号
	兰州市新华书店	刘增文、王宣文	0931-8464377/8476977	730030	兰州市通渭路10号
河北	石家庄市新华书店	王树臣、刘军考	0311-6694721/6694749	050011	建设北大街1号
	河北保定市新华书店	邸小力、吴浩	0312-2038213	071000	保定市裕华西路41号
	秦皇岛市新华书店	马丽华、韩景京	0335-3034446	066000	秦皇岛市文化北路101号
	河北海纳图书经销有限公司	邵玉强	0311-399669	050051	石家庄市友谊南大街86号图书批发市场
宁夏	宁夏银川市新华书店	郭德海、唐云璞	0951-5018005	750004	银川市解放东街47号
青海	西宁市新华书店	刘海妹、王欣	0971-8177644	810000	西宁市东关大街71号
	西宁市三田书店	谢炜、刘涛	0971-8248258	810000	西宁市南大街11号
山西	山西尔雅书店	靳小文、冯英敏	0351-7231473	030012	太原市双塔西街130号
	太原市新华书店	张锐、张艾平	0351-4047158	030002	太原市解放路111号
	山西省外文书店	马峰、李红	0351-4040393	030002	太原市解放路167号
	太原华拓文化发展有限公司	武长虹、王慧伟	0351-7067551	030006	太原市坞城路24号
	山西图书大厦	何莉、崔建平	0351-4956526	030012	太原市建设南路89号
新疆	新疆区图书批销中心	何云峰	0991-2860116	830001	乌鲁木齐市延安路4号
	乌鲁木齐市新华书店	卢艳	0991-8780455	830001	乌鲁木齐市解放南路346号
湖北	武汉财经书店	曾早霞	027-88049048	430060	武昌区武珞路126号
	湖北政博图书公司	杨华	027-50248986	430070	武汉市楚雄大道268-A13号
	湖北三新图书公司	罗彗华	027-87870493	430070	武汉楚雄大道268-515号
	武汉高文图书公司	雷明	027-87197858	430079	武汉市洪山区广埠屯广八路口
	武汉市新华书店采购中心	方亚红	027-83825252	430030	武汉市桥口区武胜路93号
贵州	贵阳西南风文化发展有限公司	文杰	0851-5933050	550002	贵阳市延安东路130号
	贵州西西弗书店	谢思晴	0851-6817599	550001	贵阳市中山西路38号
云南	昆明市新华书店	张平宏	0871-3646224	650021	昆明市南屏街56号昆明书城6楼
	昆明新知图书城	李李	0871-4184679	650032	云南省昆明市新闻路348号
	云南清华书屋	沈台平	0871-5314348	650092	昆明12.1大街158号(云南师大)
广西	南宁市新华书店	黎坚	0771-2829636	530012	南宁市新华街15号
	广西南国书店	蒲朝军	0771-2615485	530012	广西南宁市中山路99号
	南宁培育书店	刘立新	0771-5307995	530021	广西南宁市金湖路53号图书批销市场309号
	广西师范大学出版社大学书店	伍萍	0773-2824653	541001	桂林市中华路36号
广东	深圳市经济学会求知图书批发部	王新	0755-26139197	518029	深圳市福田区八卦岭512栋A14-A15
	广州海林书店	魏碧林	020-84463522	510275	广州市中山大学西门西侧六号
	珠海市新华书店	郑德军	0756-3260028	519015	珠海市吉大景乐路36号
	广东省佛山市新华书店	陈树铭	0757-83361679	528000	广东省佛山市汾江南路唐园东三街23号
	深圳市新华书店	周晓雨	0755-82073020	518008	深南东路5033号金山大厦六楼
	广州购书中心有限公司	杨毅	020-38868509	510620	广州市天河路123号
	广东新东园书社	陈伏明	020-89037122	510230	广州市海珠区新港西路91-93号
浙江	浙江省新华书店杭州市店	吴东宿	0571-87924811	310004	杭州市上城区青年路31号
	宁波市新华书店	汪文	0574-87246576	315000	宁波市中山东路99号
吉林	长春联合图书城有限公司	王琳	0431-2722248	130051	长春市宽城区芙蓉路1号
	吉林省新华书店图书批销中心	徐树武	0431-4640676	130022	长春市南关区人民大街170号
黑龙江	哈尔滨学府书店	张修龙	0451-86609508	150080	哈尔滨市南岗区学府路68号
	哈尔滨市新华书店	李榕	0451-83643855	150010	哈尔滨市奋斗路368号
	黑龙江省财经书店	刘丹君	0451-82732336	150008	哈尔滨市南岗区宣庆街40号
	哈尔滨学友书店	李彦文	0451-86413314	150001	哈尔滨市南岗区司令街11号

省份	客户名称	联系人	电话	邮编	单位地址
辽宁	新华书店北方图书城有限公司	王文永	024-23945027	110016	沈阳市沈河区青年大街270号
	辽宁北方出版物配送有限公司	赵喜今	024-24230809	110015	沈阳市东陵区文萃路249号
	大连佳诚考试辅导书店	石杰	0411-4670294	116023	大连市沙河口区黑石礁中山路661号
	东北财经大学出版社图书代办站	王兆勇	0411-4712239	116025	大连市黑石礁
	沈阳市新华书店	李迎锋	024-86617291	110032	沈阳市于洪区鸭绿江东街30号
	大连东财经济文化服务中心	黄玉鑫	0411-4712758	116225	大连黑石礁东财大内
	沈阳市理想书店图书有限公司	陈林	024-86191841	110031	辽宁省沈阳市皇姑区长江街147号
	沈阳现代教育书店	孙德萱	024-88112726	110031	沈阳市皇姑区黄河南大街12号
内蒙古	内蒙古外文书店	张少鹏	0471-6966778	010020	呼和浩特市新华大街46号
江苏	江苏省新华书店集团有限公司	高荣	025-85462063	210042	南京市花园路8号
	南京书城图书有限公司	陈容芳	025-84791669-8100	210005	南京市白下区中山东路18号2901室
	南京大学出版社书店	荣再艳	025-83320583	210093	南京市汉口路22号
	江苏南京新华联合图书	吴永振	025-83300361	210009	江苏省南京市湖南路47号
	南京唐风书社	宋宁	025-86633689	210009	南京市中山北路105号出版物市场2楼38号
安徽	合肥市考试书店	倪晓辉	0551-4233375	230061	合肥市金寨路337号
	安徽省新华书店图书音像发行有限公司	李永红	0551-5615040	230041	合肥蒙城北路84号
	合肥三人行书屋	陈群	0551-4233589	230011	安徽合肥市大市场书刊城6区1960#1961#1962#
四川	西南书局	任向宏	028-86511882	610081	成都市梨花街2号四川书市2楼7号
	四川新华书店集团文轩连锁有限责任公司	石冰	028-86527750	610017	成都市顺城大街306号凯乐大厦14层
重庆	重庆财经书店	郑懿彦	023-63868717	400013	重庆市渝中区华福巷25号
	重庆新华书店(集团)有限责任公司	常布克	023-68410778	400050	重庆市九龙坡滩子口76号
上海	上海新华书店图书配送中心(书城)	姚成刚	021-63914007	200072	上海市沪太路785号
	长春出版社上海图书发行部	李毅	021-56948228	200010	延长中路765号4号2楼
	上海财大书店	袁道枚	021-65422251	200083	上海中山北一路369号
北京	西单图书大厦	乔鹏	010-66083007	100031	西城区西长安街17号
	王府井书店	张瑜	010-66082767	100006	王府井大街218号
	中关村图书大厦	张瑜	010-66082767	100080	海淀区北四环西路6号
	北京金麦山图书公司	王长兰	010-65065315	100026	朝阳区甜水园北里16楼图书市场甲101
	图书城电子科技书店	王顺	010-62564631	100080	海淀图书城昊海楼1层
福建	福州远景书店	林蔚宇	0591-3317198	350004	福州市工业路118号东辉花园图书批发市场
	福州市新华书店	陈佳芳	0591-7552090	350000	福州市六一北路前岐村24号新华书店物流中心
	厦门对外图书交流中心	林元添	0592-5054027	361004	厦门市湖滨南路809号
	厦门市新华书店	吴丽娜	0592-5130710	361003	厦门市莲坂嘉禾路94号
	越洋图书文化城	马胜初	0591-7501498	350003	福州市古田路22号（五一广场越洋图书文化城）
	泉州市新华书店	粘忠奋	0595-2984659	360200	福建省泉州市温陵南路
江西	江西省新华书店	黄心刚	0791-8166026	330000	南昌市高新大道88号
	南昌青苑书店	万国英	0791-8592290	330000	南昌市洪都北大道图书城1号
	弘苑书店(零售店)	戴小华	0791-8593587	330000	南昌市洪都北大道213号图书市场38号
山东	山东省新华书店	展晓雯	0531-2779449	250000	济南市十六里河镇万寿路2号
	济南市新华书店	刘世勇	0531-6096310	250000	济南市泉城路185号
	青岛市新华书店	毕小艳	0532-4881294	266000	青岛市金华路38号
	济南东方学林书店	刘静一	0531-2076737	250000	济南市马鞍山路46号文化市场248号305号
河南	河南省新华书店(中原图书大厦)	孙斯军	0371-5684915	450000	郑州市郑汴路350号中原图书批销中心
	河南省新华书店郑州市店	孙宪荣	0371-6256783/6262229	450000	郑州市西太康路19号
陕西	陕西万邦文化传播有限公司	魏红建	029-2222531	710003	西安市东大街383号方汇大厦一层万邦书城
	西安市新华书店采购中心	陈力丁	029-87457438	710004	西安市解放路236号
天津	天津市新华书店连锁配送中心	王守华	022-88111943	300221	天津市河西区珠江道55号

欲获取相关教学辅导资料的教师烦请填写如下支持表，传真或 E-MAIL 给我们，以确此教辅只被教师获得。

培生教育出版集团北京办事处
Pearson Education Beijing Office
电话：8610-88817788-2301
8610-88817788-2302
8610-88816659
传真：8610-88817499
E-mail：service@pearsoned.com.cn

北京大学出版社培文教育文化公司
地址：北京市海淀区中关村 118 号
1 号楼 1209 室
邮编：100871
电话：010-58874097
传真：010-58874098
E-mail：pw@pup.pku.edu.cn

教辅资料支持表

兹证明__________大学(English Name)__________系／院______学年（学期）开设的__________课程，采用__________出版社出版的__________（英文原版／简体中文版）作为主要教材任课教师为__________，学生____个班共____人。

任课教师可选择以下所需教辅资料的 2-3 种，并注明所需资料的形式。

教学辅导资料种类

☐教师指导手册（Instructor's Manual）
☐习题解答（Solution Manual）
☐题库（Test Bank）
☐幻灯片（PowerPoint Slides）

教学辅导资料提供形式

☐E-file（须拷贝）
☐印刷手册
☐网上注册后下载 （PearsonEd 推荐使用）

英文书名（Title）：__________
版次（Edition）：__________ **作者（Author）：**__________
课程名称（Course Name）：__________
年级／程度（Year/Level）：__________ **院／系（School/Fac）：**__________
必修课（Required）： ☐是 ☐否

联系地址(Adress)：__________ 邮编(Zip Code)：__________
电话(Tel)：__________ 传真(Fax)：__________ E-mail：__________

系/院主任：　　　　（签字）

（系/院办公室章）

年　　月　　日